Computational Quantum Chemistry II

# The Group Theory Calculator

Computational Quantum Chemistry II

# The Group Theory Calculator

**Charles M. Quinn**

*Department of Chemistry, National University of Ireland, Maynooth, Maynooth, Co. Kildare, Ireland*

**Patrick W. Fowler**

*Department of Chemistry, Sheffield University, Sheffield, UK*

**David B. Redmond**

*Department of Mathematics, National University of Ireland, Maynooth, Maynooth, Co. Kildare, Ireland*

*Amsterdam • Boston • Heidelberg • London • New York • Oxford*
*Paris • San Diego • San Francisco • Singapore • Sydney • Tokyo*
Academic Press is an imprint of Elsevier

Academic Press is an imprint of Elsevier
Linacre House, Jordan Hill, Oxford OX2 8DP, UK
The Boulevard, Langford Lane, Kidlington, Oxford OX5 1GB, UK
84 Theobald's Road, London WC1X 8RR, UK
Radarweg 29, PO Box 211, 1000 AE, Amsterdam, The Netherlands
30 Corporate Drive, Suite 400, Burlington, MA 01803, USA
525 B Street, Suite 1900, San Diego, California 92101-4495, USA



**British Library Cataloguing in Publication Data**
A catalogue record for this book is available from the British Library

ISBN-13: 978-0-12-370456-6
ISBN-10: 0-12-370456-1

*Accompanying CD-ROM*
ISBN-13: 978-0-12-370457-3
ISBN-10: 0-12-370457-X

For information on all Academic Press publications
visit our web site at http://books.elsevier.com

Printed and bound in The Netherlands

05 06 07 08 09 10    10 9 8 7 6 5 4 3 2 1

# Contents

# Preface

This book describes the approach to applied group theory that we have developed in the past twenty-five years and illustrates how this approach, known as the 'Spherical Shell' method, can be applied to solve a variety of problems that benefit from a group theory analysis. To complement the theory, the book is supplied with a CDROM, on which interactive files, based on EXCEL® spreadsheet technology controlled by Visual Basic for Applications [VBA] code, can be used to perform straightforwardly group-theory analyses for direct application to the simplification of physical problems in Chemistry, Physics and even Engineering Science.

These powerful calculators are simple to use and do not require a detailed knowledge of the technicalities of group theory for their application.

The manuscript, also, has provided an opportunity to summarize our distinct methodology, which allows the direct identification of the appropriate linear combinations of atomic orbitals needed for the Qualitative Molecular Orbital theory analyses, so useful in modern Chemistry. This approach fits very naturally with the orbit-by-orbit procedure for performing the group theory, and which is at the heart of the construction of the calculator files on the CDROM.

Dr. J.G McKiernan, Department of Mathematics and Computer Science, Dundalk Institute of Technology, Ireland, wrote the original programmes for the plotting of functions as elliptical projections of their amplitudes on the unit sphere, during his studies at Maynooth for his Ph.D in Chemistry. Ms. Maryann Ryan, as a senior undergraduate, during two recent summer projects worked hard to prove the reliability and correctness of the considerable VBA programme that controls the operation of the calculators.

The 'BonusPack' on the CDROM contains two other programmes and a novelty item that may appeal to the reader and user of the calculator files. Thus, you will find novel EXCEL® spreadsheets for Hückel theory calculations for the regular orbit cages of the $I_h$ and I point groups and Extended Hückel theory calculations for hydrogen-atom cages of the regular orbits of the cubic point groups. These are applied to demonstrate the mutual orthogonality and so correctness of the icosahedral and *kubic* harmonics returned in the calculator basis function button displays. The original structure and function drawing programmes for the creation of Apianus II projections of the structure orbits and the LCAO-MO functions of Chapter 3 are included as the console executables, ApianusII.exe and FunctionPlot.exe. These folders contain sample input and output files for inspection and application to generate other diagrams of the reader's choice.

The fun item is Dr. Liu Hou's wonderful screen saver of active Archimedean polyhedra written in OpenGL.

All the programmes on the CDROM are provided for personal non-commercial use by the reader. Charles M. Quinn is the copyright holder of the EXCEL® files and Liu Hou is the copyright holder of the screen-saver programme in the BonusPack folder. You are expressly forbidden to copy and distribute any of the files, other than to create a copy of the CDROM on your own computer.

The standard installation of Microsoft Office® does not include two extra items: the 'Analysis Tool Pack', and the Frontline Systems SOLVER® macro. Since the GT_Calculator files require complex arithmetic, the 'Analysis Tool Pack' must be present. Since the EXCEL® Hückel and Extended Hückel programmes depend on optimization as required by the application of the variation principle to the LCAO-MO Hamiltonian, the SOLVER® macro, also, is needed. Both can be added to an existing installation of the OFFICE® software using the 'Add-ins' option in the TOOLS menu.

The EXCEL files on the CDROM were created on a PC running MS WINDOWS 2000, version 5, SP4 and MS EXCEL 2002, SP3. These files are not written for the MAC environment, but can be run on a MAC if the 'Virtual PC' programme is present.

To ensure that your calculator files operate as described in Chapter 1, it is necessary to lower the security level on your copy of EXCEL® to 'medium' [Tools/Options/Security menu] and to hide any external toolbars, e.g. Adobe Acrobat PDFMaker 6.0 [View/Toolbars and deselect as necessary].

**Charles M. Quinn,**
*Department of Chemistry, NUI, Maynooth,*
*Maynooth, Co. Kildare, Ireland.*

**Patrick W. Fowler,**
*Department of Chemistry, Sheffield University,*
*Sheffield S3 7HF, UK.*

**David B. Redmond,**
*Department of Mathematics, NUI, Maynooth,*
*Maynooth, Co. Kildare, Ireland.*

# 1

# Operating instructions for the Group Theory Calculator

The Group Theory Calculator [the GT_Calculator] is the set of interactive EXCEL® spreadsheet files, one for each of the main molecular point groups, on the CDROM supplied with this manual. The group theoretical calculations, which can be performed with the calculator, are rendered possible because of the enhancement of the basic spreadsheet operations and displays using *Visual Basic for Applications* code and the complex-arithmetic routines available in the Analysis Tool Pack EXCEL® 'Add-ins'. Since the *Analysis Tool Pack* is not one of the standard components loaded in a typical installation of the EXCEL® software package, it is necessary, before attempting to use the GT_Calculator, to ensure that your version of the EXCEL® programme includes this extra component. With any spreadsheet open, check the 'Add-ins' list in the TOOLS menu on the main EXCEL® toolbar and, if necessary, install this component in the usual way.

It is advisable to open and use only one EXCEL file and, especially, not to have blank 'new' spreadsheets open at the same time. While the default exit procedure in the VBA code involves the instruction '*Application.quit*', on occasions the presence of another open workbook can interfere with the closing sequence.

With the GT_Calculator you can perform a variety of standard group theory calculations simply by entering the appropriate structure details for the molecular geometry. In addition, on the various worksheets of the calculator files, it is straightforward to determine more advanced group theoretical results, such as the numbers of isomers generated for a given structure by decoration, or to calculate and decompose the symmetric and antisymmetric powers of permutation representations.

If such matters are familiar and you are comfortable with the concepts of *Group Theory* on the *Spherical Shell*, it is probably necessary only to read through the operating instructions in the remainder of this chapter before using the calculator files for your own group theory analyses. Otherwise, you are advised to study the material in the remaining chapters prior to any extensive use of the calculator.

## 1.1 Overview

The GT_Calculator can be operated directly from the CDROM. With the MS Windows EXPLORER active, your monitor screen display, when you inspect the contents of the folders on the CDROM and their contents, will be similar to the screen dumps shown in Figure 1.1.

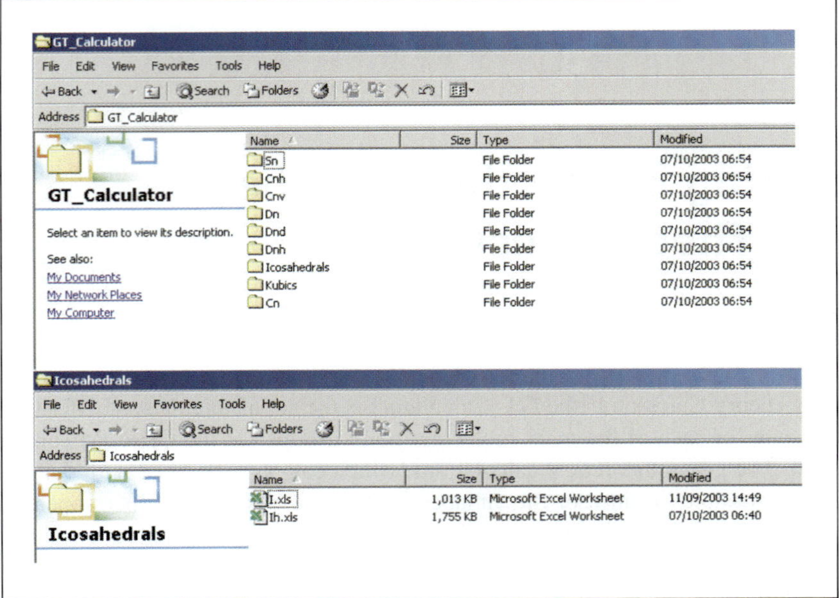


**Figure 1.1** Screen dumps of the EXPLORER trees for the files on the GT_Calculator CDROM. Each file is activated by selecting the particular icon of the tree and then activating with the mouse or by pressing the ENTER button on the keyboard.

In the second screen dump, the contents of the icosahedral folder is displayed and, as you see, this folder contains EXCEL® files for the two point groups, I and $I_h$, of icosahedral symmetry.

For example, to operate the GT_Calculator for the case of structures of $I_h$ symmetry, select Ih.xls and open this spreadsheet in the usual manner, either by pressing the ENTER button on your keyboard or with your mouse. The initial screen display will be as in the first diagram in Figure 1.2 and it is necessary to choose the 'enable macros' option in order to activate the functionality of the calculator. After a short graphic display, which can be cancelled with the ESC button, the standard logo screen for the calculator files is displayed with the centre text used to distinguish the different point groups.

This logo screen is quite different to the standard MS-EXCEL® display. There are no familiar toolbars and the spreadsheet is hidden under the Calculator logo and its dark grey background. Access and control of the GT_Calculator files are managed from the new toolbar at the top of the logo screen in Figure 1.2, which contains only the three command buttons: 'Exit'; 'Character Table' and 'Setup', with the underscore device used to identify the letters to access the actions of these command buttons using the combined keyboard sequence, for example, ALT+x [case is not significant], to exit the file.

The 'Setup' command button in this toolbar can be used to adjust the display size and position for different monitors, with the option to save these optimum settings to suit different monitor sizes. On selecting the 'Setup' button the subsidiary bar shown in Figure 1.3 is activated. The actions of the individual command buttons are from the left; 'Zoom In'; 'Zoom Out'; 'Scroll Down'; 'Scroll Up'; 'Scroll Right'; 'Scroll Left'; 'Set', to save the display size

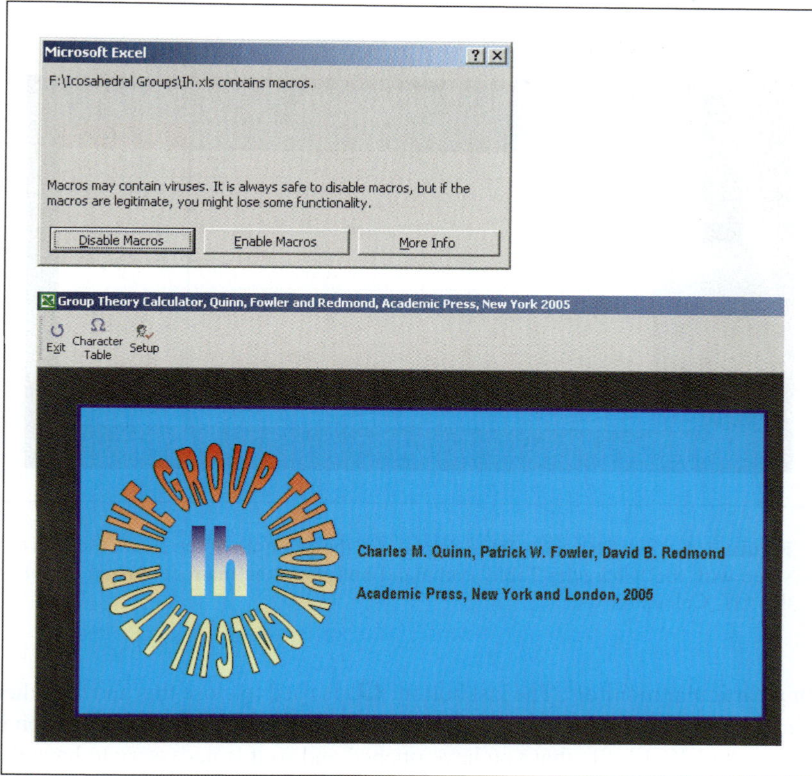


**Figure 1.2** Screen displays during the activation of the Ih.xls GT_Calculator file. Note that all the message box windows in the calculator are *modal* and so take precedence over the background display. Remember to check that the security level is set to 'medium' in a blank EXCEL® file before opening the calculator files, otherwise this opening sequence will not appear and the option to 'enable macros' is not given.

**Figure 1.3** The 'Setup' command bar accessible from the initial logo screen for each GT_Calculator file.

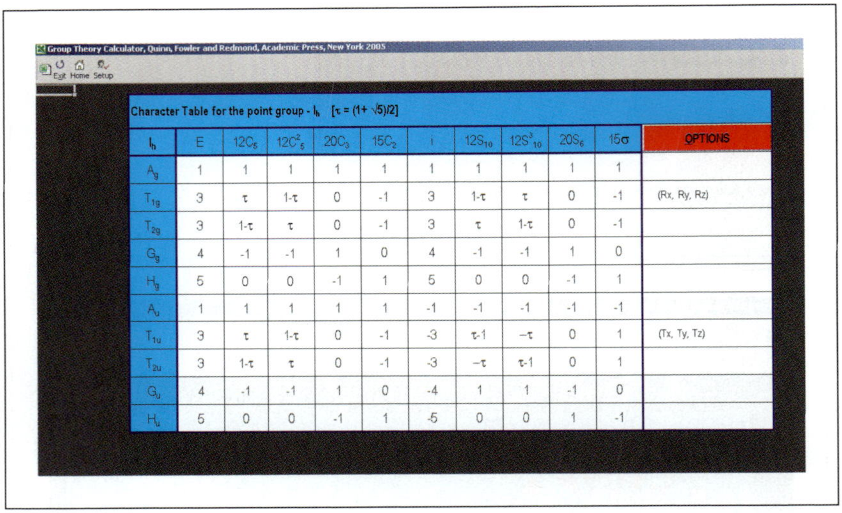

Group Theory Calculator, Quinn, Fowler and Redmond, Academic Press, New York 2005

Exit Home Setup

Character Table for the point group - $I_h$ [$\tau = (1+\sqrt{5})/2$]

| $I_h$ | E | $12C_5$ | $12C^2_5$ | $20C_3$ | $15C_2$ | i | $12S_{10}$ | $12S^3_{10}$ | $20S_6$ | $15\sigma$ | OPTIONS |
|---|---|---|---|---|---|---|---|---|---|---|---|
| $A_g$ | 1 | 1 | 1 | 1 | 1 | 1 | 1 | 1 | 1 | 1 | |
| $T_{1g}$ | 3 | $\tau$ | $1-\tau$ | 0 | -1 | 3 | $1-\tau$ | $\tau$ | 0 | -1 | (Rx, Ry, Rz) |
| $T_{2g}$ | 3 | $1-\tau$ | $\tau$ | 0 | -1 | 3 | $\tau$ | $1-\tau$ | 0 | -1 | |
| $G_g$ | 4 | -1 | -1 | 1 | 0 | 4 | -1 | -1 | 1 | 0 | |
| $H_g$ | 5 | 0 | 0 | -1 | 1 | 5 | 0 | 0 | -1 | 1 | |
| $A_u$ | 1 | 1 | 1 | 1 | 1 | -1 | -1 | -1 | -1 | -1 | |
| $T_{1u}$ | 3 | $\tau$ | $1-\tau$ | 0 | -1 | -3 | $\tau-1$ | $-\tau$ | 0 | 1 | (Tx, Ty, Tz) |
| $T_{2u}$ | 3 | $1-\tau$ | $\tau$ | 0 | -1 | -3 | $-\tau$ | $\tau-1$ | 0 | 1 | |
| $G_u$ | 4 | -1 | -1 | 1 | 0 | -4 | 1 | 1 | -1 | 0 | |
| $H_u$ | 5 | 0 | 0 | -1 | 1 | -5 | 0 | 0 | 1 | -1 | |

**Figure 1.4** The Character Table display for the example of the Ih.xls file, activated with the 'Character Table' command button on the main command bar of this GT_Calculator file.

chosen for a particular monitor[1]. The last button, ↻, is used to close this and the other 'Setup' toolbar. Note, the option to save cannot be applied directly on the CD files, its action is to save a new copy of the EXCEL file that you have opened and so it is necessary to have this file on your hard drive or other suitable medium with the transfer 'read-only' property switched off.

Selecting the 'Character Table' button on the main command bar of the GT_Calculator files activates the main display window as shown, again, for the case of the $I_h$ point group, in Figure 1.4.

For all the calculator files on the CDROM, this display comprises the standard character table for the particular group and identifies the irreducible symmetries of translations and rotations about the origin of the coordinate system for molecular structures with this point symmetry.

The OPTIONS command button [activate with your mouse or the keyboard sequence ALT+O] provides the access path to the main calculator functions. These are identified by the captions on the command buttons shown in Figure 1.5. The actions of each of these buttons are described in the remaining sections of this chapter.

Note the availability of another 'Setup' command button on the main toolbar of this worksheet: its functionality is similar to the 'Setup' button on the main logo sheet, but, as you can see in Figure 1.6, the 'SET' command bar button to save the window size is replaced by a new button, the action of which is to place particular displays on this window and the others accessible using the command buttons of Figure 1.4 as considered best for a particular monitor and the sharpness of the viewer's eyes. This second command bar is included as the 'Setup' option on the main command bar for all the worksheets accessed through the command buttons displayed in Figure 1.5.

[1]Since each Character Table is of different size, the 'Setup' macro may need to be run on several files for any particular monitor to obtain the optimum display.

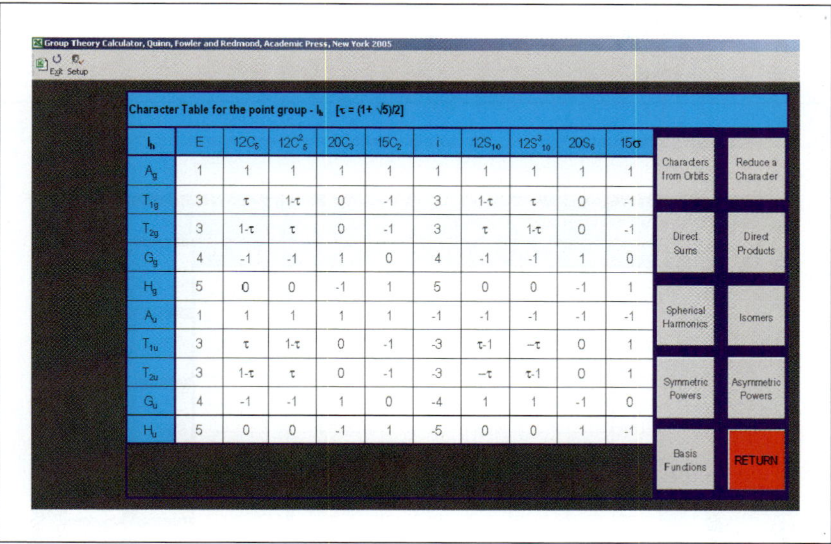


| $I_h$ | E | $12C_5$ | $12C_5^2$ | $20C_3$ | $15C_2$ | i | $12S_{10}$ | $12S_{10}^3$ | $20S_6$ | $15\sigma$ |
|---|---|---|---|---|---|---|---|---|---|---|
| $A_g$ | 1 | 1 | 1 | 1 | 1 | 1 | 1 | 1 | 1 | 1 |
| $T_{1g}$ | 3 | $\tau$ | $1-\tau$ | 0 | -1 | 3 | $1-\tau$ | $\tau$ | 0 | -1 |
| $T_{2g}$ | 3 | $1-\tau$ | $\tau$ | 0 | -1 | 3 | $\tau$ | $1-\tau$ | 0 | -1 |
| $G_g$ | 4 | -1 | -1 | 1 | 0 | 4 | -1 | -1 | 1 | 0 |
| $H_g$ | 5 | 0 | 0 | -1 | 1 | 5 | 0 | 0 | -1 | 1 |
| $A_u$ | 1 | 1 | 1 | 1 | 1 | -1 | -1 | -1 | -1 | -1 |
| $T_{1u}$ | 3 | $\tau$ | $1-\tau$ | 0 | -1 | -3 | $\tau-1$ | $-\tau$ | 0 | 1 |
| $T_{2u}$ | 3 | $1-\tau$ | $\tau$ | 0 | -1 | -3 | $-\tau$ | $\tau-1$ | 0 | 1 |
| $G_u$ | 4 | -1 | -1 | 1 | 0 | -4 | 1 | 1 | -1 | 0 |
| $H_u$ | 5 | 0 | 0 | -1 | 1 | -5 | 0 | 0 | 1 | -1 |

**Figure 1.5** The typical options window, for the example of the calculator file Ih.xls, accessible from the display of Figure 1.4 by activating the OPTIONS command button.

The display, in Figure 1.5, is the starting point for all applications of the GT_Calculator. Particular applications are accessed by activating the command buttons on the keypad. The actions of these buttons are as follows.

## 1.2 Characters from Orbits

The '*Characters from Orbits*' command button leads to the worksheet displayed in Figure 1.7 and there are several features to note about the display in the figure.

First, the character table for the $I_h$ group has been reduced in size so that more information can be displayed in the window. Secondly, the worksheet is *protected* [the general condition of all the worksheets] except for the cells used to input data, which, in all cases, are bordered in red. Thirdly, there is a series of command buttons on the right of the display, with actions described by the button labels. These observations are general for all the worksheet displays of the GT_Calculator.

There are 6 possible structure orbits [see Chapter 2, Section 2.6] of the $I_h$ symmetry group including the trivial orbit, $O_1$, of a single structure point at the coordinate origin. The others are $O_{12}$ [the icosahedron], $O_{20}$ [the dodecahedron], $O_{30}$ [the icosidodecahedron], $O_{60}$ [the truncated dodecahedron or the truncated icosahedron] and, finally, the regular orbit, $O_{120}$. This orbit describes a set of 120 points, all equivalent under $I_h$ symmetry, but with each point in a general position with respect to the symmetry elements. It can be realized by decorating each of the vertices of the icosahedron with a decagon of five-fold rotational symmetry to generate the *great* rhombicosidodecahedron.

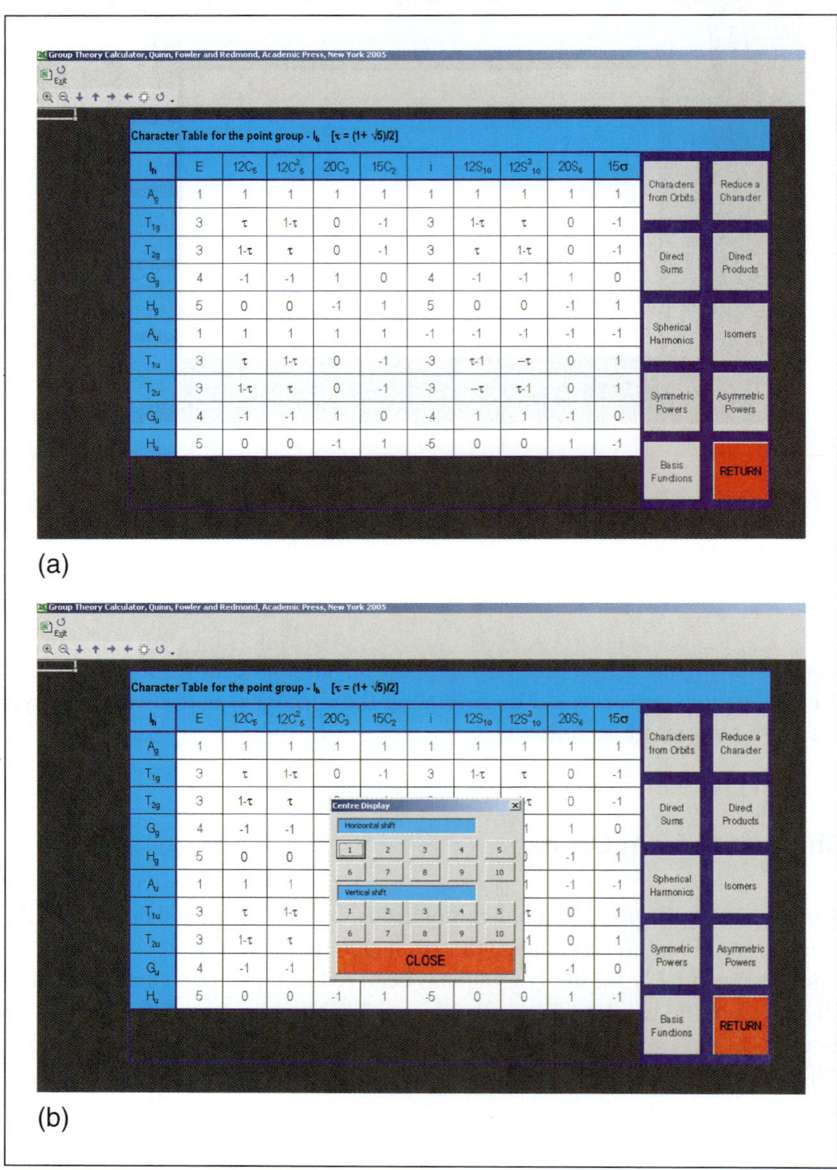


| $I_h$ | E | $12C_5$ | $12C_5^2$ | $20C_3$ | $15C_2$ | i | $12S_{10}$ | $12S_{10}^3$ | $20S_6$ | $15\sigma$ |
|---|---|---|---|---|---|---|---|---|---|---|
| $A_g$ | 1 | 1 | 1 | 1 | 1 | 1 | 1 | 1 | 1 | 1 |
| $T_{1g}$ | 3 | τ | 1-τ | 0 | -1 | 3 | 1-τ | τ | 0 | -1 |
| $T_{2g}$ | 3 | 1-τ | τ | 0 | -1 | 3 | τ | 1-τ | 0 | -1 |
| $G_g$ | 4 | -1 | -1 | 1 | 0 | 4 | -1 | -1 | 1 | 0 |
| $H_g$ | 5 | 0 | 0 | -1 | 1 | 5 | 0 | 0 | -1 | 1 |
| $A_u$ | 1 | 1 | 1 | 1 | 1 | -1 | -1 | -1 | -1 | -1 |
| $T_{1u}$ | 3 | τ | 1-τ | 0 | -1 | -3 | τ-1 | -τ | 0 | 1 |
| $T_{2u}$ | 3 | 1-τ | τ | 0 | -1 | -3 | -τ | τ-1 | 0 | 1 |
| $G_u$ | 4 | -1 | -1 | 1 | 0 | -4 | 1 | 1 | -1 | 0 |
| $H_u$ | 5 | 0 | 0 | -1 | 1 | -5 | 0 | 0 | 1 | -1 |

**Figure 1.6** The 'Setup' toolbar, Figure 1.6a, for all worksheet windows other than the initial logo screen. The actions of all the buttons are the same as previously, except for the ⌗ button, which on activation leads to the centring device shown in Figure 1.6b.

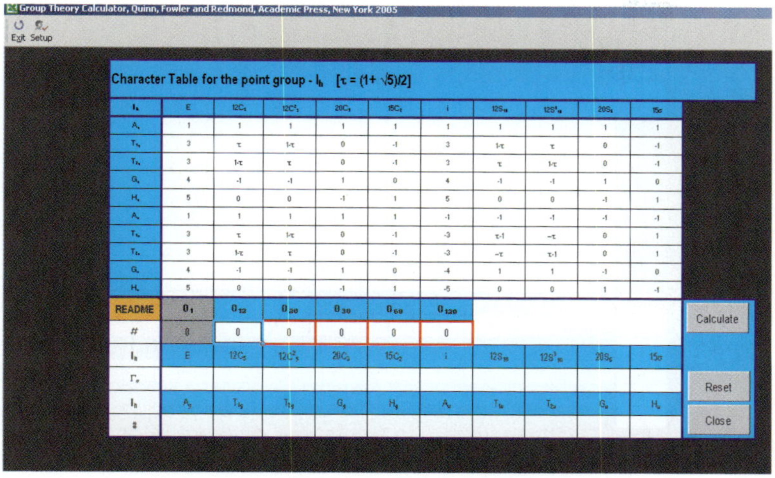

Group Theory Calculator, Quinn, Fowler and Redmond, Academic Press, New York 2005

Exit Setup

Character Table for the point group - $I_h$ [$\tau = (1+\sqrt{5})/2$]

| $I_h$ | E | $12C_5$ | $12C_5^2$ | $20C_3$ | $15C_2$ | i | $12S_{10}$ | $12S_{10}^3$ | $20S_6$ | $15\sigma$ |
|---|---|---|---|---|---|---|---|---|---|---|
| $A_g$ | 1 | 1 | 1 | 1 | 1 | 1 | 1 | 1 | 1 | 1 |
| $T_{1g}$ | 3 | $\tau$ | $1-\tau$ | 0 | -1 | 3 | $1-\tau$ | $\tau$ | 0 | -1 |
| $T_{2g}$ | 3 | $1-\tau$ | $\tau$ | 0 | -1 | 3 | $\tau$ | $1-\tau$ | 0 | -1 |
| $G_g$ | 4 | -1 | -1 | 1 | 0 | 4 | -1 | -1 | 1 | 0 |
| $H_g$ | 5 | 0 | 0 | -1 | 1 | 5 | 0 | 0 | -1 | 1 |
| $A_u$ | 1 | 1 | 1 | 1 | 1 | -1 | -1 | -1 | -1 | -1 |
| $T_{1u}$ | 3 | $\tau$ | $1-\tau$ | 0 | -1 | -3 | $\tau-1$ | $-\tau$ | 0 | 1 |
| $T_{2u}$ | 3 | $1-\tau$ | $\tau$ | 0 | -1 | -3 | $-\tau$ | $\tau-1$ | 0 | 1 |
| $G_u$ | 4 | -1 | -1 | 1 | 0 | -4 | 1 | 1 | -1 | 0 |
| $H_u$ | 5 | 0 | 0 | -1 | 1 | -5 | 0 | 0 | 1 | -1 |

| README | $O_1$ | $O_{12}$ | $O_{20}$ | $O_{30}$ | $O_{60}$ | $O_{120}$ | | | | |
|---|---|---|---|---|---|---|---|---|---|---|
| # | 0 | 0 | 0 | 0 | 0 | 0 | | | | |
| $I_h$ | E | $12C_5$ | $12C_5^2$ | $20C_3$ | $15C_2$ | i | $12S_{10}$ | $12S_{10}^3$ | $20S_6$ | $15\sigma$ |
| $\Gamma_\sigma$ | | | | | | | | | | |
| $I_h$ | $A_g$ | $T_{1g}$ | $T_{2g}$ | $G_g$ | $H_g$ | $A_u$ | $T_{1u}$ | $T_{2u}$ | $G_u$ | $H_u$ |
| # | | | | | | | | | | |

Calculate

Reset

Close

**Figure 1.7** The worksheet for the calculation of the permutation character and its direct sum components, listed as Mulliken symbols from orbit lists. This display is accessed from the 'Characters from Orbits' command button of the window shown in Figure 1.5.

In Figure 1.7, the input cell for the number of $O_1$ orbits is greyed out. This is because the $\sigma$, $\pi$ and $\delta$ classification[2], which is used to describe radial and tangential properties on the surface of the sphere does not apply to the centre. In groups where the $O_1$ orbit, if present, must lie at the centre, this input is 'locked' in the 'Characters from Orbits' and 'Isomers' [Figure 1.18] displays.

Characters and their decompositions into direct sums of irreducible characters are determined by entering an orbit list, to define the geometry of the icosahedral object under consideration. For example, in Figure 1.8, the permutation character of the regular orbit is calculated and decomposed into a direct sum of irreducible components. The result illustrates the general rule that each irreducible representation occurs in the regular representation with a number of copies equal to its dimensions. Thus, in this example, we find one copy of each 1D irreducible character, three copies of each 3D irreducible character and so on, in this decomposition of the regular character for the $I_h$ point group.

For all the spreadsheet files of the GT_Calculator the convention applied is to mark the non-zero components of a direct sum decomposition of a permutation character by green background shading. For the regular character in the example of Figure 1.8, all the irreducible symmetries contribute and so all the 'direct sum' cells exhibit green shadings.

There are two extra buttons visible on the display in Figure 1.8. The one labelled 'Print' has the obvious use that the calculation result can be printed on your active printer. The one labelled '$\pi$, $\delta$ ...' facilities the use of the permutation result to calculate the characters and direct sum components for the higher-order harmonics of the permutation character and

[2] $\sigma$, $\pi$ and $\delta$ representations of the molecular point groups, Patrick W. Fowler and Charles M. Quinn, *Theoretica Chimica Acta* (1986) 333.

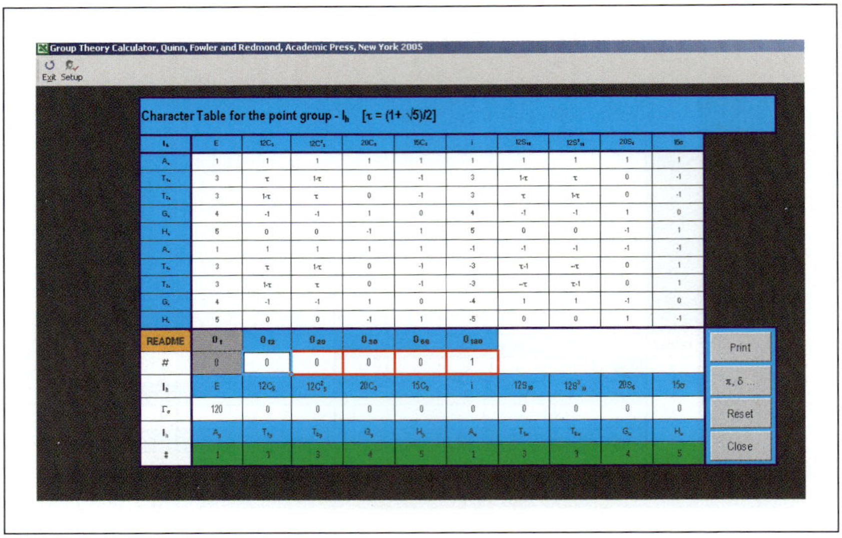

Group Theory Calculator, Quinn, Fowler and Redmond, Academic Press, New York 2005

Exit Setup

Character Table for the point group - $I_h$ [$\tau = (1+\sqrt{5})/2$]

| $I_h$ | E | $12C_5$ | $12C_5^2$ | $20C_3$ | $15C_2$ | i | $12S_{10}$ | $12S_{10}^3$ | $20S_6$ | $15\sigma$ |
|---|---|---|---|---|---|---|---|---|---|---|
| $A_g$ | 1 | 1 | 1 | 1 | 1 | 1 | 1 | 1 | 1 | 1 |
| $T_{1g}$ | 3 | τ | 1-τ | 0 | -1 | 3 | 1-τ | τ | 0 | -1 |
| $T_{2g}$ | 3 | 1-τ | τ | 0 | -1 | 3 | τ | 1-τ | 0 | -1 |
| $G_g$ | 4 | -1 | -1 | 1 | 0 | 4 | -1 | -1 | 1 | 0 |
| $H_g$ | 5 | 0 | 0 | -1 | 1 | 5 | 0 | 0 | -1 | 1 |
| $A_u$ | 1 | 1 | 1 | 1 | 1 | -1 | -1 | -1 | -1 | -1 |
| $T_{1u}$ | 3 | τ | 1-τ | 0 | -1 | -3 | τ-1 | −τ | 0 | 1 |
| $T_{2u}$ | 3 | 1-τ | τ | 0 | -1 | -3 | −τ | τ-1 | 0 | 1 |
| $G_u$ | 4 | -1 | -1 | 1 | 0 | -4 | 1 | 1 | -1 | 0 |
| $H_u$ | 5 | 0 | 0 | -1 | 1 | -5 | 0 | 0 | 1 | -1 |

| README | $O_1$ | $O_{12}$ | $O_{20}$ | $O_{30}$ | $O_{60}$ | $O_{120}$ |
|---|---|---|---|---|---|---|
| # | 0 | 0 | 0 | 0 | 0 | 1 |

| $I_h$ | E | $12C_5$ | $12C_5^2$ | $20C_3$ | $15C_2$ | i | $12S_{10}$ | $12S_{10}^3$ | $20S_6$ | $15\sigma$ |
|---|---|---|---|---|---|---|---|---|---|---|
| $\Gamma_\sigma$ | 120 | 0 | 0 | 0 | 0 | 0 | 0 | 0 | 0 | 0 |
| $I_h$ | $A_g$ | $T_{1g}$ | $T_{2g}$ | $G_g$ | $H_g$ | $A_u$ | $T_{1u}$ | $T_{2u}$ | $G_u$ | $H_u$ |
| # | 1 | 3 | 3 | 4 | 5 | 1 | 3 | 3 | 4 | 5 |

Print

π, δ ...

Reset

Close

**Figure 1.8** Calculation of the permutation character generated on the regular orbit of the $I_h$ point group geometry by the actions of the symmetry operation and its decomposition into the direct sum components identified by their Mulliken symbols. Note the appearance of extra option command buttons on the right of the display.

for the calculation of the normal modes of vibration of the structure defined by the orbit list[2]. In MO theory, for example, the permutation character corresponds to the symmetry spanned by a set of $\sigma$ basis functions on the vertices of the orbit, the $\pi$ character to the symmetry of a set of pairs of tangential p-orbitals and the $\delta$ character to a set of pairs of tangential d-orbitals and so on through higher harmonics[3].

Two examples of this kind of extra calculation are displayed in Figure 1.9 and Figure 1.10. In Figure 1.9, the action of the '$\pi$, $\delta$ ...' command button is used to calculate the $\delta$ character of the regular orbit permutation character, in LCAO-MO theory, the character generated by the actions of the symmetry operations of the $I_h$ point group on local pairs of d-orbitals at each of the 120 vertices of the regular orbit. The 'answer box' entry, in this calculation, is limited to a maximum value of 60. In Figure 1.10, the action of the '$\pi$, $\delta$ ...' command button is used to calculate the vibration character for the 120-vertex regular-orbit cage of the icosahedral structure, which character is returned for the entry '100' in the 'answer box'.

## 1.3 Error Traps

All the worksheets in the GT_Calculator files are supported by error detection macros, which trap and delete incorrect data entered into the input cells. For the case of the 'Characters from Orbits' command button worksheet, only integer numbers of orbits may be input for a calculation of the permutation character and its direct sum components though for certain

[3]The complete set of local d-orbitals on a set of vertices generates $\sigma$, $\pi$ and $\delta$ characters under actions of the symmetry operations of the point group. The set of local f-orbitals generates $\sigma$, $\pi$ and $\delta$ and $\phi$ characters, and so on.

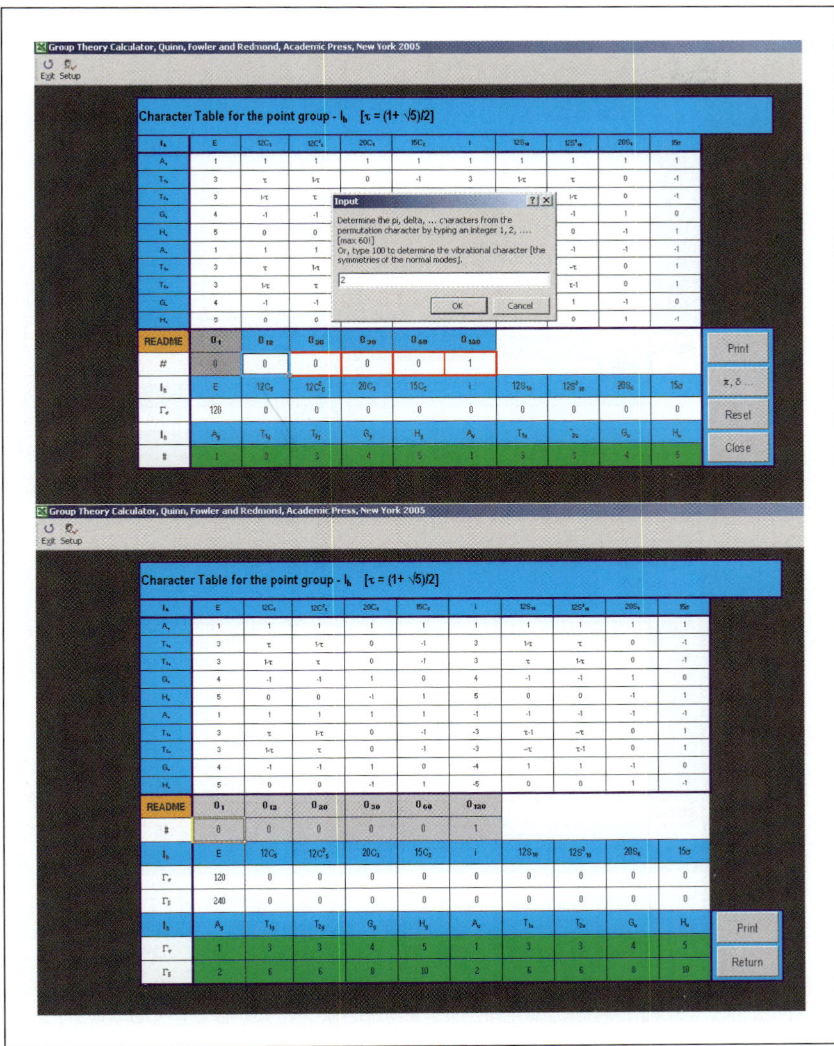

**Figure 1.9** Calculation of the $\delta$ character and its direct sum irreducible components for the regular orbit permutation character of Figure 1.8. The 'Print' command button generates a hard copy of the result on your local printer, while the 'Return' button restores the screen for the permutation character result.

applications both positive and negative integers may be used. However, any entries involving non-integer numbers of orbits or non-numeric characters return error flags and restore the display to its initial state, Figure 1.11.

Similar traps and warning messages are included in the macros driving the calculation options on the worksheets accessed through the other command buttons of the main display of Figure 1.5. Explicit illustration of their actions is not necessary in this section. But, it is appropriate, perhaps, to mention that traps are provided, also, to detect impossible character inputs and the omission of necessary input components in, for example, calculations of the characters of direct products and the corresponding direct-sum components.

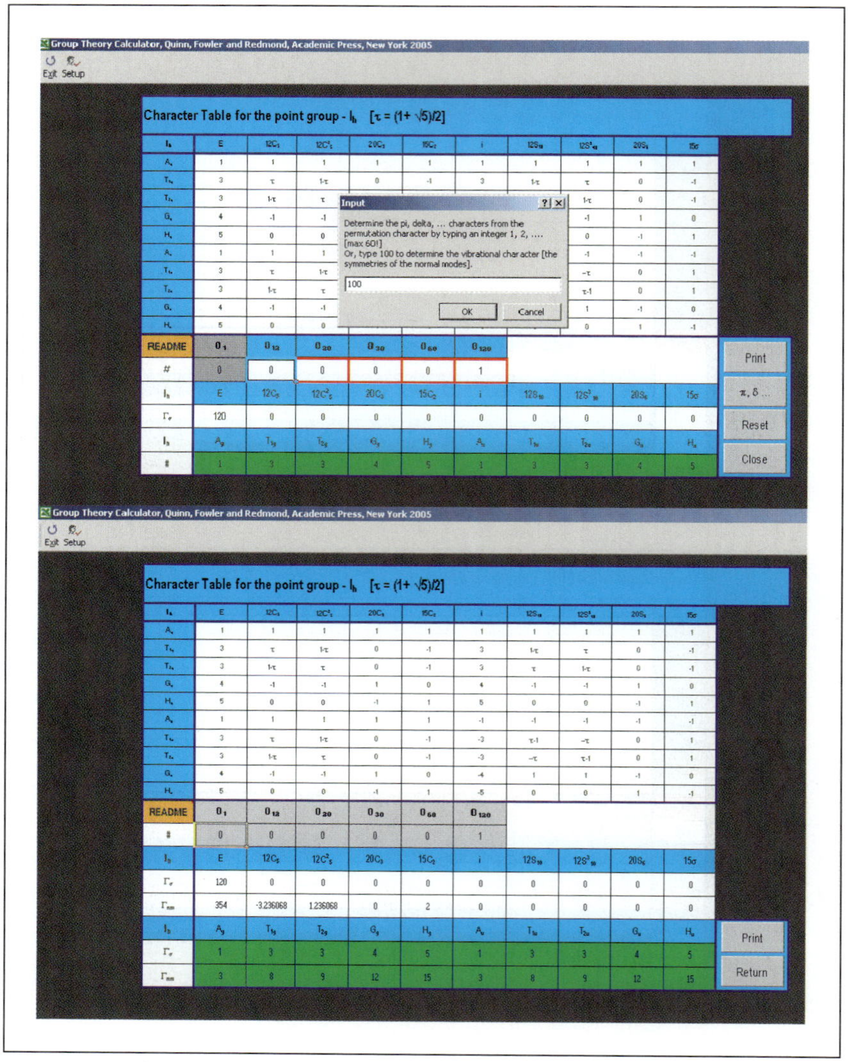

**Figure 1.10** Calculation of the character and direct sum irreducible components of the normal modes of vibration for the regular orbit of 120 vertices on the unit sphere.

As mentioned already, the limit on the integer input for the determination of the higher order harmonics from a given permutation character is 60. This is also the limit on the input on the worksheet activated with the 'Spherical Harmonics' command button of the main options display, Figure 1.5.

For technical reasons, your keyboard 'BACKSPACE' button should not be used to correct errors made during data input. The default state of the input cells on all the calculator worksheets is that they contain the number zero.

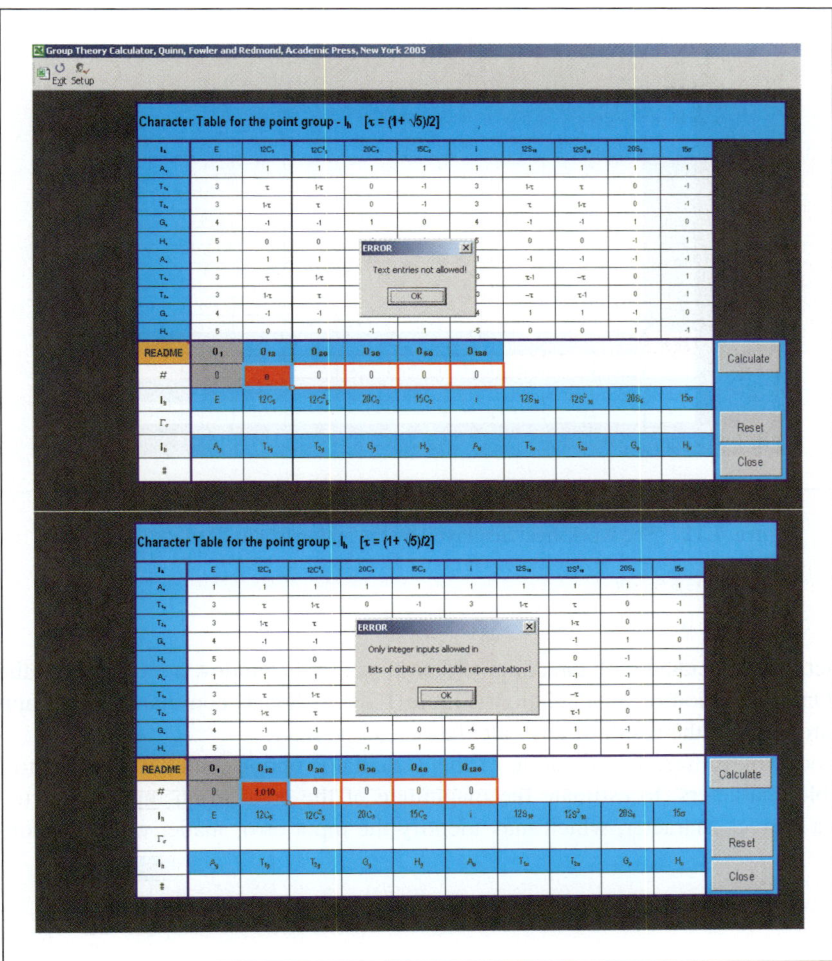


**Figure 1.11** Demonstration of the error trapping routines in the code controlling worksheet calculations activated through the 'Character from Orbits' command button. Note that the second screen dump has been enhanced to emphasise that a non-integer entry was made, because the default format for the '# of orbits' input cells is integer.

## 1.4 Reduce a Character

The worksheet displayed by the GT_Calculator files by the action of the 'Reduce a Character' command button, Figure 1.5, is shown in Figure 1.12. This worksheet takes as input the character, Γ, normally a reducible representation, i.e. a set of traces of the matrix representatives of the operators of the group and returns the direct sum components of this character, identified by Mulliken symbols. This input is entered in the 'red-bordered' cells and the direct sum components are returned as numbers of Mulliken symbols in the last row of the display.

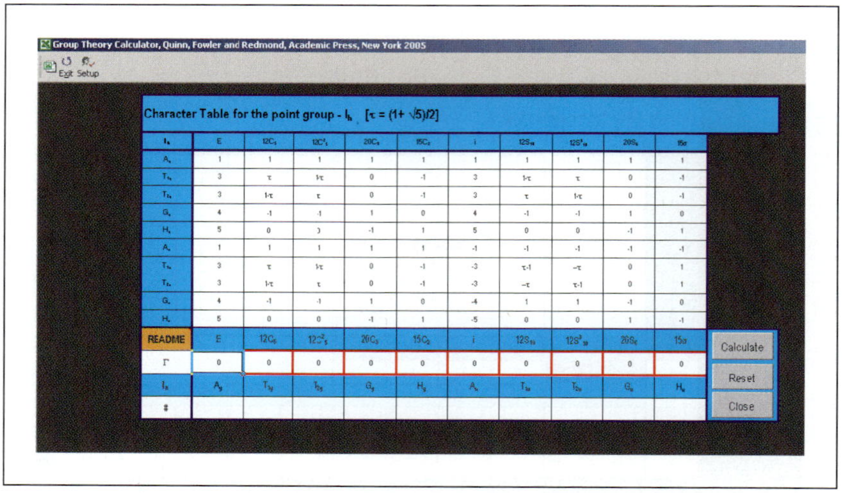

**Figure 1.12** The worksheet displayed by the action of the 'Reduce a Character' button.

The actions associated with the command buttons of this window are clear from the button labels. On completion of a calculation (Figure 1.13), a 'Print' command button appears so that a hard copy of the result of any calculation can be made.

The code controlling the operation of this Calculator option includes routines to identify impossible characters, to estimate the magnitude of the error in an input character and to suggest a correct character, which may identify the input error made, when the correct and

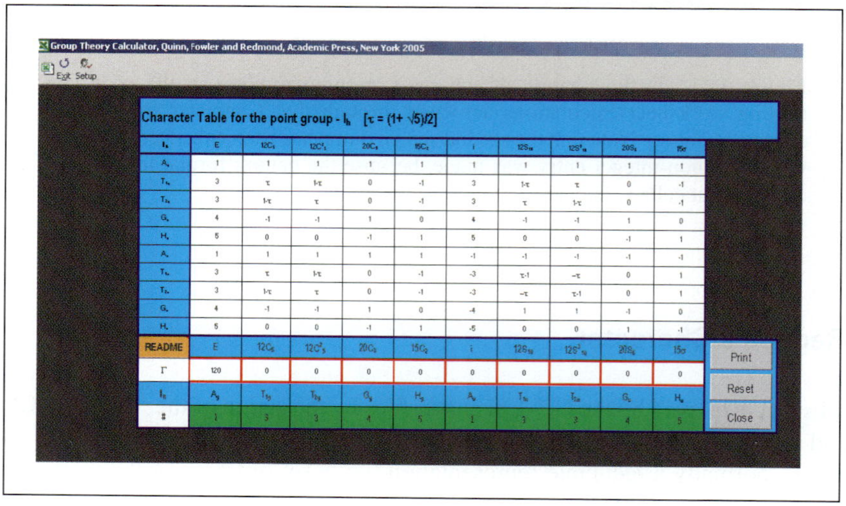

**Figure 1.13** Demonstration of the operation of the calculator for a correct character input.

incorrect characters are not too different, Figure 1.14. The sequence leading from the incorrect input to a possible and, often, a correct one is illustrated in the displays of Figure 1.14.

In Figure 1.14a, the regular character has been spoiled by setting the trace to be 1 under the inversion operation of the $I_h$ point group. This error is trapped and marked by the display in Figure 1.14b and the user then is offered the option to review the input as shown in Figure 1.15. This review is conducted by setting the number of decimal places counted in the coefficients of the direct sum components. As you can see in Figure 1.14e, setting two places of decimal as the 'integer' allows inaccuracies in the calculation in the third decimal place to be insignificant and then truncation leads to the final guess, which, in this case, corresponds to a correction of the input, rather than an alternative input!

## 1.5 Direct Sums

The 'Direct Sum' command button in all the GT_Calculator files leads to the worksheet display shown in Figure 1.14. Direct Sums, as reducible characters, are returned on input of the appropriate integer numbers of Mulliken symbols for the particular point group and starting the calculation using the *Calculate* command button. A *Print* command button becomes available as one of the actions of the *Calculate* command button.

## 1.6 Direct Products

The 'Direct Product' command button in all the GT_Calculator files leads to the worksheet display shown in Figure 1.16, in which reducible characters determined from the input of direct sums are multiplied together. The direct product result is returned both as a reducible character and as a list of direct sum components of this reducible character, identified, again, by their Mulliken symbols.

The usual error traps are present in the code controlling the direct-product calculations. In addition, you will be prompted should you forget to specify either or both components of the direct sums required to perform the calculation. A 'Print' command button is available.

## 1.7 Spherical Harmonics

The 'Spherical Harmonics' command button leads to the display shown in Figure 1.17. The calculation leads to the representations (usually reducible) spanned by a set of harmonics of given $\ell$-value, placed at the origin of the coordinates in an object exhibiting the particular point group symmetry output as the direct sum components of the central harmonics character for a given angular momentum quantum number as input. Thus the initial condition shown in Figure 1.17 for the case of the Ih.xls file of the GT_Calculator is simply that $\ell = 0$ central harmonic transforms as $A_g$ and exhibits the totally symmetric character shown in the last row of the display.

In general, this display shows how the set of harmonic polynomials for given $\ell$-value splits on *Descent in Symmetry* from the spherical group to the particular point group in question. The calculation is limited to $\ell$-values less than or equal to 60.

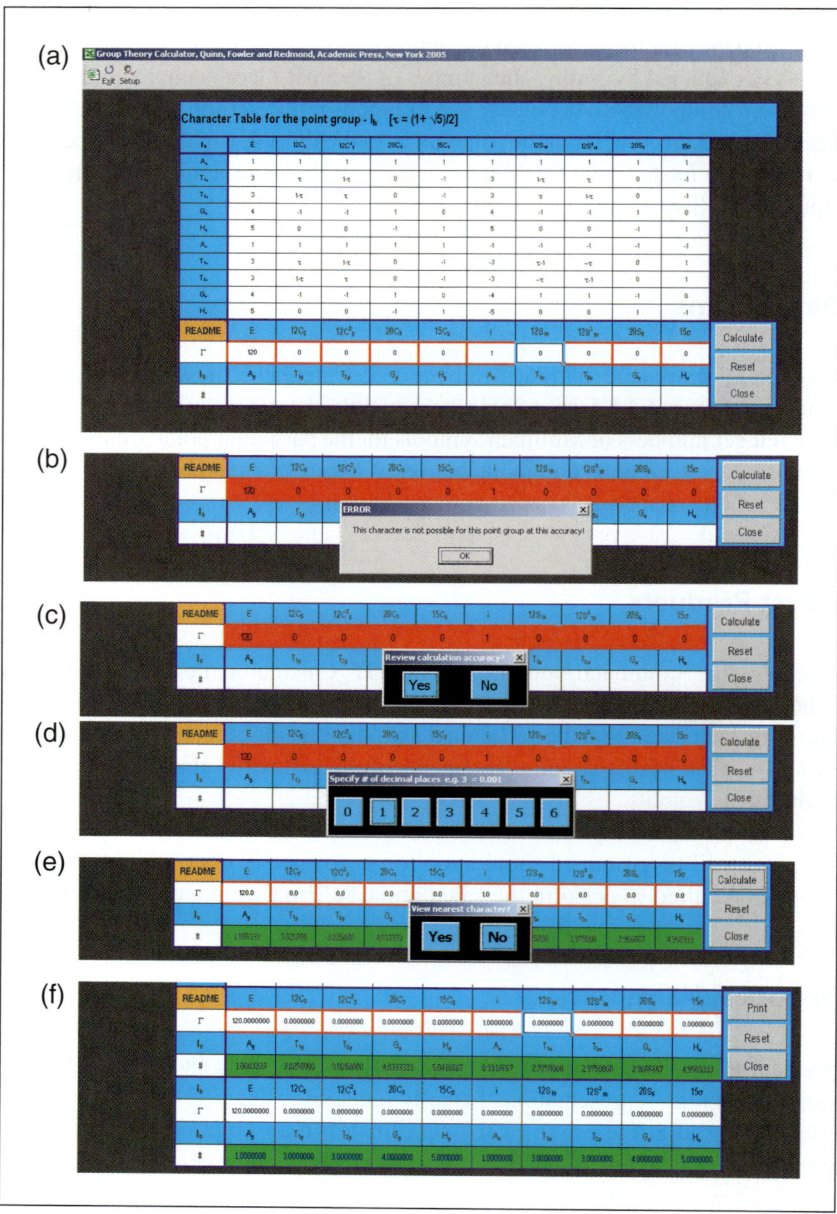


**Figure 1.14** Demonstration of the operation of the calculator for an incorrect character input.

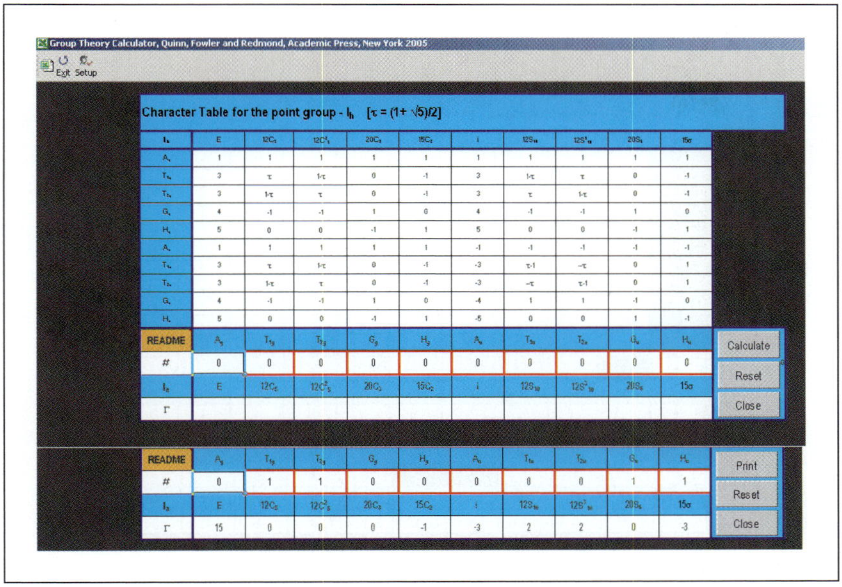

Group Theory Calculator, Quinn, Fowler and Redmond, Academic Press, New York 2005

Exit Setup

Character Table for the point group - $I_h$ [$\tau = (1+\sqrt{5})/2$]

| $I_h$ | E | $12C_5$ | $12C_5^2$ | $20C_3$ | $15C_2$ | i | $12S_{10}$ | $12S_{10}^3$ | $20S_6$ | $15\sigma$ |
|---|---|---|---|---|---|---|---|---|---|---|
| $A_g$ | 1 | 1 | 1 | 1 | 1 | 1 | 1 | 1 | 1 | 1 |
| $T_{1g}$ | 3 | τ | 1-τ | 0 | -1 | 3 | 1-τ | τ | 0 | -1 |
| $T_{2g}$ | 3 | 1-τ | τ | 0 | -1 | 3 | τ | 1-τ | 0 | -1 |
| $G_g$ | 4 | -1 | -1 | 1 | 0 | 4 | -1 | -1 | 1 | 0 |
| $H_g$ | 5 | 0 | 0 | -1 | 1 | 5 | 0 | 0 | -1 | 1 |
| $A_u$ | 1 | 1 | 1 | 1 | 1 | -1 | -1 | -1 | -1 | -1 |
| $T_{1u}$ | 3 | τ | 1-τ | 0 | -1 | -3 | τ-1 | -τ | 0 | 1 |
| $T_{2u}$ | 3 | 1-τ | τ | 0 | -1 | -3 | -τ | τ-1 | 0 | 1 |
| $G_u$ | 4 | -1 | -1 | 1 | 0 | -4 | 1 | 1 | -1 | 0 |
| $H_u$ | 5 | 0 | 0 | -1 | 1 | -5 | 0 | 0 | 1 | -1 |

| README | $A_g$ | $T_{1g}$ | $T_{2g}$ | $G_g$ | $H_g$ | $A_u$ | $T_{1u}$ | $T_{2u}$ | $G_u$ | $H_u$ |
|---|---|---|---|---|---|---|---|---|---|---|
| # | 0 | 0 | 0 | 0 | 0 | 0 | 0 | 0 | 0 | 0 |
| $I_h$ | E | $12C_5$ | $12C_5^2$ | $20C_3$ | $15C_2$ | i | $12S_{10}$ | $12S_{10}^3$ | $20S_6$ | $15\sigma$ |
| Γ | | | | | | | | | | |

Calculate Reset Close

| README | $A_g$ | $T_{1g}$ | $T_{2g}$ | $G_g$ | $H_g$ | $A_u$ | $T_{1u}$ | $T_{2u}$ | $G_u$ | $H_u$ |
|---|---|---|---|---|---|---|---|---|---|---|
| # | 0 | 1 | 1 | 0 | 0 | 0 | 0 | 0 | 1 | 1 |
| $I_h$ | E | $12C_5$ | $12C_5^2$ | $20C_3$ | $15C_2$ | i | $12S_{10}$ | $12S_{10}^3$ | $20S_6$ | $15\sigma$ |
| Γ | 15 | 0 | 0 | 0 | -1 | -3 | 2 | 2 | 0 | -3 |

Print Reset Close

**Figure 1.15** Calculation of the reducible character of a direct sum.

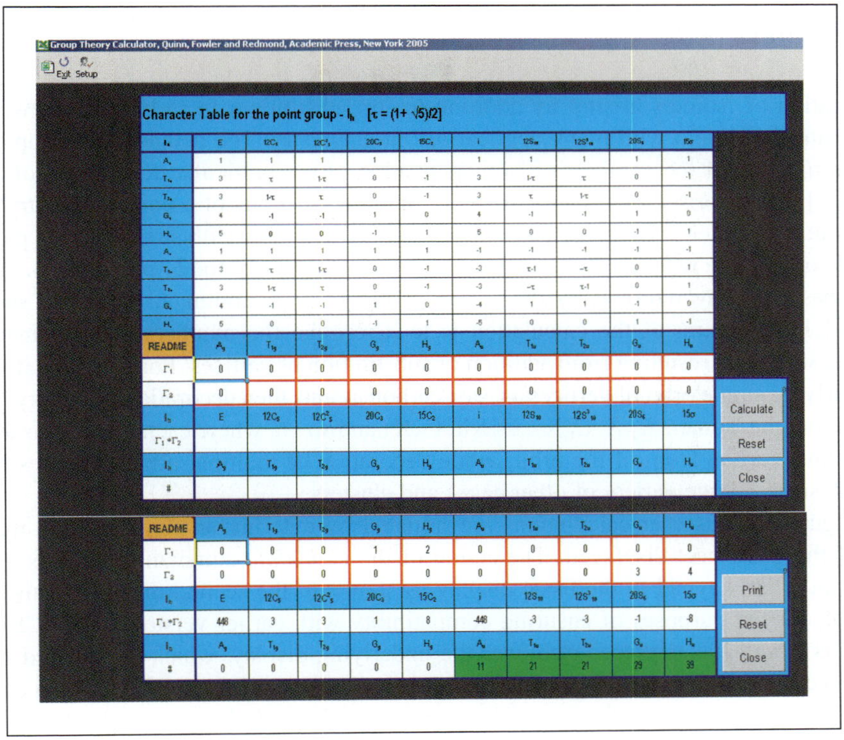

Group Theory Calculator, Quinn, Fowler and Redmond, Academic Press, New York 2005

Exit Setup

Character Table for the point group - $I_h$ [$\tau = (1+\sqrt{5})/2$]

| $I_h$ | E | $12C_5$ | $12C_5^2$ | $20C_3$ | $15C_2$ | i | $12S_{10}$ | $12S_{10}^3$ | $20S_6$ | $15\sigma$ |
|---|---|---|---|---|---|---|---|---|---|---|
| $A_g$ | 1 | 1 | 1 | 1 | 1 | 1 | 1 | 1 | 1 | 1 |
| $T_{1g}$ | 3 | τ | 1-τ | 0 | -1 | 3 | 1-τ | τ | 0 | -1 |
| $T_{2g}$ | 3 | 1-τ | τ | 0 | -1 | 3 | τ | 1-τ | 0 | -1 |
| $G_g$ | 4 | -1 | -1 | 1 | 0 | 4 | -1 | -1 | 1 | 0 |
| $H_g$ | 5 | 0 | 0 | -1 | 1 | 5 | 0 | 0 | -1 | 1 |
| $A_u$ | 1 | 1 | 1 | 1 | 1 | -1 | -1 | -1 | -1 | -1 |
| $T_{1u}$ | 3 | τ | 1-τ | 0 | -1 | -3 | τ-1 | -τ | 0 | 1 |
| $T_{2u}$ | 3 | 1-τ | τ | 0 | -1 | -3 | -τ | τ-1 | 0 | 1 |
| $G_u$ | 4 | -1 | -1 | 1 | 0 | -4 | 1 | 1 | -1 | 0 |
| $H_u$ | 5 | 0 | 0 | -1 | 1 | -5 | 0 | 0 | 1 | -1 |

| README | $A_g$ | $T_{1g}$ | $T_{2g}$ | $G_g$ | $H_g$ | $A_u$ | $T_{1u}$ | $T_{2u}$ | $G_u$ | $H_u$ |
|---|---|---|---|---|---|---|---|---|---|---|
| $\Gamma_1$ | 0 | 0 | 0 | 0 | 0 | 0 | 0 | 0 | 0 | 0 |
| $\Gamma_2$ | 0 | 0 | 0 | 0 | 0 | 0 | 0 | 0 | 0 | 0 |
| $I_h$ | E | $12C_5$ | $12C_5^2$ | $20C_3$ | $15C_2$ | i | $12S_{10}$ | $12S_{10}^3$ | $20S_6$ | $15\sigma$ |
| $\Gamma_1 * \Gamma_2$ | | | | | | | | | | |
| $I_h$ | $A_g$ | $T_{1g}$ | $T_{2g}$ | $G_g$ | $H_g$ | $A_u$ | $T_{1u}$ | $T_{2u}$ | $G_u$ | $H_u$ |
| # | | | | | | | | | | |

Calculate Reset Close

| README | $A_g$ | $T_{1g}$ | $T_{2g}$ | $G_g$ | $H_g$ | $A_u$ | $T_{1u}$ | $T_{2u}$ | $G_u$ | $H_u$ |
|---|---|---|---|---|---|---|---|---|---|---|
| $\Gamma_1$ | 0 | 0 | 0 | 1 | 2 | 0 | 0 | 0 | 0 | 0 |
| $\Gamma_2$ | 0 | 0 | 0 | 0 | 0 | 0 | 0 | 0 | 3 | 4 |
| $I_h$ | E | $12C_5$ | $12C_5^2$ | $20C_3$ | $15C_2$ | i | $12S_{10}$ | $12S_{10}^3$ | $20S_6$ | $15\sigma$ |
| $\Gamma_1 * \Gamma_2$ | 448 | 3 | 3 | 1 | 8 | -448 | -3 | -3 | -1 | -8 |
| $I_h$ | $A_g$ | $T_{1g}$ | $T_{2g}$ | $G_g$ | $H_g$ | $A_u$ | $T_{1u}$ | $T_{2u}$ | $G_u$ | $H_u$ |
| # | 0 | 0 | 0 | 0 | 0 | 11 | 21 | 21 | 29 | 39 |

Print Reset Close

**Figure 1.16** Calculation of the *Direct Product* of the reducible characters of two direct sums.

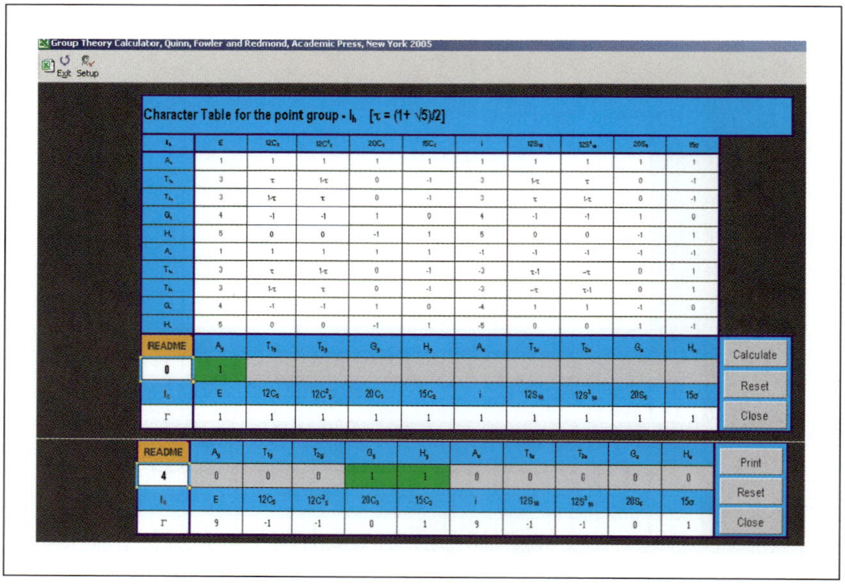

Group Theory Calculator, Quinn, Fowler and Redmond, Academic Press, New York 2005

Exit Setup

Character Table for the point group - $I_h$ [$\tau = (1+\sqrt{5})/2$]

| $I_h$ | E | $12C_5$ | $12C_5^2$ | $20C_3$ | $15C_2$ | i | $12S_{10}$ | $12S_{10}^3$ | $20S_6$ | $15\sigma$ |
|---|---|---|---|---|---|---|---|---|---|---|
| $A_g$ | 1 | 1 | 1 | 1 | 1 | 1 | 1 | 1 | 1 | 1 |
| $T_{1g}$ | 3 | $\tau$ | $1-\tau$ | 0 | -1 | 3 | $1-\tau$ | $\tau$ | 0 | -1 |
| $T_{2g}$ | 3 | $1-\tau$ | $\tau$ | 0 | -1 | 3 | $\tau$ | $1-\tau$ | 0 | -1 |
| $G_g$ | 4 | -1 | -1 | 1 | 0 | 4 | -1 | -1 | 1 | 0 |
| $H_g$ | 5 | 0 | 0 | -1 | 1 | 5 | 0 | 0 | -1 | 1 |
| $A_u$ | 1 | 1 | 1 | 1 | 1 | -1 | -1 | -1 | -1 | -1 |
| $T_{1u}$ | 3 | $\tau$ | $1-\tau$ | 0 | -1 | -3 | $\tau-1$ | $-\tau$ | 0 | 1 |
| $T_{2u}$ | 3 | $1-\tau$ | $\tau$ | 0 | -1 | -3 | $-\tau$ | $\tau-1$ | 0 | 1 |
| $G_u$ | 4 | -1 | -1 | 1 | 0 | -4 | 1 | 1 | -1 | 0 |
| $H_u$ | 5 | 0 | 0 | -1 | 1 | -5 | 0 | 0 | 1 | -1 |

| README | $A_g$ | $T_{1g}$ | $T_{2g}$ | $G_g$ | $H_g$ | $A_u$ | $T_{1u}$ | $T_{2u}$ | $G_u$ | $H_u$ |
|---|---|---|---|---|---|---|---|---|---|---|
| 0 | 1 | | | | | | | | | |
| $I_h$ | E | $12C_5$ | $12C_5^2$ | $20C_3$ | $15C_2$ | i | $12S_{10}$ | $12S_{10}^3$ | $20S_6$ | $15\sigma$ |
| Γ | 1 | 1 | 1 | 1 | 1 | 1 | 1 | 1 | 1 | 1 |

Calculate Reset Close

| README | $A_g$ | $T_{1g}$ | $T_{2g}$ | $G_g$ | $H_g$ | $A_u$ | $T_{1u}$ | $T_{2u}$ | $G_u$ | $H_u$ |
|---|---|---|---|---|---|---|---|---|---|---|
| 4 | 0 | 0 | 0 | 1 | 1 | 0 | 0 | 0 | 0 | 0 |
| $I_h$ | E | $12C_5$ | $12C_5^2$ | $20C_3$ | $15C_2$ | i | $12S_{10}$ | $12S_{10}^3$ | $20S_6$ | $15\sigma$ |
| Γ | 9 | -1 | -1 | 0 | 1 | 9 | -1 | -1 | 0 | 1 |

Print Reset Close

**Figure 1.17** An example of the use of the *Spherical Harmonics* command button. The result shows that the central 'g' harmonics transform as $G_g$ and $H_g$ in $I_h$ symmetry.

## 1.8 Isomers

Enumeration of isomers arising by addition to, or substitution in, a basis framework is a mathematical problem with many applications in chemistry. The details of our approach to the identification of the numbers of isomers resulting from particular decorations of a given structure [i.e. comprised of a single orbit or a number of orbits of a point symmetry group] are set out in Chapter 4. Here, the basic operating instructions are illustrated for the particular example of a single regular orbit and then a mixture of orbits of the $I_h$ point group.

The basic display returned by the action of the *Isomers* command button is shown in Figure 1.18. The aim of the calculation is to determine the numbers and symmetries of isomeric structures resulting from addition or substitution of vertices. For example, if we take an underlying single orbit of vertices as, say, carbon atoms, then decorations x, xx, xy, xxx . . . tell us about the isomers $C_nX$, $C_nX_2$, $C_nXY$, $C_nX_3$ and so on, where each X, Y, . . . is attached to a distinct carbon atom of the orbit, while decoration of multiple-orbit structures extends the analysis to a large variety of other cages and clusters.

In Figure 1.18, isomer calculations are initiated by entering an orbit list into the input cells of the worksheet displayed. The action of the *Calculate* command button is to add a 'Decoration list' keypad to the basic display as shown in the second diagram in the figure. As usual, for the purposes of ensuring compatibility with earlier versions of EXCEL®, the keypad is *modal* [takes precedence over the underlying display, which is rendered inactive and inaccessible until control is transferred back from the active modal form] and a choice of decoration must be made before control is returned to the underlying worksheet.

Figure 1.19 displays the remaining sequence leading to the final result window of the isomer calculation for simple decorations or a relatively small number of vertices for which a

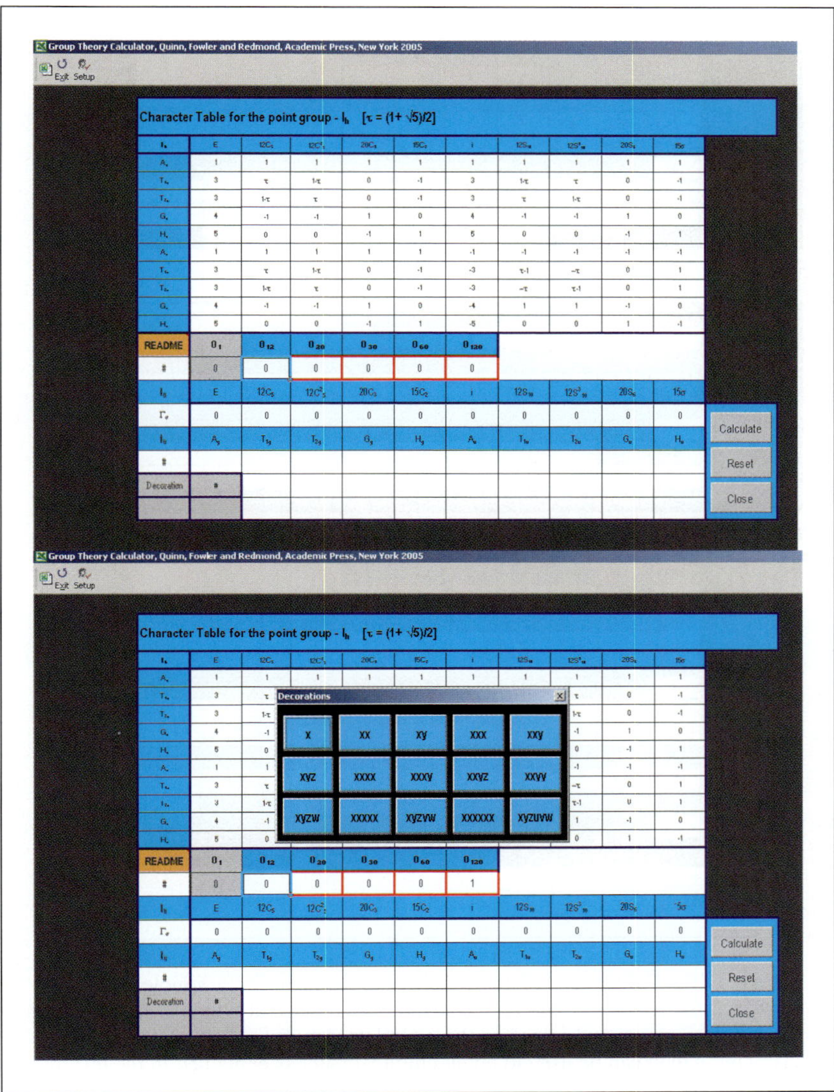


**Figure 1.18** The initial display returned by the action of the *Isomers* command button of the GT_Calculator followed by the interrogation sequence in which the decoration pattern is chosen, here for the 120-vertex cage, the great rhombicosadodecahedron.

decoration has been specified. In Figure 1.19a, the result for the choice of *triple replacements* in a single regular orbit of the $I_h$ point group is shown. The regular orbit character is displayed together with the list of its direct sum components identified by their Mulliken symbols. For the choice of the 'xxx' decoration pattern, the number of isomers is returned as 2347. This result indicates that this decoration of a $C_{120}$ cage, with three addends on distinct centres, leads to a total of 2347 different $C_{120}X_3$ structures, counting each enantiomeric pair as two structures.

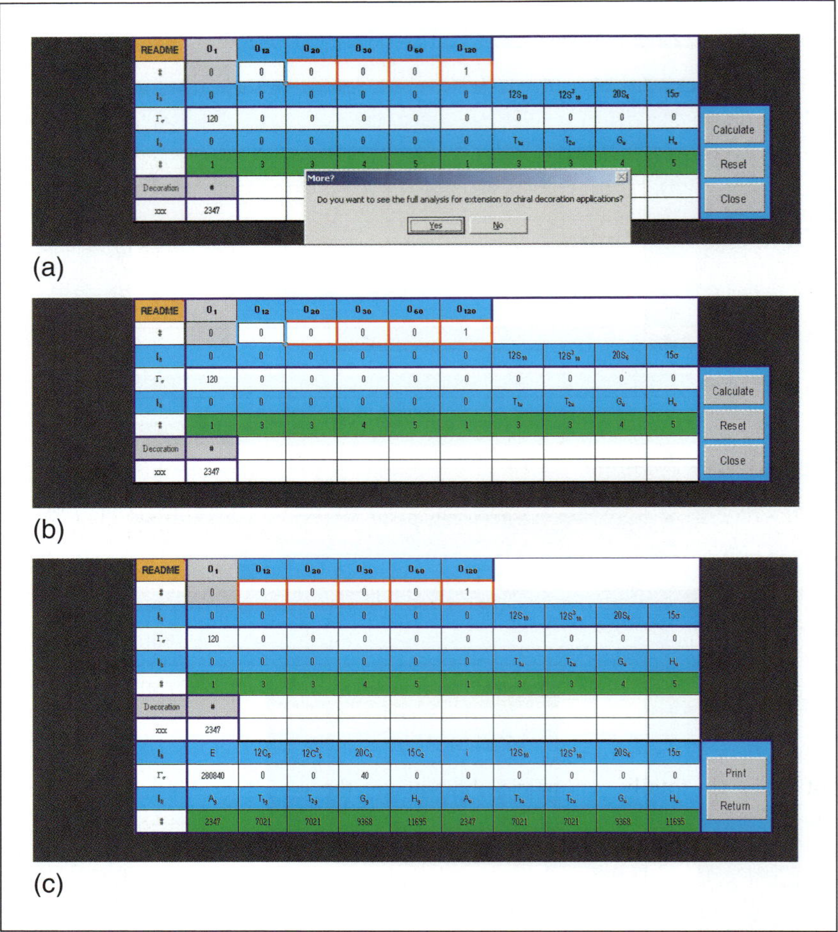


**Figure 1.19** Calculation of the numbers of isomers for a simple decoration for which the result still can be displayed on the original window, in (b) as the number of isomers and in (c) as the complete decomposition of the reducible character, which can be used to count chiral isomers as explained in the text and in Chapter 4.

There is further potential in this calculation. You can choose to see the whole of the direct sum, Figure 1.19c, corresponding to the character of the representation generated by the decoration, and so distinguish between the numbers of chiral and achiral structures resulting from a given decoration of the vertices of the original structure.

The number of chiral pairs, for a given decoration pattern, the number of times the determinantally antisymmetric character[4] appears in the decoration character. From this result the number of achiral structures is obtained as the difference between the total number of decoration isomers possible and the number of chiral pairs. The total number of isomers is the

[4]The totally antisymmetric character has the trace +1 for all proper rotations and trace −1 for all improper rotations.

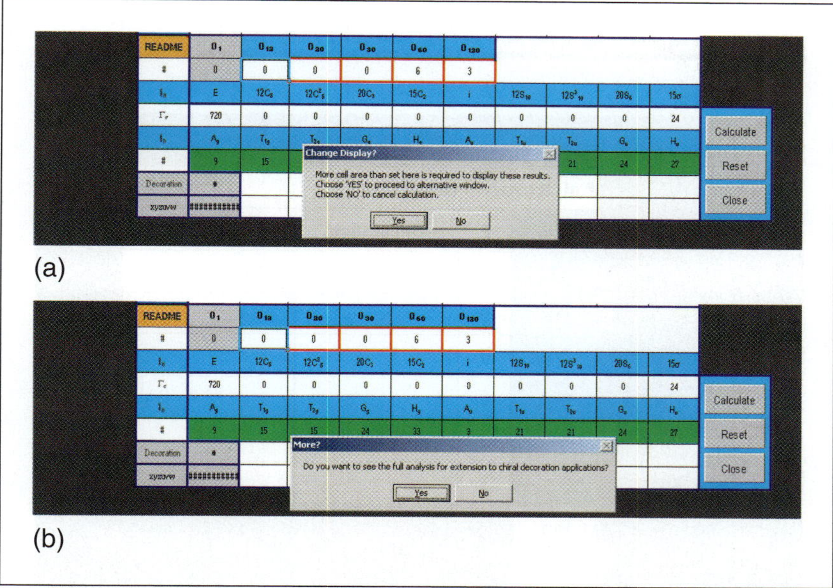


**Figure 1.20** For calculations leading to large numbers of decorations a different display is appropriate. This sequence can lead to a final display of the total number of isomers for a given decoration choice and input orbit list as in Figure 1.20. The alternative full analysis is presented in Figure 1.21.

number of times the totally symmetric character of the group appears and this is the result visible, Figure 1.19b display, when the response 'no' is offered on that display.

Because very large numbers of possible isomers arise for complicated decorations of even basic structures exhibiting relative few vertices or for any decoration of structures with a large number of vertices, the display formats of Figures 1.18 and 1.19 are not appropriate. A different format is used to display the results when very large integer numbers arise.

Figure 1.20 illustrates the alternative sequence of displays for such cases, the example of a $C_{720}$ fullerene cage formed as $6 \times C_{60}$ and $3 \times C_{120}$ orbits with a decoration pattern corresponding to six different decoration elements xyzuvw i.e. six different atoms on six distinct positions on the $C_{720}$ fullerene cage.

Then, in Figure 1.21 the large-format result is displayed as the number of totally symmetric components of the decoration direct sum, while Figure 1.22 is the more detailed display for chiral applications, in this example, the full list of direct sum components in the decoration character of the $C_{720}$-vertexed structure, of $I_h$ point symmetry, subject to xyzuvw decoration.

## 1.9 Symmetric and Antisymmetric Powers

The symmetric and antisymmetric powers of group representations have been identified as important in the analysis of several physical problems subject to group theoretical algebra since the appearance of the classic paper by Tisza[5].

[5] L. Tisza, *Z. Physik,* **82** (1933) 48.

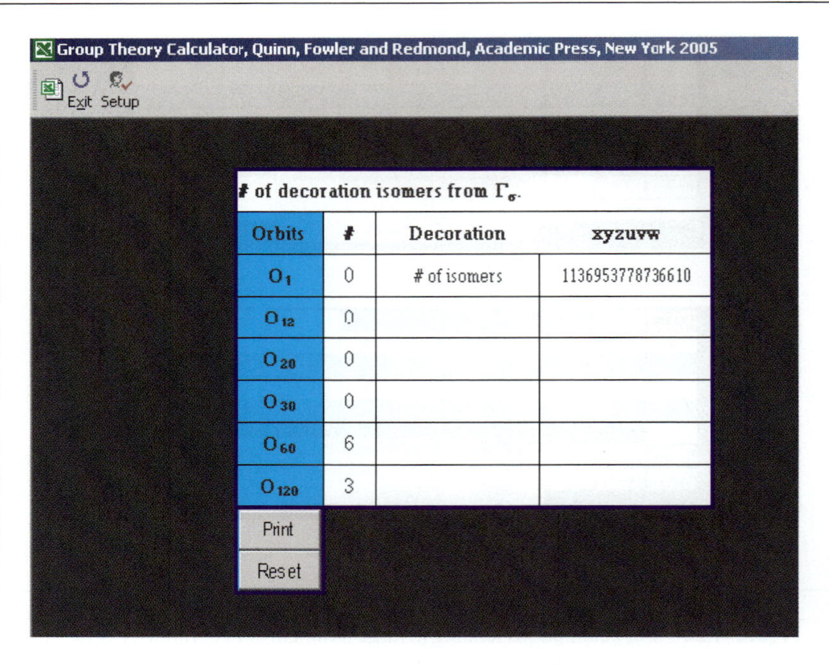


| Orbits | # | Decoration | xyzuvw |
|---|---|---|---|
| $O_1$ | 0 | # of isomers | 1136953778736610 |
| $O_{12}$ | 0 | | |
| $O_{20}$ | 0 | | |
| $O_{30}$ | 0 | | |
| $O_{60}$ | 6 | | |
| $O_{120}$ | 3 | | |

**Figure 1.21** For decoration calculations in which a large number of isomers are found, this alternative window is used to display the result in a larger format.

The most familiar example is of symmetric and antisymmetric squares of Cartesian coordinates x, y and z. The symmetric square of (x, y, z) is the set of products xx, yy, zz, xz, yz and xy, which describe, for example, polarizabilities. The antisymmetric components [yz′ − zy′], [xz′ − zx′], [xy′ − yx′], describe components of rotations $R_x$, $R_y$ and $R_z$.

The GT_Calculator includes options for the calculation of the symmetric and antisymmetric powers in the range 1 to 6 for a character input as a direct sum. Since the operation of the calculator for both applications is identical, only the instructions for the determination of symmetric powers is given in this section.

The initial display returned by the *Symmetric Powers* command button is shown in Figure 1.23. Input to initiate a calculation is in the form of a direct sum characterized by the numbers of irreducible components identified by their Mulliken symbols.

For the standard test case of the $I_h$ point group and the GT_Calculator file Ih.xls, Figure 1.23a displays the input needed to carry out the powers calculations for $1 \times H_g$. The symmetric square is calculated according to the character relation (see equation 4.9),

$$\chi_{[2]}(R) = \frac{1}{2}\left\{\chi^2(R) + \chi(R^2)\right\}$$

with $R^2$ a symmetry operation of the icosahedral group. Thus, in Figure 1.23b, we find that the reducible character of the symmetric square of $h_g$ is the direct sum character of $1A_g$, $1G_g$ and $2H_g$.

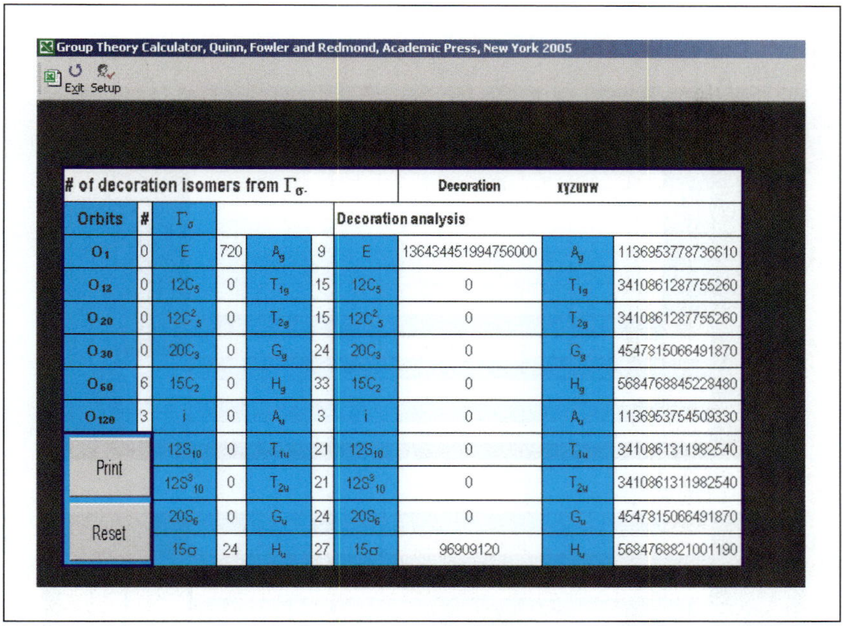

Group Theory Calculator, Quinn, Fowler and Redmond, Academic Press, New York 2005

Exit Setup

| # of decoration isomers from $\Gamma_\sigma$. | | | | | | Decoration | xyzuvw | | |
|---|---|---|---|---|---|---|---|---|---|
| Orbits | # | $\Gamma_\sigma$ | | | | Decoration analysis | | | |
| $O_1$ | 0 | E | 720 | $A_g$ | 9 | E | 136434451994756000 | $A_g$ | 1136953778736610 |
| $O_{12}$ | 0 | $12C_5$ | 0 | $T_{1g}$ | 15 | $12C_5$ | 0 | $T_{1g}$ | 3410861287755260 |
| $O_{20}$ | 0 | $12C^2_5$ | 0 | $T_{2g}$ | 15 | $12C^2_5$ | 0 | $T_{2g}$ | 3410861287755260 |
| $O_{30}$ | 0 | $20C_3$ | 0 | $G_g$ | 24 | $20C_3$ | 0 | $G_g$ | 4547815066491870 |
| $O_{60}$ | 6 | $15C_2$ | 0 | $H_g$ | 33 | $15C_2$ | 0 | $H_g$ | 5684768845228480 |
| $O_{120}$ | 3 | i | 0 | $A_u$ | 3 | i | 0 | $A_u$ | 1136953754509330 |
| Print | | $12S_{10}$ | 0 | $T_{1u}$ | 21 | $12S_{10}$ | 0 | $T_{1u}$ | 3410861311982540 |
| | | $12S^3_{10}$ | 0 | $T_{2u}$ | 21 | $12S^3_{10}$ | 0 | $T_{2u}$ | 3410861311982540 |
| Reset | | $20S_6$ | 0 | $G_u$ | 24 | $20S_6$ | 0 | $G_u$ | 4547815066491870 |
| | | $15\sigma$ | 24 | $H_u$ | 27 | $15\sigma$ | 96909120 | $H_u$ | 5684768821001190 |

**Figure 1.22** The complete list of components in the direct sum of the decoration character for a large structure, $C_{720}$, subject to a six-fold decoration, xyzuvw.

The traces of the antisymmetric square character for the same input follow from the relation (see equation 4.13),

$$\chi_{\{2\}}(R) = \frac{1}{2}\left\{\chi^2(R) - \chi(R^2)\right\}$$

and this result is presented in Figure 1.22c, together with the direct sum decomposition and we note that the sum of the dimensions of the symmetric and antisymmetric squares is 25, which is the expected result of the taking of the square.

The other relations leading to higher-order symmetric and antisymmetric squares are discussed in Chapter 4. For example, for the regular character of $I_h$, the sixth symmetric power, Figure 1.24, describes the symmetry spanned by all the symmetric sixth powers of a set of objects transforming as the regular orbit, i.e. of the 120 radial displacements of the 120 vertices. An application of this calculation would occur in setting up a model sextic force field for a $C_{120}$ molecule.

It is presumed in the calculations for symmetric and antisymmetric powers that sufficient cell areas are available to display the results of most calculations of this kind without the need for alternative displays. Difficulties with resolution can be remedied using the Zoom command, accessible via the *Setup* subsidiary command bar.

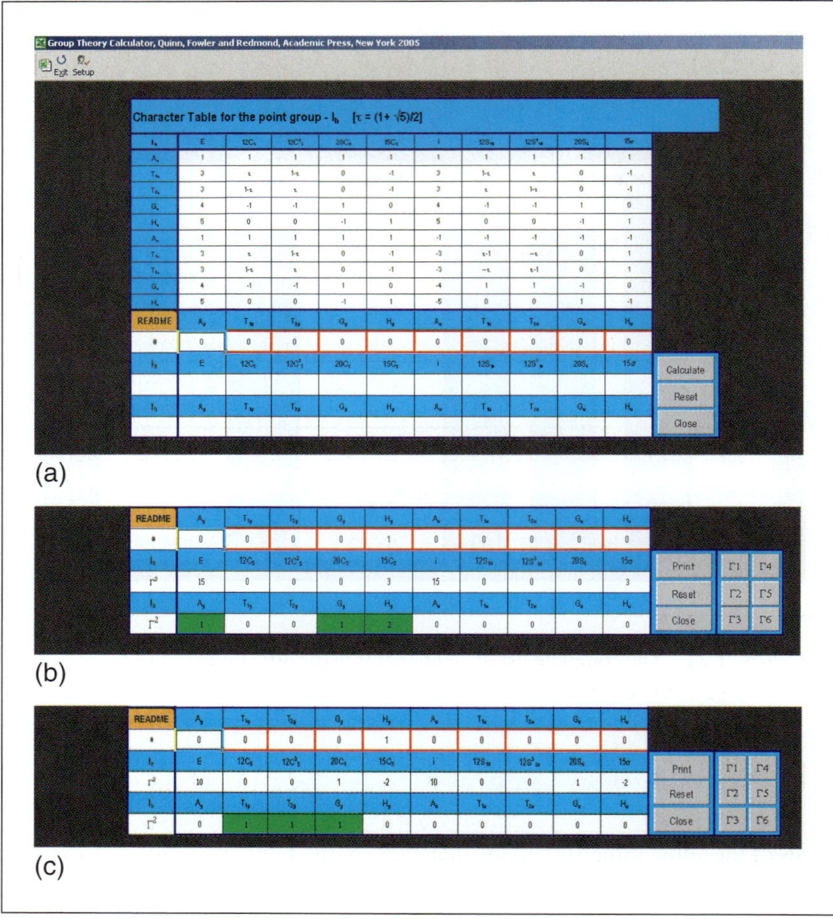

**Figure 1.23** The sequence of worksheet displays in the calculation of the square symmetric and antisymmetric powers of the $1H_g$ irreducible character of $I_h$. Note that Figure 1.23c is generated from the action of the antisymmetric powers command button in the options menu displayed in Figure 1.6b.

## 1.10 Basis Functions

The command button labelled *Basis Functions* leads, for each GT_Calculator file, to a keypad from which all the polynomial functions required to provide basis functions for all the irreducible components of the regular representation of each point group can be displayed. For the example of the Ih.xls file, the action of the *Basis Function* command button is shown in Figure 1.25.

As Chapter 3 describes, the basis functions for each irreducible representation are limited in number. Basis functions are polynomial functions with specific behaviour under symmetry operations. Thus, in $I_h$, the set x, y, z transforms as components of the $T_{1u}$ representation. It is useful to have explicit basis functions to display the properties of the representations,

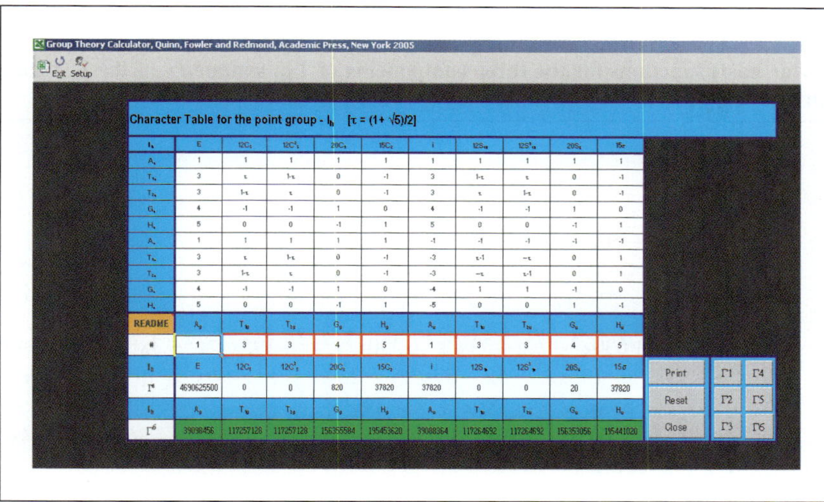


**Figure 1.24** The Calculator result for the 6[th] symmetric power of the regular representation for the 120-vertex cage of $I_h$ point symmetry.

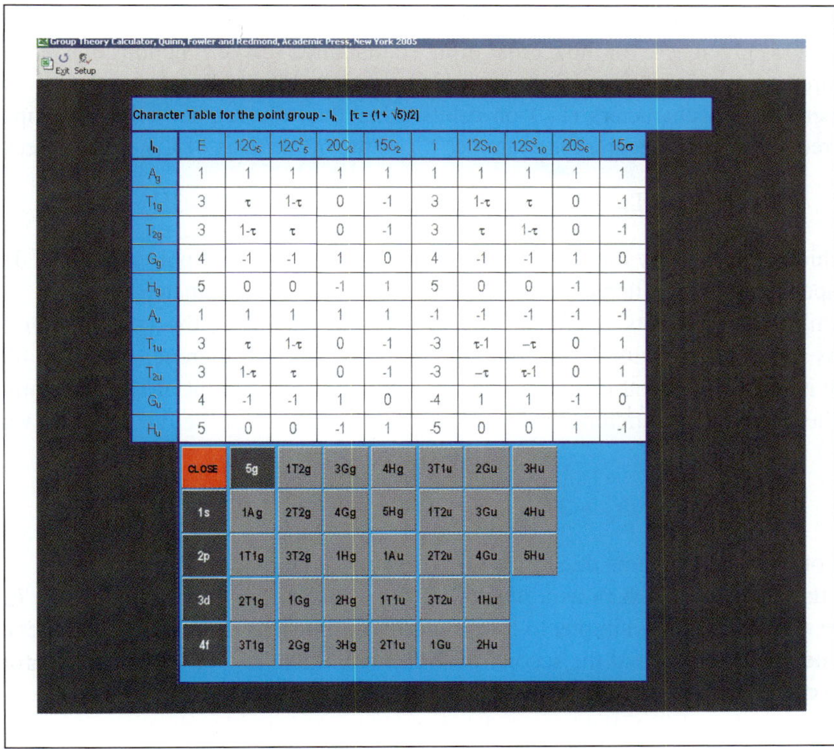


**Figure 1.25** The keypad displayed by the action of the *Basis Function* command button for the example of the Ih.xls file. The regular character of a group decomposes to a direct sum of 1 of every 1D character, 2 of every 2D character and so on.

**Table 1.1 The three sets of mutually-orthonormal polynomial functions required to provide basis functions for the three components of $T_{1u}$ symmetry in the $I_h$ regular representation.**

| | |
|---|---|
| $1T_{1u}$ | x<br>y<br>z |
| $2T_{1u}$ | $126z^5 - 140z^3 + 30z + 7x^5 - 70x^3y^2 + 35xy^4$<br>$12x^2z^3 - 12y^2z^3 - 4x^2z + 9x^3z^2 - 27xy^2z^2 - x^3 + 3xy^2$<br>$24xyz^3 - 8xyz - 27x^2yz^2 + 9y^3z^2 + 3x^2y - y^3$ |
| $3T_{1u}$ | $858z^7 - 1386z^5 + 630z^3 - 70z + 39x^5z^2 - 390x^3y^2z^2 + 195xy^4z^2 - 3x^5 + 30x^3y^2 - 15xy^4$<br>$429xz^6 - 495xz^4 + 135xz^2 - 5x - 39x^4z^3 + 234x^2y^2z^3 - 39y^4z^3 + 9x^4z - 54x^2y^2z + 9y^4z + 39x^6z - 585x^4y^2z + 585x^2y^4z - 39y^6z$<br>$429yz^6 - 495yz^4 + 135yz^2 - 5y + 156x^3yz^3 - 156xy^3z^3 - 36x^3yz + 36xy^3z + 234x^5yz - 780x^3y^3z + 234xy^5z$ |

illustrate nodal properties of molecular orbitals, etc. It turns out that the maximum number of independent functions needed in any problem is g sets of g components per orbit, where g is the degeneracy of the irreducible representation. A complete enumeration is therefore possible and is given here for the regular orbit, so that three sets of functions of $T_{1u}$ symmetry are required, {x, y, z} as $1T_{1u}$ and the two other sets listed in Table 1.1.

The permutation characters ($\Gamma_\sigma$) on orbits other than the regular orbit decompose into fewer irreducible components, for example, $\Gamma_\sigma$ of the 60 vertices of $C_{60}$ is the direct sum

$$\Gamma_\sigma = A_g + T_{1g} + 2T_{1u} + T_{2g} + 2T_{2u} + 2G_g + 2G_u + 3H_g + 2H_u$$

result which requires only the first two of the possible three sets of polynomials of Table 1.1 for the specification of suitable basis functions exhibiting $T_{1u}$ symmetry.

The utility of such lists is emphasised in a further example, the construction of basis functions transforming with $T_{1u}$ symmetry for the two orbit fullerene, $C_{80}$. The 80-vertex structure is formed by combination of the $O_{20}$ orbit and an $O_{60}$ orbit of $I_h$ point symmetry to realize the three-valent fullerene. The permutation character over the full set of 80 vertices is the direct sum

$$\Gamma_\sigma = 2A_g + T_{1g} + 3T_{1u} + T_{2g} + 3T_{2u} + 3G_g + 3G_u + 4H_g + 2H_u$$

because of extra components due to the $O_{20}$ orbit permutation character.

Over the 80 vertices, taken as a whole, all three sets of the basis functions in Table 1.1 would be required, but it is simpler to carry out a two-orbit analysis, taking two copies of $1T_{1u}$, the first over 60 vertices and the second over 20, with the third $T_{1u}$ set being the projections of $2T_{1u}$ on the vertices of $O_{60}$.

In operation of the GT_Calculator, a shorthand notation is used to display polynomial functions for all of the basis functions required by the regular representations of the point groups. This notation was suggested by Elert[6]. The polynomial $C_{mnp}x^my^nz^p$ is written $C_{mnp}(mnp)$,

[6] W. Elert, *Z. Physik*, **51** (1928) 8.

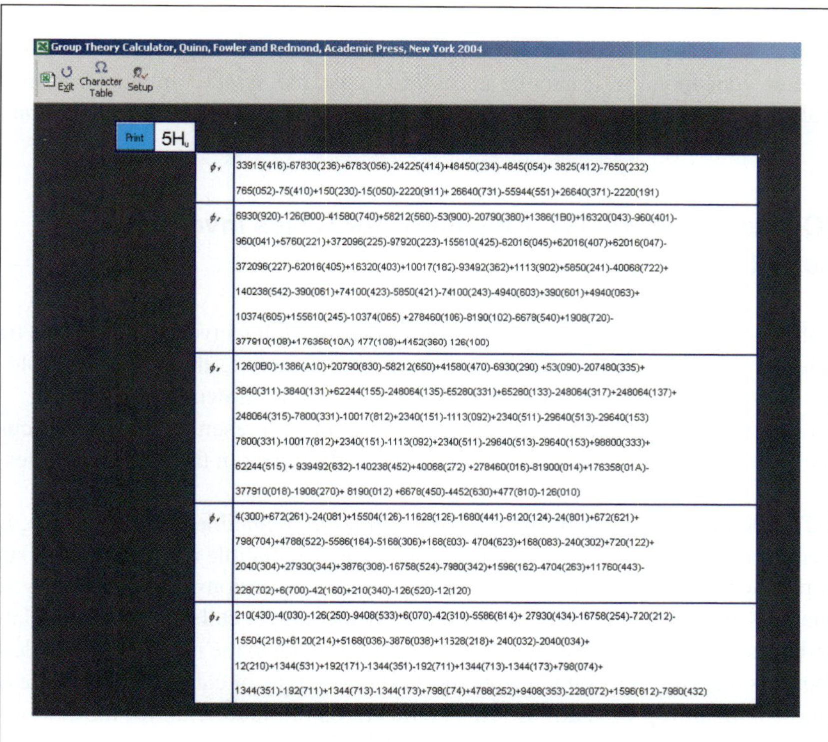


**Figure 1.26** The central polynomial functions of $5H_u$ symmetry, which are orthonormal on the unit sphere and, which along with $1H_u$, $2H_u$, $3H_u$ and $5H_u$ are required to make basis functions for all of the 25 functions of $H_u$ symmetry in the regular representation of $I_h$.

with the additional rule that letter codes are used for powers greater than 9, i.e.

$$m, n, p = 0, 1, 2, 3, 4, 5, 6, 7, 8, 9, A, B, C, \ldots$$

For example, the 5Hu command button on the keypad leads to the display, Figure 1.26, of the $5^{th}$ copy of five polynomial functions, which, on the unit sphere, are mutually orthonormal to one another and to the 20 other polynomial functions of this irreducible symmetry forming the $1^{st}$, $2^{nd}$, $3^{rd}$ and $4^{th}$ sets of functions also of this symmetry, displayed when the other buttons on the keypad labelled with this symmetry are selected by a mouse click. The third function of $5h_u$ irreducible symmetry in Figure 1.26 has leading polynomial terms

$$\phi_3 = 126(0B0) - 1386(A10) + \cdots$$

i.e.

$$\phi_3 = 126y^{11} - 1386x^{10}y + \cdots$$

In the display in Figure 1.26, as usual, a *Print* command button is available so that hard copies of the polynomial functions can be generated on your local printer. Depending on the screen size of your monitor, it may be necessary to use the *Setup* command button, in order to bring all of the basis functions into view using the *Zoom*, scroll and centring controls.

## 1.11 Operation of the GT_Calculator for Cases Involving Complex Algebra

Several of the molecular point groups exhibit character tables involving complex traces for certain classes of rotational symmetry operation. For example, the character table display [*compare Figure 1.4*] for the GT_Calculator file, C3h.xls, is shown in Figure 1.27.

As you can see, the group exhibits separably degenerate representations and it is customary to specify these explicitly, so that complex exponentials appear in the character tables of such groups.

The GT_Calculator has been designed to deal with *real* reducible representations, in which both the components of any given separably degenerate irreducible representations, such as E′ in $C_{2h}$ appear with equal weight. Data input and output conventions ensure that only such real representations are used. Thus, all the relevant worksheet input cells for these calculations are linked and the second member of each linked pair is set equal to the first. For such calculations, the red-borders marking cells on the worksheets for each type of calculation straddle only one each of these pairs of representations. This is illustrated in Figure 1.28 for the example of the construction of the direct sum character by combination of the irreducible characters of the point group $C_{3h}$.

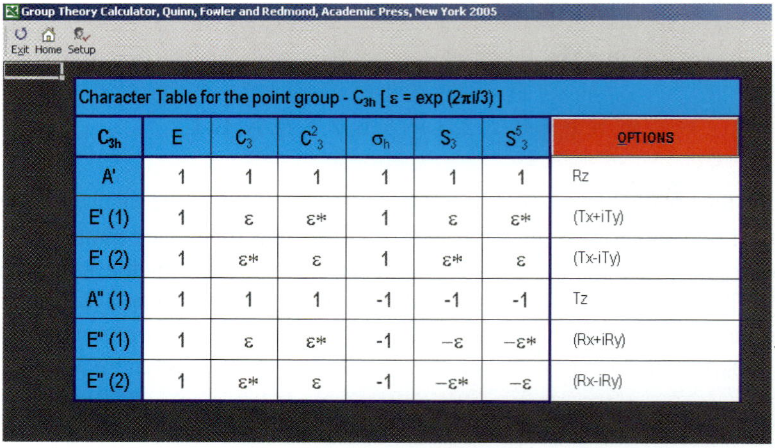

Group Theory Calculator, Quinn, Fowler and Redmond, Academic Press, New York 2005

Exit Home Setup

Character Table for the point group - $C_{3h}$ [ $\varepsilon = \exp(2\pi i/3)$ ]

| $C_{3h}$ | E | $C_3$ | $C_3^2$ | $\sigma_h$ | $S_3$ | $S_3^5$ | OPTIONS |
|---|---|---|---|---|---|---|---|
| A' | 1 | 1 | 1 | 1 | 1 | 1 | Rz |
| E' (1) | 1 | $\varepsilon$ | $\varepsilon^*$ | 1 | $\varepsilon$ | $\varepsilon^*$ | (Tx+iTy) |
| E' (2) | 1 | $\varepsilon^*$ | $\varepsilon$ | 1 | $\varepsilon^*$ | $\varepsilon$ | (Tx-iTy) |
| A" (1) | 1 | 1 | 1 | -1 | -1 | -1 | Tz |
| E" (1) | 1 | $\varepsilon$ | $\varepsilon^*$ | -1 | $-\varepsilon$ | $-\varepsilon^*$ | (Rx+iRy) |
| E" (2) | 1 | $\varepsilon^*$ | $\varepsilon$ | -1 | $-\varepsilon^*$ | $-\varepsilon$ | (Rx-iRy) |

**Figure 1.27** Character table for the point group $C_{3h}$ in which separably degenerate representations occur, with the result that the character table exhibits traces under the three-fold rotations that are complex as indicated in the header title.

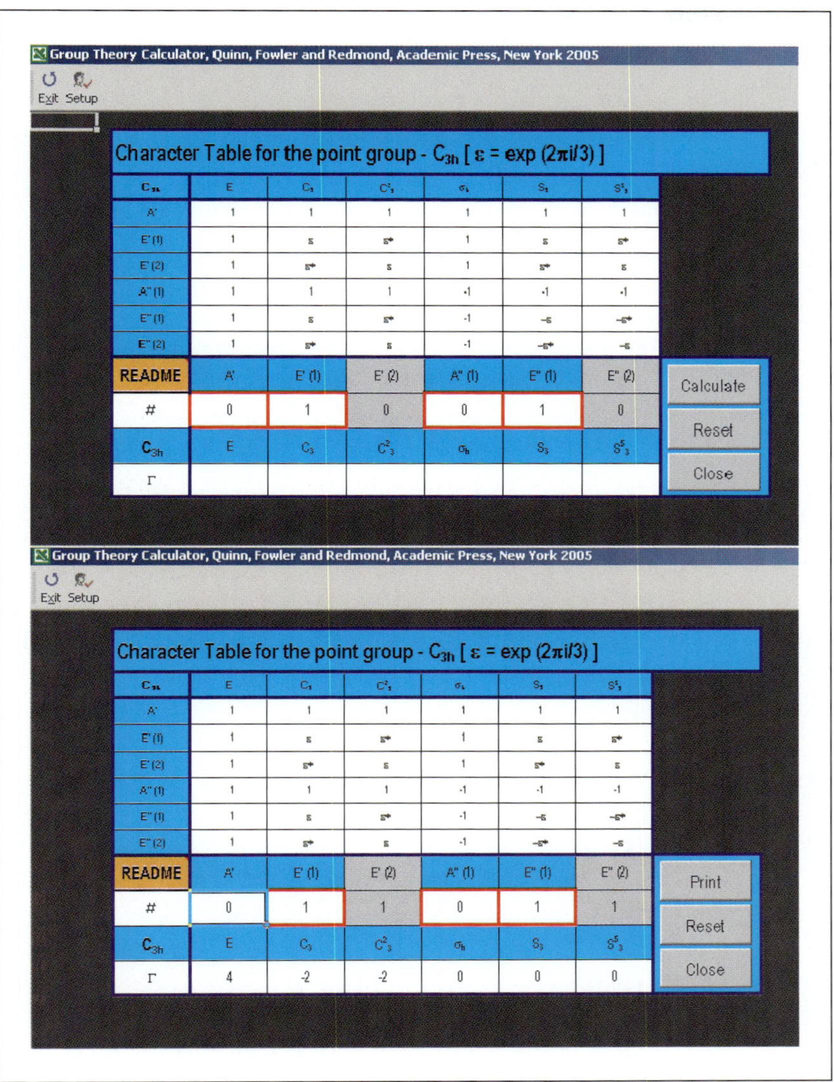

Group Theory Calculator, Quinn, Fowler and Redmond, Academic Press, New York 2005

Exit Setup

Character Table for the point group - $C_{3h}$ [ $\varepsilon = \exp(2\pi i/3)$ ]

| $C_{3h}$ | E | $C_3$ | $C_3^2$ | $\sigma_h$ | $S_3$ | $S_3^5$ |
|---|---|---|---|---|---|---|
| A' | 1 | 1 | 1 | 1 | 1 | 1 |
| E'(1) | 1 | $\varepsilon$ | $\varepsilon^*$ | 1 | $\varepsilon$ | $\varepsilon^*$ |
| E'(2) | 1 | $\varepsilon^*$ | $\varepsilon$ | 1 | $\varepsilon^*$ | $\varepsilon$ |
| A"(1) | 1 | 1 | 1 | -1 | -1 | -1 |
| E"(1) | 1 | $\varepsilon$ | $\varepsilon^*$ | -1 | $-\varepsilon$ | $-\varepsilon^*$ |
| E"(2) | 1 | $\varepsilon^*$ | $\varepsilon$ | -1 | $-\varepsilon^*$ | $-\varepsilon$ |
| README | A' | E'(1) | E'(2) | A"(1) | E"(1) | E"(2) |
| # | 0 | 1 | 0 | 0 | 1 | 0 |
| $C_{3h}$ | E | $C_3$ | $C_3^2$ | $\sigma_h$ | $S_3$ | $S_3^5$ |
| Γ | | | | | | |

Calculate

Reset

Close

Group Theory Calculator, Quinn, Fowler and Redmond, Academic Press, New York 2005

Exit Setup

Character Table for the point group - $C_{3h}$ [ $\varepsilon = \exp(2\pi i/3)$ ]

| $C_{3h}$ | E | $C_3$ | $C_3^2$ | $\sigma_h$ | $S_3$ | $S_3^5$ |
|---|---|---|---|---|---|---|
| A' | 1 | 1 | 1 | 1 | 1 | 1 |
| E'(1) | 1 | $\varepsilon$ | $\varepsilon^*$ | 1 | $\varepsilon$ | $\varepsilon^*$ |
| E'(2) | 1 | $\varepsilon^*$ | $\varepsilon$ | 1 | $\varepsilon^*$ | $\varepsilon$ |
| A"(1) | 1 | 1 | 1 | -1 | -1 | -1 |
| E"(1) | 1 | $\varepsilon$ | $\varepsilon^*$ | -1 | $-\varepsilon$ | $-\varepsilon^*$ |
| E"(2) | 1 | $\varepsilon^*$ | $\varepsilon$ | -1 | $-\varepsilon^*$ | $-\varepsilon$ |
| README | A' | E'(1) | E'(2) | A"(1) | E"(1) | E"(2) |
| # | 0 | 1 | 1 | 0 | 1 | 1 |
| $C_{3h}$ | E | $C_3$ | $C_3^2$ | $\sigma_h$ | $S_3$ | $S_3^5$ |
| Γ | 4 | -2 | -2 | 0 | 0 | 0 |

Print

Reset

Close

**Figure 1.28** An example of the GT_Calculator calculation of direct sums for the case of the irreducible characters of the point group $C_{3h}$, which exhibit complex values. On input only one member of the each pair of separably degenerate irreducible representations can be chosen as it is required that the other is present in equal numbers. This restriction is emphasised in the grey background chosen for these input cells, which are locked to prevent user input. In Figure 1.28b, the input is updated to take account of this and shown in the final display.

The input of the direct sum components is restricted so that integers can be entered into one member of each pair of separably degenerate characters as indicated in the display by the grey shading for the E′(2) and E″(2) possible components of a direct sum. Thus, as in the second worksheet display, the result of a calculation involving the separably degenerate characters of $C_{3h}$ is shown. Both partners of the E′ and E″ representations are involved automatically in the calculation.

# 2

# Geometry, orbits and decorations

It is important to understand the notion of an orbit of structure, for an object exhibiting a particular point symmetry, in order to operate the GT_Calculator.

In this chapter, you will learn:

1. how to recognize the orbits of the different point groups;
2. how these orbits fit into the general geometry of polygons and polyhedra;
3. how the idea of 'decorations' connects all the orbits of one group with the regular orbit of that group and provides a general basis for the understanding of structural chemistry ranging from simple molecules to crystals and complex large molecules.

## 2.1 Structure Orbits

In the present context, we are dealing with molecular and other structures in which sets of points [atoms] are distributed in fixed positions, which definition includes the structures of regular solids and indeed polymers of defined point symmetry. In cases of high symmetry, such structures exhibit a natural centre and the sets of fixed points lie on spheres about this centre.

For such objects, a *structural orbit* [an orbit] is a set of positions related by symmetry operations of a group, G. The effect of any symmetry operation on a single point of the orbit is either to leave it in place or to shift it to another position within the set. Thus, the effect of the operation on the set as a whole is that of a permutation of labels[1]. Where the point group defines a fixed centre, all points of a single orbit lie at the same radial distance from this centre, but the inverse is not true: points at the same radial distance from the centre are not necessarily members of the same orbit[2].

The regular orbit of a point symmetry group is the set of positions for which the only operation that leaves each vertex label invariant is the identity; all other operations permute

[1] We adopt this convention to preserve the notion that a symmetry operation on a structure, within a fixed frame of reference, returns the structure as if no action had been performed. That is to say all the vertices of the structure remain in their initial positions and the action of the symmetry operation is described by the suitable interchange of the labels of the vertices.

[2] For the case of primitive cubic lattice symmetry, distinct orbits can be identified by the solution of the '3-squares' problem that $m^2 + n^2 + p^2$ should be $R^2$, with m, n and p integers. For large R, there are many sets of solutions as m, n and p sets and so there are many distinct orbits of the point group $O_h$ on the single spherical shell defined by R.

labels within the set. From this definition it is easy to see that the size of the regular orbit is equal to the number of operations in the group G. The size of any other orbit is a divisor of the order of G.

The *site group* of a point is the subgroup, H, of operations within G that leave the point invariant. For example, a point may lie on an axis or plane of symmetry and thus be unshifted by the corresponding rotations and reflections. It may lie on all symmetry elements simultaneously and have H = G. In the case of the regular orbit, each point is shifted by all but the identity and so has site group $C_1$. Apart from the trivial single-point orbit, orbits exhibit site groups of type $C_{nv}$ or a subgroup. The product of the size of the orbit and the order of its site group is equal to the order of G.

In the GT_Calculator, orbits of a group are labelled $O_n$, where n is the number of points in the orbit. Where necessary, to distinguish orbits of the same order, a label is added to denote the type of symmetry element on which the points lie. Thus $O_{4h}$, $O_{4d}$, $O_{4v}$ would be orbits of size 4, with their points lying respectively on $\sigma_h$, $\sigma_d$ and $\sigma_v$ mirror planes.

A simple example provides useful illustration of these observations. Consider the $C_{3v}$ point symmetry of the ammonia molecule. The group character table is

| $C_{3v}$ | E | $2C_3$ | $3\sigma_v$ |
|---|---|---|---|
| $A_1$ | 1 | 1 | 1 |
| $A_2$ | 1 | 1 | −1 |
| E | 2 | −1 | 0 |

Using the labels 1, 2 and 3, in white circles, to identify the hydrogen atoms and N in a blue circle to identify the nitrogen, within a fixed frame of reference, then the actions of the group symmetry operations on these labels for the atoms in ammonia are to generate the permutation matrix representation shown in Figure 2.1. The column vector in Figure 2.1 refers to fixed positions. Thus, its 1st row element is the label attaching to the position of N after each symmetry operation; the 2nd row element is the label attaching to the position of the first hydrogen after the each symmetry operation and so on.

The set of matrices are written out again in Table 2.1 in order to emphasize that from the elementary properties of matrices the 4 × 4 permutation matrix representation is simply the direct sum of a 1 × 1 matrix representation on the label of the nitrogen atom alone and a 3 × 3 matrix representation on the labels for the hydrogen atoms taken together. The analysis has identified two *orbits*, the trivial $O_1$ orbit, which is possible in any point group, and the $O_3$ orbit of the point group $C_{3v}$. Only one other orbit is possible for $C_{3v}$ point symmetry structures. This is the regular orbit, $O_6$, generated from a point in a general position (i.e. not lying on any symmetry plane or axis). The six points of the orbit lie at the vertices of a planar hexagon with alternating edge lengths; their plane is not a symmetry plane of $C_{3v}$, and thus the site symmetry is $C_1$.

For any permutation representation the character is list of the traces of the permutation matrices with each trace equal to the number of labels *unshifted* under the corresponding symmetry operation of the group[3]. Thus, for the matrices in Table 2.1, the traces are 4, 1, 1, 2, 2, and 2 respectively, indicating that *every* operation leaves the nitrogen label unshifted

[3] Usually, in chemistry texts, the individual character values are identified as $\chi(R)$ for the character under operation $R$ and the $\chi(R)$ are the traces [*the sums of the diagonal matrix elements*] of the matrix representatives of the symmetry operations.

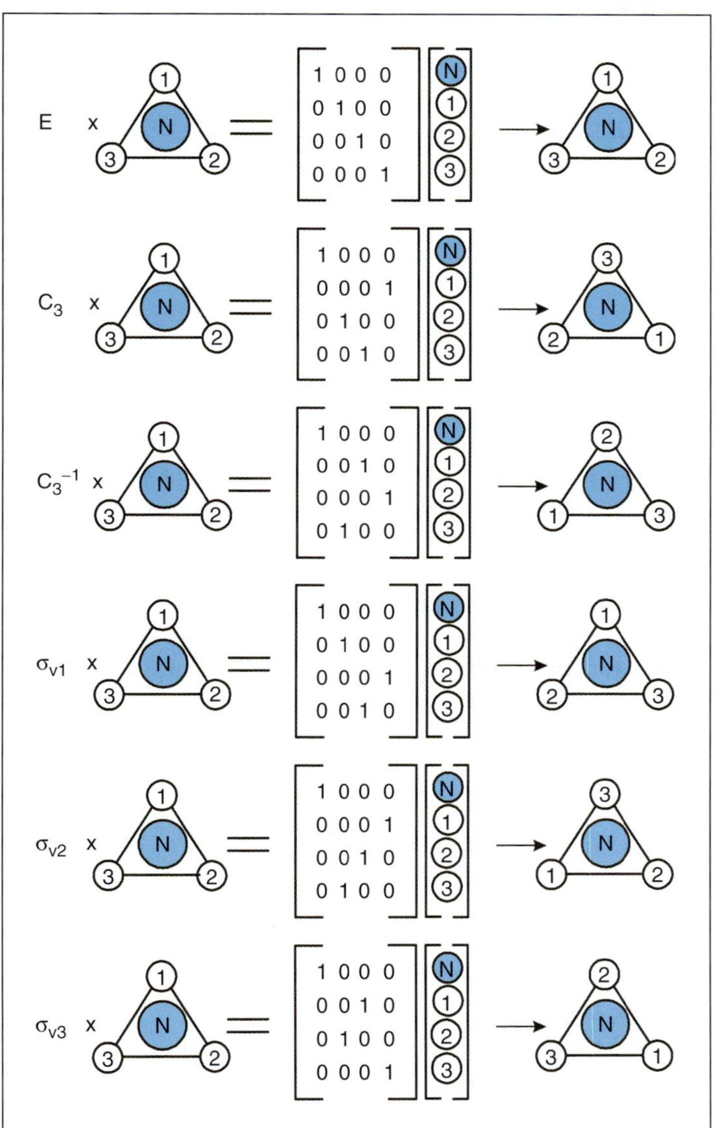


**Figure 2.1** The actions of symmetry operations of the point group, $C_{3v}$, in the structure of the ammonia molecule giving rise to the permutation representation based on the matrices in the second column of the figure. The principal rotational axis, $C_3$, is normal to the plane of the paper.

**Table 2.1 Division of the permutation matrices of Figure 2.1 into their direct sum components as involving the one-dimensional irreducible set exhibiting the invariance of the nitrogen label [or position on the principal rotational axis] and the three-dimensional reducible set over the hydrogen atom labels [or positions of the hydrogen atoms].**

| E | $C_3$ | $C_3^{-1}$ | $\sigma_{v1}$ | $\sigma_{v2}$ | $\sigma_{v3}$ |
|---|---|---|---|---|---|
| $\left[\begin{array}{c\|ccc}1&0&0&0\\\hline 0&1&0&0\\0&0&1&0\\0&0&0&1\end{array}\right]$ | $\left[\begin{array}{c\|ccc}1&0&0&0\\\hline 0&0&0&1\\0&1&0&0\\0&0&1&0\end{array}\right]$ | $\left[\begin{array}{c\|ccc}1&0&0&0\\\hline 0&0&1&0\\0&0&0&1\\0&1&0&0\end{array}\right]$ | $\left[\begin{array}{c\|ccc}1&0&0&0\\\hline 0&1&0&0\\0&0&0&1\\0&0&1&0\end{array}\right]$ | $\left[\begin{array}{c\|ccc}1&0&0&0\\\hline 0&0&0&1\\0&0&1&0\\0&1&0&0\end{array}\right]$ | $\left[\begin{array}{c\|ccc}1&0&0&0\\\hline 0&0&1&0\\0&1&0&0\\0&0&0&1\end{array}\right]$ |

and on the position of the nitrogen atom, while the $C_3$ rotations shift all the labels attaching to the positions of the hydrogen atoms in the structure and the $\sigma_v$ reflections interchange two hydrogen labels across the mirror plane upon which the third atom and its label reside.

For the regular orbit, the permutation character has the group order, $|G|$, under the identity and 0 under all other symmetry operations. This character of the regular representation, $\Gamma_{regular} = (|G|, 0, 0, 0, \ldots)$ has a simple reduction: when expressed as a sum, $\Gamma_{regular}$ contains every irreducible character in a number of copies equal to its degeneracy, thus one copy for non-degenerate irreducible character or each component of a separably degenerate character, two for doubly degenerate characters and so on.

This observation is confirmed straightforwardly, by applying the reduction formula and the result is summarized for the example of the regular character of $C_{3v}$ in the table

| $C_{3v}$ | E | $2C_3$ | $3\sigma_v$ |
|---|---|---|---|
| $A_1$ | 1 | 1 | 1 |
| $A_2$ | 1 | 1 | −1 |
| E | 2 | −1 | 0 |
| E | 2 | −1 | 0 |
| $\Gamma_6$ | 6 | 0 | 0 |

Similarly, we can derive the result that the permutation characters of the $O_1$ and $O_3$ orbits are $A_1$ and $A_1 + E$, respectively.

The notion of orbits is a device to reduce the structures of objects to their essentials. Thus, $NH_3$, and any other $C_{3v}$ $XY_3$ molecule or ion, has symmetry properties that arise from the property that each complete structure divides into a copy of $O_1$ and a copy of $O_3$. $C_{3v}$ structures may also contain multiple copies of the possible orbits, for example, $CH_3F$ comprises $2 \times O_1$ and $1 \times O_3$ orbits, but the important point is that all $C_{3v}$ molecular structures have an orbit formula

$$m_1O_1 + m_3O_3 + m_6O_6$$

describing the distribution of the vertices. In the operation of the GT_Calculator, specific integer numbers $m_1$, $m_3$ and $m_6$ are entered in order to calculate the permutation character of the structure points, from which all the other results can be obtained as described in Chapter 1.

## 2.2 Orbits and Geometry

The orbits of $C_{3v}$ correspond to simple geometric shapes: $O_1$ is a point; $O_3$ is the set of vertices of an equilateral triangle, while $O_6$ is a planar hexagon with alternating edge lengths. The same geometrical motifs can occur in many groups and all orbits can be represented, sometimes in several different ways, as vertices of polygons and polyhedra.

Associated with any orbit is its number of degrees of freedom; the number of geometric parameters needed to specify the orbit polygon or polyhedron. The regular orbit of a group has $C_1$ site symmetry and is constructed from a starting vertex in the general position, so it has three geometric degrees of freedom (or two if we confine the vertices to the unit sphere). An orbit with $C_s$ symmetry has vertices that lie in a reflection plane of the group and so has two degrees of freedom (or one on the unit sphere). The vertices of an orbit with $C_n$ or $C_{nv}$ site symmetry ($n > 1$) are constrained to lie at poles of rotational axes and so have only one (radial) degree of freedom (or no freedom at all on the unit sphere). $O_1$ orbits of single points have the full symmetry of the group, and if this is not a $C_{nv}$ symmetry are constrained to lie at the centre of the sphere, and have no residual geometric freedom. The degrees of freedom of a multi-orbit structure can be counted by adding up the contributions from orbits, or by counting the number of copies of $\Gamma_0$ in the vibrational character, or equivalently the number of whole copies of $\Gamma_{xyz}$ in the total permutational character for the whole structure. The relevance of these geometric freedoms to our pictorial method is that where the vertices of an orbit can be moved around on the sphere, there may be several polyhedra that describe the orbit equally well. The Archimedean solids show this 'polymorphism' in two cases: in the $O_h$ point group, the truncated cube and the small rhombicuboctahedron are instances of the same 24-vertex orbit, and in $I_h$, the truncated dodecahedron, the truncated icosahedron and the small rhombicosidodecahedron are all instances of the unique 60-vertex orbit of this group.

As an example, Figure 2.2 lists the orbits of the group $D_{3h}$ for which the character table is

| $D_{3h}$ | E | $2C_3$ | $3C_2$ | $\sigma_h$ | $2S_3$ | $3\sigma_v$ |
|---|---|---|---|---|---|---|
| $A_1'$ | 1 | 1 | 1 | 1 | 1 | 1 |
| $A_2'$ | 1 | 1 | −1 | 1 | 1 | −1 |
| $E'$ | 2 | −1 | 0 | 2 | −1 | 0 |
| $A_1''$ | 1 | 1 | 1 | −1 | −1 | −1 |
| $A_1''$ | 1 | 1 | −1 | −1 | −1 | 1 |
| $E''$ | 2 | −1 | 0 | −2 | 1 | 0 |

These are identified easily by exhaustion, taking the possible intersections of the symmetry elements and occur as $O_1$, $O_2$, $O_3$, $O_{6h}$, $O_{6v}$ and $O_{12}$.

We have the point, $O_1$, the triangle, $O_3$ and the hexagon, $O_{6[h]}$, orbits found in $C_{3v}$ structures, but, now, the $\sigma_h$ symmetry element of the $D_{3h}$ group introduces the possibility to double each motif of $C_{3v}$ structures. Thus, for $D_{3h}$, the orbit list includes the 2-point orbit $O_2$,

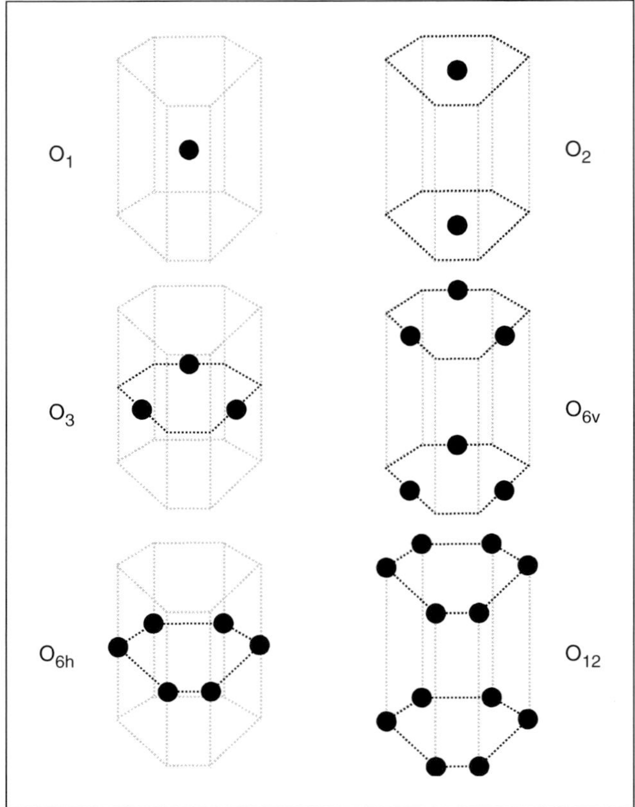


**Figure 2.2** The six orbits for structures exhibiting $D_{3h}$ point symmetry. In rows, these are $O_1$, $O_2$; $O_3$, $O_{6v}$; $O_{6h}$ and $O_{12}$.

the trigonal prism, $O_6$ and the three-fold symmetric hexagonal prism of $O_{12}$. The permutation characters on these orbits of structure and the site groups of the vertices are specific to the overall point group symmetry of the complete object. This is shown in Table 2.2, in which are listed the permutation representations of the common orbits for the groups $C_{3v}$ and $D_{3h}$ as direct sums over the irreducible representations of the groups.

The same geometrical object can appear as an orbit of different groups for which the description of its permutation character, as a list of Mulliken symbols [a sum of irreducible characters], can be different. For example, the $O_6$ and $O_{6'}$ orbits in the group $D_{6h}$ are regular hexagons with permutation characters $A_{1g} + E_{2g} + B_{1u} + E_{1u}$ and $A_{1g} + E_{2g} + B_{2u} + E_{1u}$. Whereas in $D_{3h}$ and $C_{3v}$ the $O_{6v}/O_6$ orbit is a planar hexagon with two distinct edge lengths and has permutation character $A_1' + E' + A_2'' + E''/A_1 + A_2 + 2E$.

These observations establish our approach to the role of point group orbits in the characterization of molecular geometry and in point group theory. The specific point symmetry of an orbit is limited by its environment, the surrounding molecular skeleton. Thus, we find triangles of atoms, squares of atoms and so on, identified as distinct orbits in different point

**Table 2.2 Comparisons of direct sum components of the permutation characters on orbits in the point groups $C_{3v}$ and $D_{3h}$.**

| $C_{3v}$ | Orbit | Site group | Direct Sums |
|---|---|---|---|
| | $O_1$ | $C_{3v}$ | $A_1$ |
| | $O_3$ | $C_s$ | $A_1 + E$ |
| | $O_6$ | $C_1$ | $A_1 + A_2 + 2E$ |
| $D_{3h}$ | | | |
| | $O_1$ | $D_{3h}$ | $A_1'$ |
| | $O_2$ | $C_{3v}$ | $A_1' + A_2''$ |
| | $O_3$ | $C_{2v}$ | $A_1' + E'$ |
| | $O_{6v}$ | $C_s$ | $A_1' + E' + A_2'' + E''$ |
| | $O_{6h}$ | $C_s$ | $A_1' + A_2' + 2E'$ |
| | $O_{12}$ | $C_1$ | $A_1' + A_2' + 2E' + A_1'' + A_2'' + 2E''$ |

groups although only some of the intrinsic point symmetry of these objects may be sampled by the symmetry operations present in the particular point group of the molecular structure.

## 2.3 The Platonic Solids, the Archimedean Polyhedra and General Orbits

The five fundamental solids, the tetrahedron, the octahedron, the icosahedron and the dodecahedron were known to the Ancient Greeks. Constructions based on isosceles triangles are described for the first four by Plato in his *Dialogue Timaeus*, where he associated them with fire, earth, air, water and noted the existence of the fifth, the dodecahedron, standing for the Universe as a whole. These five objects are now known as the Platonic solids — defined as the convex polyhedra because they exhibit equivalent convex regular polygonal faces.

A polyhedron is a volume bounded by polygons, with each edge shared by exactly two polygons. The five Platonic solids, Figure 2.3, are characterized by the fact that each such solid exhibits faces all of the same kind, which are regular polygons, that is to say, straight-sided figures of equal sides and angles.

That there cannot be other than five such objects, enclosing volume, is not difficult to demonstrate. Each vertex of any of such solids must be a common vertex of at least three faces. The sum of the angles of the faces coinciding at a vertex must be less than 360°, otherwise there would be overlap and the faces would not fit together. These conditions are met for the cases of equilateral triangular faces (interior angles of 60°), which means that three, four or five such faces could meet at a vertex. This observation accounts for the occurrence of the tetrahedron, the octahedron and the icosahedron. Then, three squares can fit together in a solid object, but four would reduce it to a tessellation of the plane, so the construction on square faces leads only to the cube as a possibility. Finally, there is the possibility to take three regular pentagons and form a vertex. The interior pentagonal angle is 108° and the object formed by this construction is the fifth Platonic solid, the dodecahedron, shown as the last drawing in Figure 2.3. Attempts to join hexagonal faces of interior angle 120°, with

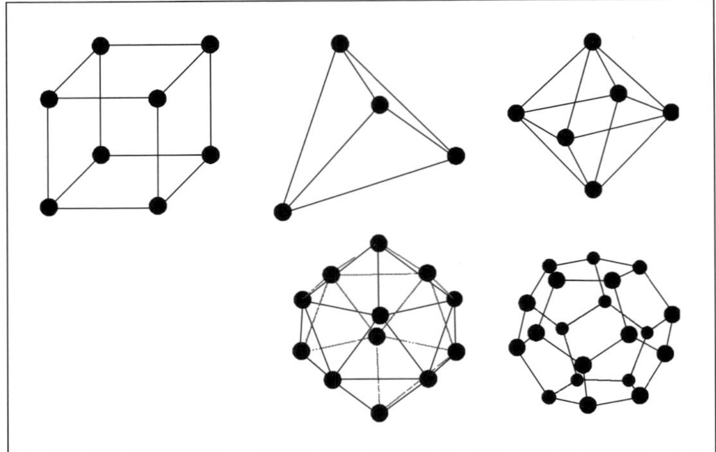

**Figure 2.3** The five Platonic solids upon which can be built the non-trivial structure orbits of Cubic and Icosahedral point symmetries.

the minimum requirement to join three, return a flat object. No regular polygon with more than six sides can be used since the requirement to join at least three must lead to overlap.

Molecular frameworks with the shapes of each Platonic solid are known. In symmetry terms, the Platonic solids split into two families: the tetrahedron, the cube and the octahedron, which have cubic symmetry, and the icosahedron and the dodecahedron, which have icosahedral symmetry.

A further set of semi-regular polyhedra, many of which are also important in Chemistry, follows on relaxation of the requirement for equivalent polygonal faces. The thirteen Archimedean polyhedra have equivalent vertex figures on all vertices and all faces remain planar and equilateral, but are of two or three distinct kinds. In orbit terms, the Platonic and Archimedean polyhedra all have single orbits of vertices, but, whereas, the face centres of Platonic solids also fall into single orbits, those of Archimedean solids span either two or three. Again, the Archimedean solids fall into cubic and icosahedral families.

In analysing polyhedra and the relationships between them, a useful formula is Euler's relation

$$\mathrm{v} + \mathrm{f} = \mathrm{e} + 2$$

which holds for polyhedra without holes or handles (i.e. spherical polyhedra) and relates the numbers of vertices (v), faces (f) and edges (e). This formula has a symmetry equivalent[4]

$$\Gamma_\sigma(\mathrm{v}) + \Gamma_\sigma(\mathrm{f})\mathrm{x}\Gamma_\varepsilon = \Gamma_{||}(\mathrm{e}) + \Gamma_0 + \Gamma_\varepsilon$$

$$\Gamma_\sigma(\mathrm{v})\mathrm{x}\Gamma_\varepsilon + \Gamma_\sigma(\mathrm{f}) = \Gamma_\perp(\mathrm{e}) + \Gamma_0 + \Gamma_\varepsilon$$

where $\Gamma_\sigma$ is a permutation character for vertices or faces and $\Gamma_{||}$ and $\Gamma_\perp$ are characters for sets of tangential vectors respectively along and perpendicular to the edges. $\Gamma_0$ has value $+1$ under all symmetry operations. $\Gamma_\varepsilon$ has value $+1$ under all proper symmetry operations and $-1$ under all improper symmetry operations.

[4] A. Ceulemans and P.W. Fowler, *Nature*, **353** (1991) 52.

Figure 2.4 shows the 13 Archimedean polyhedra with their names, which indicate some of the family relationships. Each of the polyhedra is conventionally regarded as derived by a process of truncation on vertices or edges from a Platonic parent — thus the truncated cube has 8 triangular faces formed by cutting off the vertices of the cube, the truncated octahedron has 6 square faces, derived by cutting off vertices of an octahedral parent, and so on.

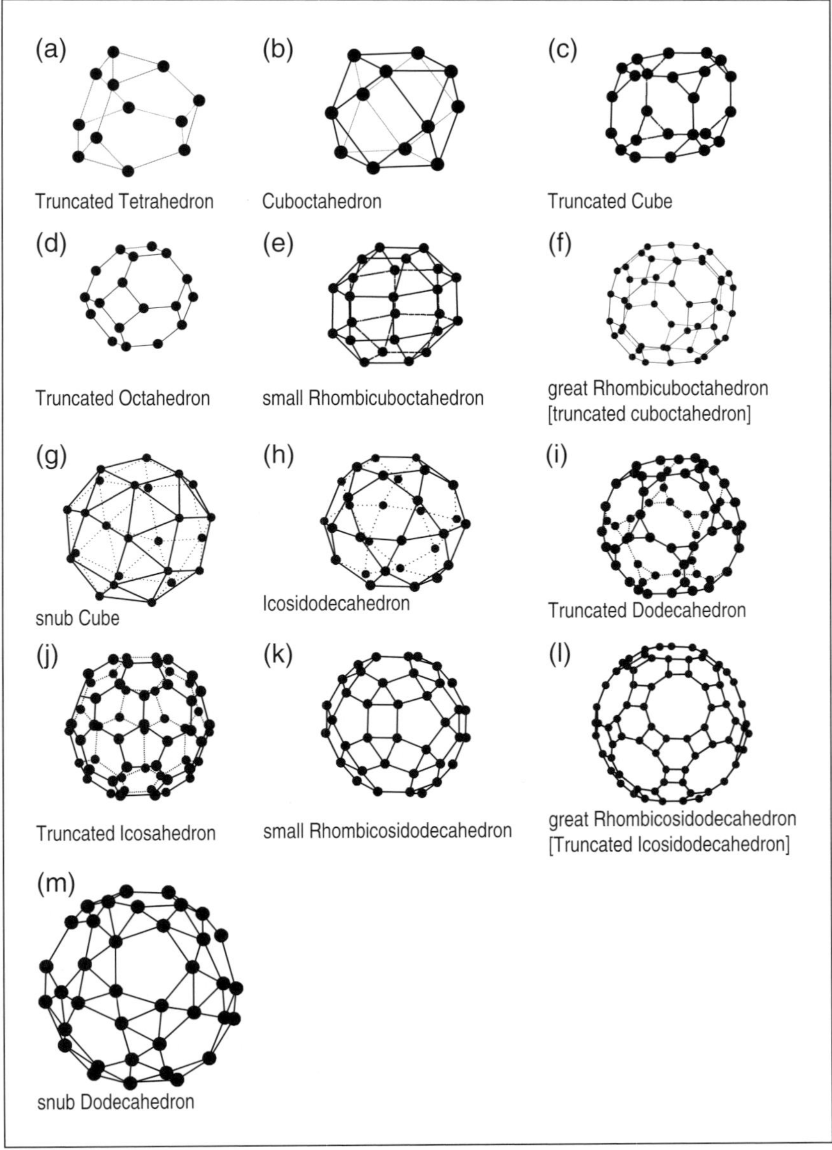


**Figure 2.4** The 13 Archimedean polyhedra, which can be constructed from the Platonic solids by relaxation of the requirement that all polygonal faces of the polyhedron be equivalent. Archimedean polyhedra have equivalent vertex figures on all vertices. All faces remain planar and equilateral, but are of two or three distinct kinds.

A more unified perspective can be developed by starting with the polyhedra derived from the regular orbits in the two parent groups, cubic ($O_h$) and icosahedral ($I_h$). The various Archimedean and Platonic solids then follow by a process of collapsing vertices of these regular-orbit polyhedra. Each polyhedron appears as the realization of an orbit of $O_h/I_h$ or a subgroup.

## 2.4 Polyhedral Orbits in $O_h$ Point Symmetry

The character table for the $O_h$ point group is

| $O_h$ | E | $8C_3$ | $6C_2$ | $6C_4$ | $3C_2[C_4^2]$ | i | $6S_4$ | $8S_6$ | $3\sigma_h$ | $6\sigma_d$ |
|---|---|---|---|---|---|---|---|---|---|---|
| $A_{1g}$ | 1 | 1 | 1 | 1 | 1 | 1 | 1 | 1 | 1 | 1 |
| $A_{2g}$ | 1 | 1 | −1 | −1 | 1 | 1 | −1 | 1 | 1 | −1 |
| $E_g$ | 2 | −1 | 0 | 0 | 2 | 2 | 0 | −1 | 2 | 0 |
| $T_{1g}$ | 3 | 0 | −1 | 1 | −1 | 3 | 1 | 0 | −1 | −1 |
| $T_{2g}$ | 3 | 0 | 1 | −1 | −1 | 3 | −1 | 0 | −1 | 1 |
| $A_{1u}$ | 1 | 1 | 1 | 1 | 1 | −1 | −1 | −1 | −1 | −1 |
| $A_{2u}$ | 1 | 1 | −1 | −1 | 1 | −1 | 1 | −1 | −1 | 1 |
| $E_u$ | 2 | −1 | 0 | 0 | 2 | −2 | 0 | 1 | −2 | 0 |
| $T_{1u}$ | 3 | 0 | −1 | 1 | −1 | −3 | −1 | 0 | 1 | 1 |
| $T_{2u}$ | 3 | 0 | 1 | −1 | −1 | −3 | 1 | 0 | 1 | −1 |

Apart from the trivial orbit, $O_1$, of a single atomic site at the centre of a molecular structure exhibiting $O_h$ point symmetry, there are the six orbits $O_6$, $O_8$, $O_{12}$, $O_{24d}$, $O_{24h}$ and the regular orbit $O_{48}$. This regular orbit of 48 vertices exhibits the structure of the great rhombicuboctahedron, Figure 2.4f. A perspective drawing of the geometrical structure of this Archimedean polyhedron with all edges of equal length is shown in Figure 2.5a.

The colour-codings of vertices of the great rhombicuboctahedron displayed in the other diagrams of Figures 2.5b–d demonstrate how this polyhedron is formed through the union of a square, an octagon and a hexagon at each vertex. The $|O_h| = 48$ vertices occupy sites of $C_1$ symmetry even in the equilateral version of the regular orbit polyhedron, but modifications in which the number of vertices is reduced, with concomitant increase in site symmetry, realize five of the Archimedean polyhedra. A useful tool for visualizing these transformations is the cartographic device known as the Apianus II projection[5] for the display of 3D detail on the plane, since in this way all the vertices and the rotational symmetry elements in a given structure can be shown on a equal footing.

[5]We believe J.W. Linnett, in his Methuen Monograph, *Wave Mechanics and Valency Theory*, 1956, was the first to use elliptical projections to display the phases of the spherical harmonics on the unit sphere. We adopt a projection in which both $\theta$ and $\phi$ coordinates are plotted on linear scales on the minor and major axes of a 30° ellipse of eccentricity $\sqrt{3/2}$. This cartographic device is the one proposed by Apianus in 1524, and known as the Apianus II projection. In our early work on the Spherical Shell method we called this a modified Mollweide projection, reversing the historical sequence.

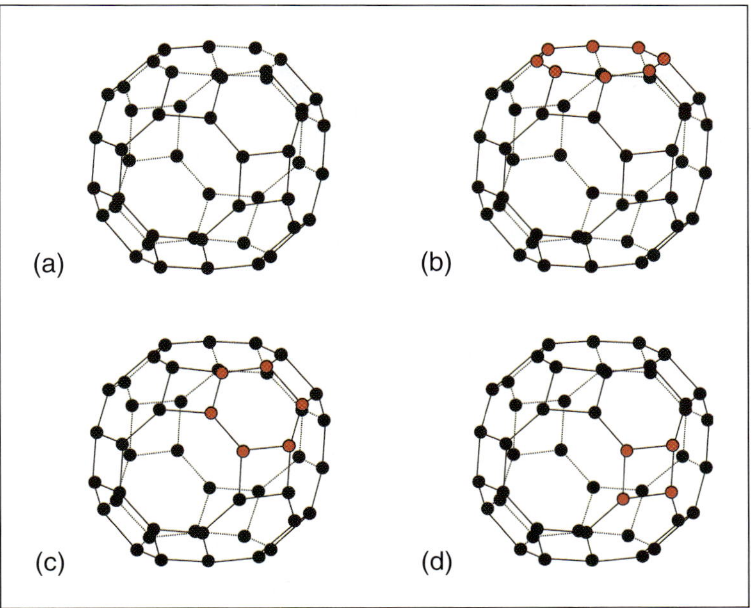


**Figure 2.5** The great rhombicuboctahedron of 48 vertices and 72 edges, formed as the union of 12 square faces, 6 octagons and 8 hexagons. Vertices of the polyhedron are coloured in sets of 8, 6 and 4 to identify the face types present, the centres of which correspond to axes of proper rotation of the point.

This elliptical projection of the unit sphere is shown in Figure 2.6. One imagines the sphere to split along the great circle from the North Pole $[+\hat{e}_z, \theta = 0]$ to the South Pole $[-\hat{e}_z, \theta = \pi]$ containing the point $[-\hat{e}_x, \theta = \pi/2, \phi = \pi)]$. Thus, along the equatorial line of the projection this point appears as the left and right extreme points of the elliptical boundary, while the circular meridian shown intersects the equatorial meridian line at the West Pole $[-\hat{e}_y, (\pi/2, -3\pi/2)]$ and the East Pole $[+\hat{e}_y, (\pi/2, \pi/2)]$. The positions of the vertices of the polyhedra corresponding to the distinct geometric orbits of each point symmetry group can be displayed as sets of points on the plane, on and within the elliptical boundary, determined by polar ($\theta$) and azimuthal ($\phi$) angles.

For a particular $\theta$ and $\phi$, the corresponding x and y coordinates on a projection are related by the identities,

$$y = (1 - 2\theta/\pi)$$

$$x = (2\phi/\pi)\sqrt{(1 - y^2)} \text{ for } \phi \text{ less than } \pi, \text{ otherwise}$$

$$x = ((2\phi - 4\pi)/\pi)\sqrt{(1 - y^2)}$$

with both results multiplied by a scale factor, to set the length of the major and minor axes of the ellipse.

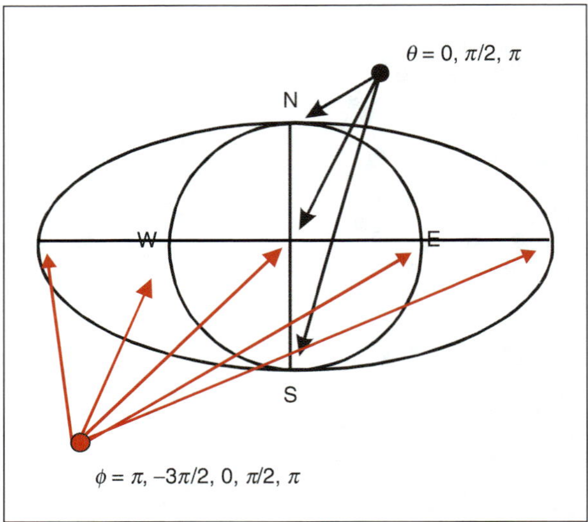


**Figure 2.6** The unit sphere as an elliptical projection. The split to form the elliptical boundary is along the great semi-circle joining $+\hat{e}_Z$ to $-\hat{e}_Z$ including the point $-\hat{e}_X$ on the unit sphere. Thus, the polar angles ($\theta$) are as marked on the primary longitude line, while the azimuthal angle ($\phi$) is measured along the central line of latitude [*the equator*] with the origin (90,0) being the central point of the projection.

In Figure 2.7, the utility of the projection is demonstrated by displaying the fully decorated regular orbit cage of the $O_h$ point group. It is straightforward to mark all the poles of rotational axes on the projection. Moreover, sets of vertices around these poles can be identified and used to demonstrate the relations between the various polyhedra that can be formed by coalescing these vertices onto these fixed points. Thus, in Figure 2.7a, octagons about the $C_4$ poles on the unit sphere are identified by the different colours, while in Figures 2.7b and c, the 48 vertices of the regular orbit are divided into sets of four and sets of six about the poles of the two- and three-fold axes.

The regular orbit displayed in Figure 2.7, is the geometry on the unit sphere such that the 'bond length', the Euclidean distance between adjacent vertices, is constant. This restriction is not necessary from a symmetry viewpoint: it may be relaxed subject only to the requirement that the local four, three and two-fold symmetries are maintained. One important example of such a relaxation occurs for the regular orbit of the $O_h$ *Crystallographic* point group. In the simplest model crystal of $O_h$ point symmetry, the primitive cubic array, for example, as in *cubium*, lattice points are distributed as dictated by the lattice vector $R_{mnp}$ such that

$$R_{mnp} = m\mathbf{a} + n\mathbf{b} + p\mathbf{c}$$

with $|\mathbf{a}| = |\mathbf{b}| = |\mathbf{c}|$ the unit cell edges mutually disposed at 90° and m, n and p integer coefficients. For $|\mathbf{a}| = |\mathbf{b}| = |\mathbf{c}|$ the first regular orbit of the $O_h$ group, about the origin of the

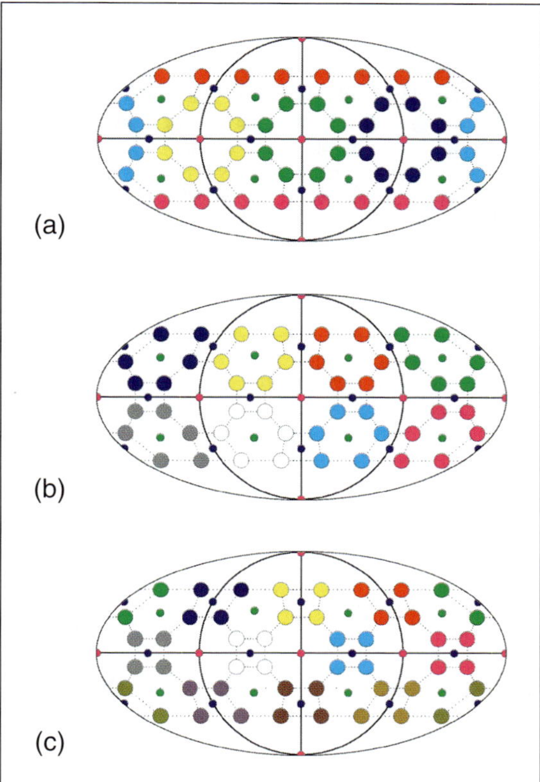


**Figure 2.7** Projection of the great rhombicuboctahedron 48-vertex regular orbit cage of the $O_h$ point group. The rotational axes are distinguished as ● [$C_4$], ● [$C_3$] and ● [$C_2$] coloured circles on the unit sphere. The 48 vertices are divided into sets of 8 (a), 6 (b) and 4 (c) about these axes using the colour coding displayed in the diagrams.

crystal, occurs for m, n and p taking the values {3,2,1} and this leads to the 48-vertex cage shown in Figure 2.8, in which, two bond lengths can be identified.

A continuous range of equisymmetric structures is permitted. Within that range, the other Archimedean polyhedra based on cubic geometry appear, in turn, as the local sets of vertices about the principal rotational axes are allowed to coalesce.

Figure 2.9 shows how contractions of local sets of 8, 6 and 4 vertices onto the poles of $C_4$, $C_3$ and $C_2$ axes, respectively, recover the octahedron, cube and cuboctahedron and thereby identify the $O_6$, $O_8$ and $O_{12}$ orbits of $O_h$ symmetry.

Pairwise contractions to superimpose two vertices of the regular structure at each vertex of the lower order 24-vertex cages identify the other Archimedean polyhedra of $O_h$ symmetry. These contractions are shown in Figures 2.10 and 2.11.

Thus, in Figure 2.10 there are two choices of pairwise contractions of the coloured sets about the $C_4$ axes of the regular orbit. In the first column, the contractions lead to the truncated

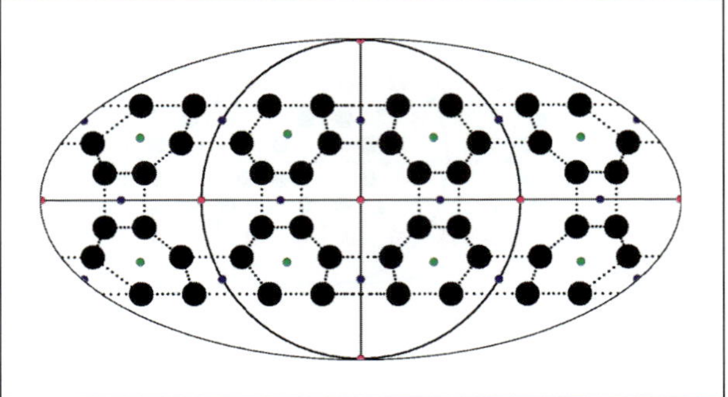

**Figure 2.8** The regular orbit of the $O_h$ point group constructed as the first 48-membered shell for *Cubium*. Two distinct bond lengths are evident in the projected structure, especially about the two-fold axes through the equator, but the structure remains of $O_h$ point symmetry.

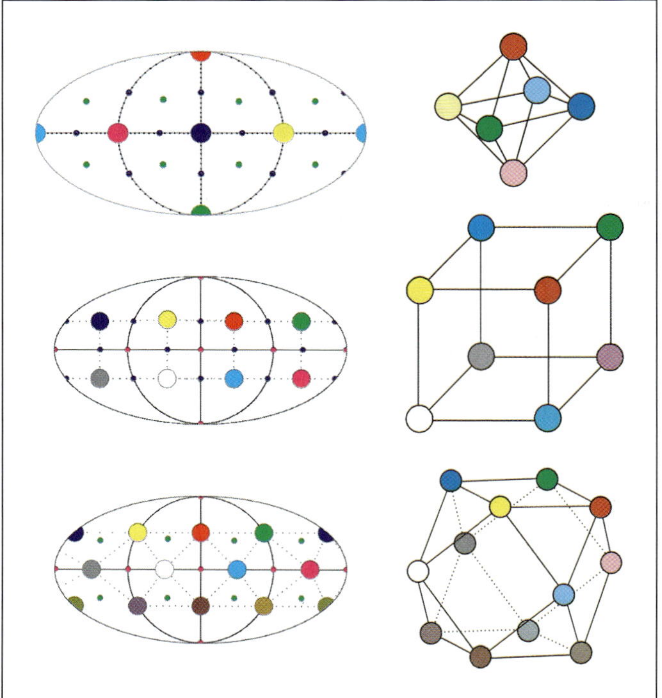

**Figure 2.9** The results of applying contractions to the coloured sets of vertices of the regular orbit cage of $O_h$ symmetry to give the octahedron, the cube and the cuboctahedron. The colour codings refer to those in Figure 2.7 and so the first row displays the results for contractions onto the $C_4$ poles, the second row the results for contractions onto the $C_3$ poles and in the third row, contractions onto the $C_2$ axes lead to the cuboctahedron.

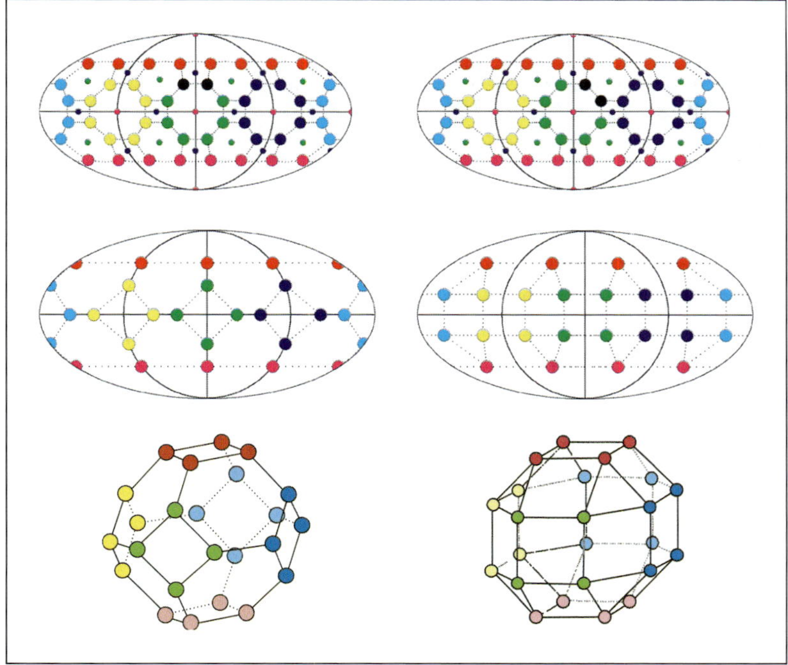

**Figure 2.10** The results of pairwise contractions of the vertices of the octagons surrounding the four-fold axes of the regular orbit cage, using the colour coding of Figure 2.7. The first choice of pairs to be contracted leads to the formation of the truncated octahedron, while the alternative choice gives rise to the small rhombicuboctahedron.

octahedron and, in the second column, to the small rhombicuboctahedron. The vertices of the truncated octahedron span the $O_{24h}$ since they lie on the three $\sigma_d$ symmetry planes, while the vertices of the small rhombicuboctahedron, which lie on the six $\sigma_d$ planes, span the $O_{24v}$ orbit of $O_h$.

The small rhombicuboctahedron is one of two Archimedean polyhedra that correspond to the same $O_{24v}$ orbit. The other polyhedron is found by carrying out pairwise contraction of the hexagonal sets of vertices about the poles of the $C_3$ axes in the regular orbit geometry as shown in Figure 2.11. There are two choices for the pairings. In the first column of Figure 2.11, the pairing choice leads once again to the small rhombicuboctahedron. In the second column the truncated cube results from the alternative pairing. Examination of the site symmetry of each vertex shows that both 24-vertex polyhedra span the same orbit.

This observation is emphasised in Figure 2.12, where the alternative geometrical constructions of enlarging and contracting the triangular truncations about the vertices of the cube are seen to lead to the interconversion of the geometry of the truncated cube and that of the small rhombioctahedron through the intermediate geometry of the cube and *vice versa*.

The second distinct orbit of 24 vertices and a cage exhibiting $O_h$ symmetry is found when the pairwise contraction procedure is applied to the square or rectangle polyhedron vertices about the $C_2$ axes of the $O_h$ regular orbit cage. The results are shown in Figure 2.13.

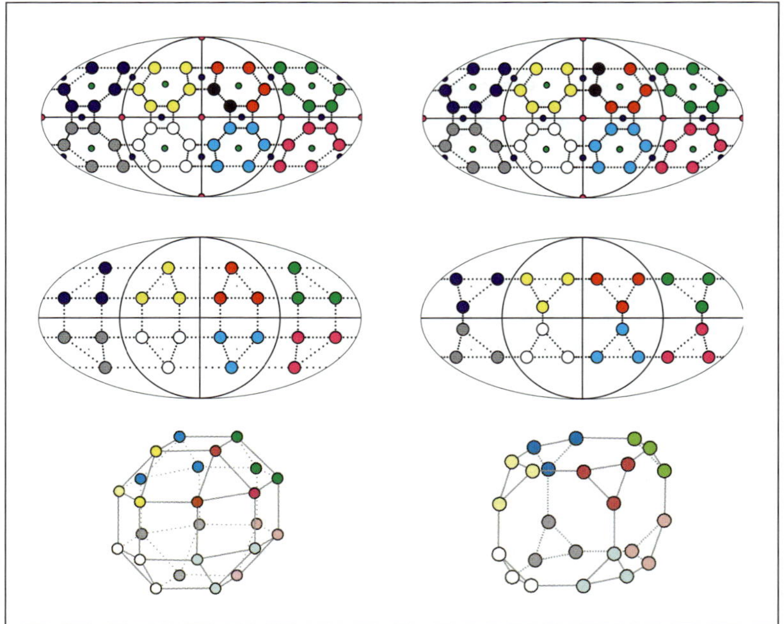

**Figure 2.11** The results of pairwise contractions of the vertices of the hexagons of the regular orbit cage of $O_h$ symmetry, again with the colour coding as in Figure 2.7. The first pairing choice, column 1, leads to the 24-vertex cage of the truncated cube, while the second choice leads to a further copy of the small rhombicuboctahedron.

## 2.5 Polyhedral Orbits of Cubic Symmetry Lower than $O_h$

Four other polyhedra based on 24-vertex cages exhibiting cubic symmetry can be formed from the regular orbit structure of the previous section. Two of these are chiral pairs, the *dextro* snub cube and its chiral partner, the *laevo* snub cube; the third is the regular orbit of $T_d$ point symmetry, while the fourth is the regular orbit of $T_h$ symmetry.

In Figure 2.14a, the 48-vertex structure of the great rhombicuboctahedron is divided into two sets of 24 points, coloured to distinguish two sets related by the inversion operation. Each set of 24 vertices now exhibit O symmetry and are examples of the chiral polyhedra based on the snub cube structure, displayed as the *d*-isomer as a projection in Figure 2.14b and as a perspective drawing in Figure 2.14b.

| **O** | E | $8C_3$ | $3C_2$ | $6C_4$ | $6C_2'$ |
|---|---|---|---|---|---|
| $A_1$ | 1 | 1 | 1 | 1 | 1 |
| $A_2$ | 1 | 1 | 1 | −1 | −1 |
| E | 2 | −1 | 2 | 0 | 0 |
| $T_1$ | 3 | 0 | −1 | 1 | −1 |
| $T_2$ | 3 | 0 | −1 | −1 | 1 |

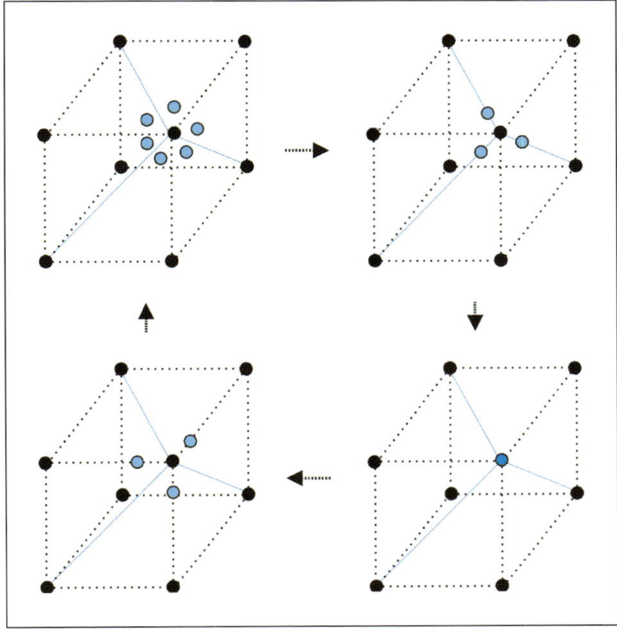

**Figure 2.12** Recovery of two Archimedean polyhedra by pairwise collapse and re-expansion of hexagonal motifs of the regular orbit of $O_h$.

The various rotational axes can be identified by examination of the *snub* cube structure, Figure 2.14c, which spans the regular orbit of O. As the lower orbits, $O_6$, $O_8$ and $O_{12}$, Figures 2.14d–f, are all intrinsically achiral, any object of O symmetry must contain at least one copy of the chiral regular orbit, it is allowed that the lower order orbits $O_6$, $O_{12}$ and $O_8$ of $O_h$ symmetry can be formed by coalescing appropriate local sets of vertices of the regular orbit onto the poles of the rotational axes as shown in Figures 2.15d–f.

The regular orbit of $T_d$ symmetry is realized by the division of the 48-point orbit of $O_h$ into two sets of 24 vertices, Figure 2.15a, grouped in sets of six about alternate vertices of the cube. For one choice of four cube vertices, the corresponding $O_{24}$ object is drawn as a projection in Figure 2.15b. For a general choice of geometrical parameters, superimposition of the $O_{24}$ orbit of $T_d$ and its inversion partner recovers a copy of the regular orbit of $O_h$. However, the $O_{24}$ orbit of $T_d$ can be converted by continuous deformation to the $O_{24h}$ orbit of $O_h$ as shown in Figure 2.15c.

The character table for $T_d$ is shown below

| $T_d$ | E | $8C_3$ | $3C_2$ | $6S_4$ | $6\sigma_d$ |
|---|---|---|---|---|---|
| $A_1$ | 1 | 1 | 1 | 1 | 1 |
| $A_2$ | 1 | 1 | 1 | −1 | −1 |
| E | 2 | −1 | 2 | 0 | 0 |
| $T_1$ | 3 | 0 | −1 | 1 | −1 |
| $T_2$ | 3 | 0 | −1 | −1 | 1 |

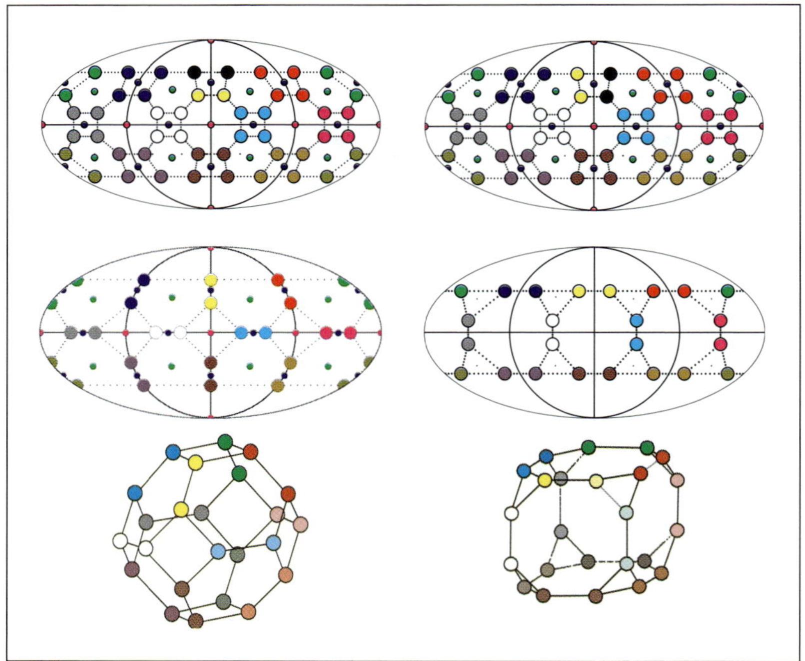

**Figure 2.13** The *pairwise contraction* procedure applied to the 4-sided polygons about the $C_2$ axes of the regular orbit cage of $O_h$ symmetry. Again, the colour coding is as in Figure 2.7. In the first column, the contraction procedure leads to another occurrence of the $O_{24h}$ 24-vertex orbit of $O_h$ symmetry, the truncated octahedron. In the second column, the pairwise contractions lead to a further copy of the truncated cube $[O_{24v}]$. These drawings differ from the ones in Figures 2.10 and 2.11 only with regard to the colours of the vertices.

Contraction and coalescence of the $O_{24}$ regular orbit vertices onto poles of the $C_2$, $C_4$ and $C_3$ axes returns the $O_{12}$, $O_6$ and $O_4$ orbits, Figures 2.15d–f. Of course, $T_d$ structures can contain at most one copy of the trivial $O_1$, e.g. in $CH_4$.

This most symmetrical, equilateral, 24-vertex orbit can be invoked to model the growth of silicon crystallites, imagined to form about a single atom since tetrahedral symmetry can be maintained by attaching to either set of 4 of the cube vertices using the intermediate $O_{12}$ orbit. Thus, in the bulk crystal there is no difference between the sets of four vertices describing the basic tetrahedron and this regular orbit and the regular geometry of the orbit $O_{12}$ of $T_d$ symmetry are found as repeating sets of nearest neighbours in the crystallography of the tetrahedral space groups. Moreover, this regular structure accounts, also, for the occurrence of the $O_6$ orbit in tetrahedral symmetry, even though there are no proper four-fold axes present. The dodecahedral holes, which are found in the silicon crystal lattice, provide an example of this $O_6$ orbit in a structure of overall $T_d$ point symmetry.

From the 24-vertex structure, of $T_d$ symmetry, the $O_{12}$ orbit follows by coalescing, in the manner of Figure 2.10 the points of each hexagon, about the alternating $C_3$ axes points on

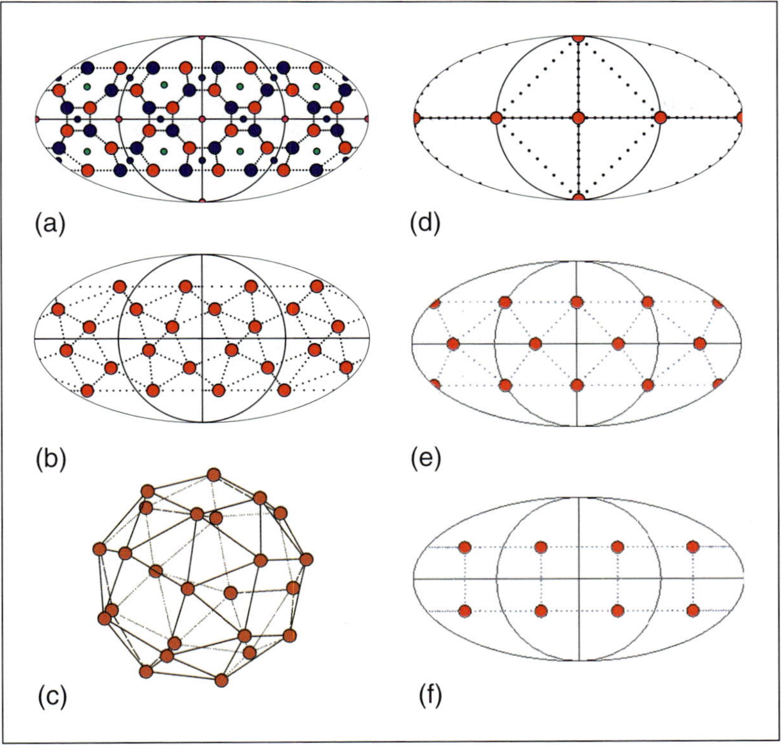


**Figure 2.14** Division of the regular 48-vertex orbit of $O_h$ symmetry (a) into the two 24-point sets of the *chiral* snub cube structure. The *dextro* projection (b) is also drawn in perspective (c)[6]. Coalescing sets of vertices onto the rotational poles leads to the other orbits $O_6$, $O_{12}$ and $O_8$ of the O symmetry group. These are achiral; they are also orbits of $O_h$.

the unit sphere, into 4 sets of three vertices. Again, there are two Archimedean polyhedra as shown in Figures 2.16a and b corresponding to the same orbit.

The character table for the T group

| **T** | E | $4C_3$ | $4C_3^2$ | $3C_2$ |
|---|---|---|---|---|
| A | 1 | 1 | 1 | 1 |
| E(1) | 1 | $\varepsilon$ | $\varepsilon^*$ | 1 |
| E(2) | 1 | $\varepsilon^*$ | $\varepsilon$ | 1 |
| $T_2$ | 3 | 0 | 0 | −1 |

includes the separably degenerate characters E(1) and E(2) with $\varepsilon = e^{(2\pi i/3)}$. This group is of order 12, but there are no reflection planes and hence the three-fold clockwise and anticlockwise rotations about each body diagonal of the inscribing cube fall into separate classes.

[6] H.M. Cundy and A.P. Rollett, *Mathematical Models* [OUP, Oxford, 1961 and Tarquin Publications].

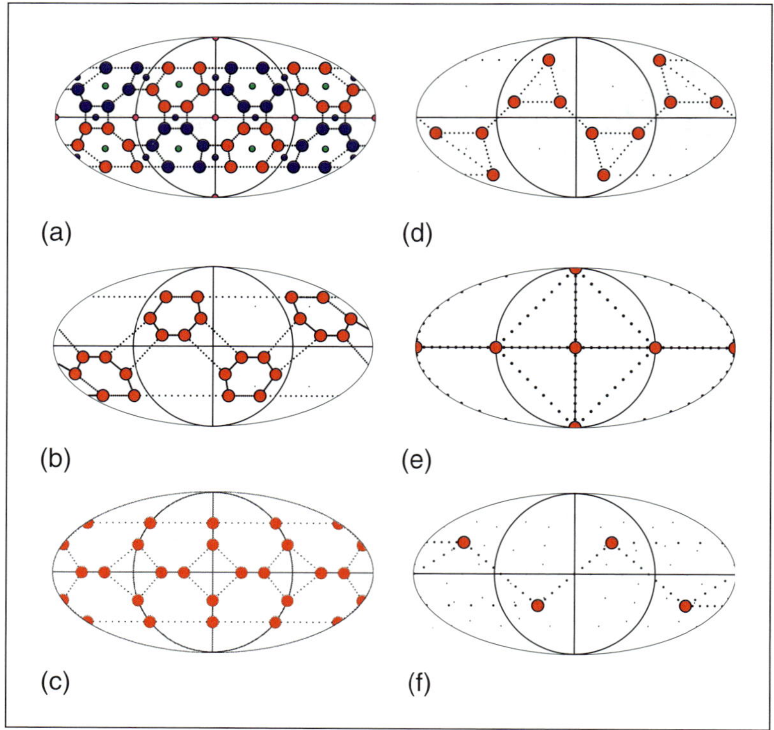


**Figure 2.15** Identification of the regular orbit of the tetrahedron by discarding 24 of the vertices of the regular orbit of $O_h$ as distinguished by the colour coding (a) to realize (b). On the unit sphere this 24 orbit of $T_d$ has two remaining degree of geometric freedom and can be continuously deformed to the $O_h$-symmetric truncated octahedron (c). The remaining projections d, e and f identify the $O_{12}$, $O_6$ and $O_4$ orbits of the group. The $O_{12}$ orbit (d) follows from the regular orbit by pairwise truncation, which on further pairwise contraction forms the $O_6$ orbit (e). The $O_4$ orbit is realized by contraction of the regular orbit onto the $C_3$ poles on the unit sphere.

The general object with 12 vertices, the regular orbit of the T group is shown in Figure 2.17a as an elliptical projection and in perspective, as the distorted truncated tetrahedron in Figure 2.17b. The only other non-trivial orbit for structures of T symmetry, is the simple tetrahedron, realized collapse of the local sets onto the poles of the three-fold rotational axes.

The last category of objects based on cubic symmetry is found for the orbits of the group $T_h$. The $T_h$ group enlarges the rotations of the tetrahedron by adding inversion symmetry. In the character table, it can be seen that there are 24 possible symmetry operations and so the regular orbit corresponds to a polyhedron of 24 vertices. Again, as with the T character table there are separably degenerate representations and the complex trace $\varepsilon = e^{(2\pi i/3)}$ and its complex conjugate appear in the $E_g$ and $E_u$ characters of these representations. The other structure orbits possible for $T_h$ symmetry are $O_{24}$, $O_{12}$, $O_8$ and $O_6$. In $T_h$ symmetry

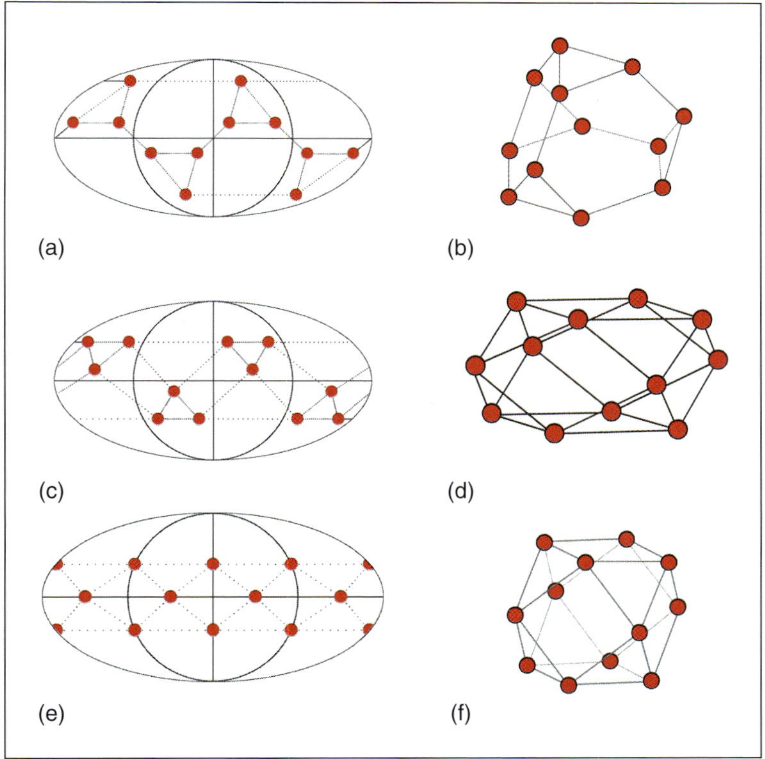


**Figure 2.16** Different geometric structures corresponding to the $O_{12}$ orbit of the tetrahedral point group $T_d$. Local inversions as in Figure 2.12 interchange the orientations of the triangular truncations and lead from the truncated tetrahedral arrangement, Figures 2.16a and b, to the distorted cuboctahedron, Figures 2.16c and d. Figures 2.16e and f show the special case in which the cuboctahedron has become regular giving the $O_{12}$ orbit common to both $T_d$ and $O_h$ groups.

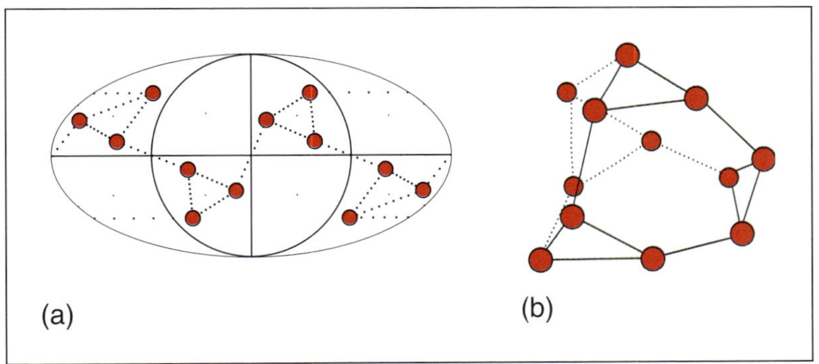


**Figure 2.17** The polyhedron of 12 vertices forming the regular geometric orbit of the T symmetry group: (a) as an elliptical projection; and (b) in perspective.

there is no orbit of 4 vertices: such an orbit would not exhibit the required inversion symmetry.

| $T_h$ | E | $4C_3$ | $4C_3^2$ | $3C_2$ | i | $4S_6$ | $4S_6^5$ | $3\sigma_d$ |
|---|---|---|---|---|---|---|---|---|
| $A_g$ | 1 | 1 | 1 | 1 | 1 | 1 | 1 | 1 |
| $E_g$ (1) | 1 | $\varepsilon$ | $\varepsilon^*$ | 1 | 1 | $\varepsilon$ | $\varepsilon^*$ | 1 |
| $E_g$ (2) | 1 | $\varepsilon^*$ | $\varepsilon$ | 1 | 1 | $\varepsilon^*$ | $\varepsilon$ | 1 |
| $T_g$ | 3 | 0 | 0 | −1 | 3 | 0 | 0 | −1 |
| $A_u$ | 1 | 1 | 1 | 1 | −1 | −1 | −1 | −1 |
| $E_u$ (1) | 1 | $\varepsilon$ | $\varepsilon^*$ | 1 | −1 | $-\varepsilon$ | $-\varepsilon^*$ | −1 |
| $E_u$ (2) | 1 | $\varepsilon^*$ | $\varepsilon$ | 1 | −1 | $-\varepsilon^*$ | $-\varepsilon$ | −1 |
| $T_u$ | 3 | 0 | 0 | −1 | −3 | 0 | 0 | 1 |

The regular orbit is displayed in Figure 2.18.

In the presence of the regular orbit, structures exhibiting $T_h$ overall symmetry can include the lower orbits $O_{12}$, $O_8$ and $O_6$ as in Figure 2.12 by coalescing appropriate local sets of vertices onto the poles on the unit inscribing sphere of the proper axes of the parent $O_h$ regular orbit.

## 2.6 Orbits and Polyhedra in $I_h$ Point Symmetry

The character table for the $I_h$ point group is

| $I_h$ | E | $12C_5$ | $12C_5^2$ | $20C_3$ | $15C_2$ | i | $12S_{10}$ | $12S_{10}^3$ | $20S_6$ | $15\sigma$ |
|---|---|---|---|---|---|---|---|---|---|---|
| $A_g$ | 1 | 1 | 1 | 1 | 1 | 1 | 1 | 1 | 1 | 1 |
| $T_{1g}$ | 3 | $\tau$ | $1-\tau$ | 0 | −1 | 3 | $1-\tau$ | $\tau$ | 0 | −1 |
| $T_{2g}$ | 3 | $1-\tau$ | $\tau$ | 0 | −1 | 3 | $\tau$ | $1-\tau$ | 0 | −1 |
| $G_g$ | 4 | −1 | −1 | 1 | 0 | 4 | −1 | −1 | 1 | 0 |
| $H_g$ | 5 | 0 | 0 | −1 | 1 | 5 | 0 | 0 | −1 | 1 |
| $A_u$ | 1 | 1 | 1 | 1 | 1 | −1 | −1 | −1 | −1 | −1 |
| $T_{1u}$ | 3 | $\tau$ | $1-\tau$ | 0 | −1 | −3 | $\tau-1$ | $-\tau$ | 0 | 1 |
| $T_{2u}$ | 3 | $1-\tau$ | $\tau$ | 0 | −1 | −3 | $-\tau$ | $\tau-1$ | 0 | 1 |
| $G_u$ | 4 | −1 | −1 | 1 | 0 | −4 | 1 | 1 | −1 | 0 |
| $H_u$ | 5 | 0 | 0 | −1 | 1 | −5 | 0 | 0 | 1 | −1 |

wherein $\tau$, the *golden section* of classical antiquity, is of value $(1+\sqrt{5})/2$. The group is of order 120 and so the regular orbit is the Archimedean polyhedron shown in Figure 2.41 and known either as the great rhombicosidodecahedron or the truncated icosidodecahedron. The polyhedron has hexagonal, square and decagonal faces centred on the poles of the three-, four- and five-fold proper axes of the $I_h$ group.

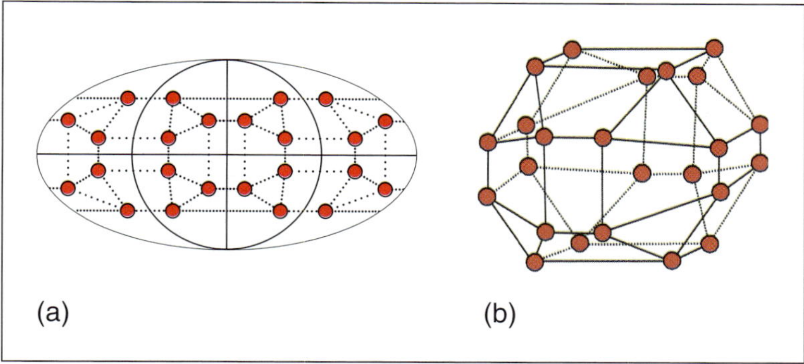


**Figure 2.18** (a) The regular orbit of $T_h$ symmetry as an elliptical projection and (b) in perspective displaying the structure as a modified small rhombicuboctahedron, in which there is no reflection symmetry across face diagonals of the inscribing cube.

These structural details are emphasized in Figure 2.19, with, in the second and later rows Figures 2.19b–d, the local sets of 10, 6 and 4 vertices of the great rhombicosidodecahedron identified about a representative pole position on a face and, then, in the second column of diagrams, as fully decorated elliptical projections of the 120-vertex cage.

There are 6 five-fold rotational symmetry elements in an object of $I_h$ point symmetry. Thus, in Figure 2.19b the 120 vertices of the great rhombicosidodecahedron are arranged in sets of 10 about the poles of these axes on the unit sphere. That construction emphasises that uniform contractions of these sets about these axes points will return the 12-vertex Platonic solid, the icosahedron, in which each vertex has $C_{5v}$ site symmetry. There are 10 three-fold rotational axes and, so, in Figure 2.19c the decoration pattern is arranged to divide the 120 vertices into sets of 6 about the 20 poles of these axes on the unit sphere. Again, uniform contraction of these subsets of vertices onto these positions on the unit sphere generates the fifth Platonic solid, the dodecahedron, and the site symmetry each vertex is $C_{3v}$.

There are 15 two-fold axes in an object of $I_h$ point symmetry. Sets of four vertices about the 30 poles of these axes are shown in Figure 2.19c. The 30-vertex cage of the icosidodecahedron to result on contraction of these local sets onto the pole positions on the unit sphere.

These transformations are displayed in Figure 2.20, with the vertices of the smaller polyhedra and their elliptical projections coloured to identify their parent decorations in Figure 2.19.

The examination of coordinate transformations as local contractions and expansions of decorations about the poles of the principal rotational axes on the unit sphere for objects of $O_h$ symmetry leads to intermediate geometries corresponding to particular Archimedean polyhedra related to the cube. In a similar manner, partial contractions and expansions of the decorations of the regular orbit of $I_h$ point symmetry, i.e. the vertices of the great rhombicosidodecahedron, leads to the remaining polyhedra within the icosahedral family of Archimedean structures and orbits of $I_h$.

The remaining lower orbits of the $I_h$ point group can be identified as effective pairwise coalescences of the various decorated regular orbit cage structures of Figure 2.19. Thus, in Figure 2.21, there are two possible pairing choices as shown in the columns of the figure.

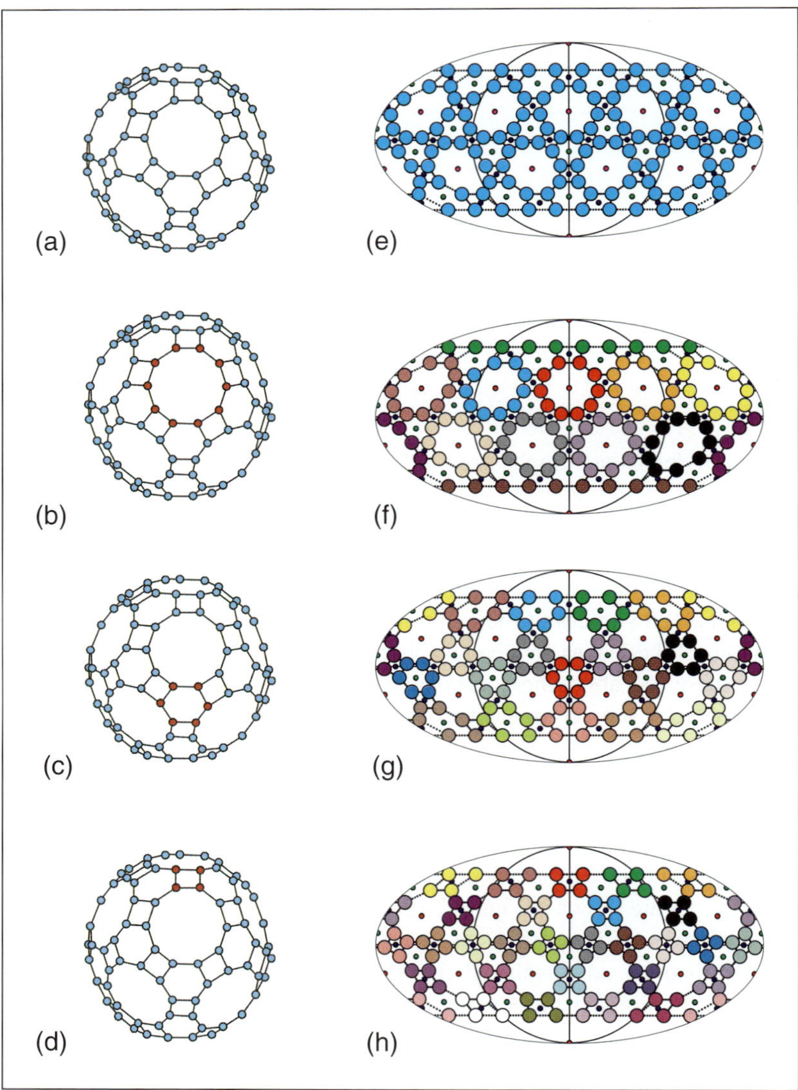


**Figure 2.19** The divisions of the vertices of the regular orbit of $I_h$ point symmetry, defining the great rhombicosidodecahedron, into *decoration* sets about the rotational axes points ● [$C_5$], row b, ● [$C_3$], row c and ● [$C_2$], row d, on the unit sphere.

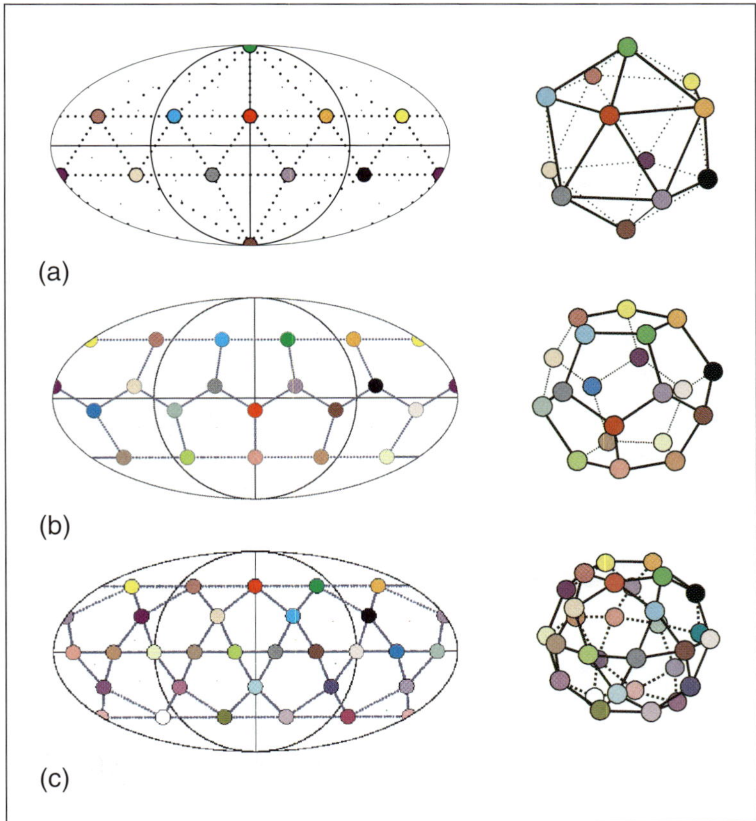


**Figure 2.20** Formation of the lower orbits of $I_h$ symmetry: $O_{12}$, the icosahedron [row a]; $O_{20}$, the dodecahedron, [row b]; and $O_{30}$, the icosidodecahedron, [row c] of Figure 2.4 by coalescing the local sets of vertices of the great rhombicosidodecahedron onto the poles of the $C_5$, $C_3$ and $C_2$ rotational axes with colour codings as in Figure 2.19.

The first pairwise contraction leads to the formation of the 3-valent structure of pentagons and hexagons, which is the cage structure of the $C_{60}$ fullerene molecule and, with equal edge lengths, is the geometrical structure known as the truncated icosahedron.

The choice of the alternative pairs of the $I_h$ regular orbit vertices as in the second column of Figure 2.21, leads to a second Archimedean polyhedron spanning the $O_{60}$ orbit. This is small rhombicosidodecahedron.

Figures 2.22 and 2.23 display the results of pairwise coalescences on the decorated regular orbit structures of Figure 2.19 corresponding to the local sets of vertices about the poles of the $C_3$ and $C_2$ rotational axes. Both of the 60-vertex structures identified in Figure 2.21 arise for particular pairwise condensations of the $O_{120}$ orbit cage, but, in addition, a third Archimedean polyhedron, in this case also 3-valent, is found. This is the truncated dodecahedral cage, Figure 2.4i, which results on applying the pairwise contractions of the regular orbit cage set out in the second column of Figure 2.22 and the first column of Figure 2.23.

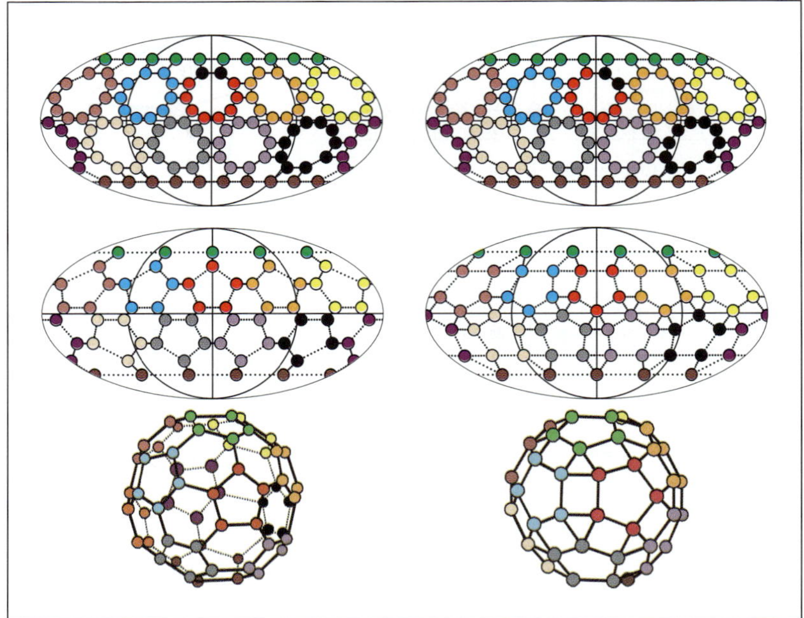

**Figure 2.21** Pairwise coalescence of the decorated decagons of the regular orbit of $I_h$ from Figure 2.19 leading to the truncated icosahedron [column a] and the small rhombicosidodecahedron [column b]. Note that the truncated icosahedral structure is 3-valent and is the archetypal $C_{60}$ cage of fullerene chemistry.

These polyhedra are the $O_{60}$ orbit realized in two ways, the truncated dodecahedron and the small rhombicosidodecahedron, which structures can be formally interconverted in the manner of Figure 2.11.

## 2.7 The Orbits of Structures Exhibiting I Symmetry

There is only one other point group for structures exhibiting icosahedral symmetry. The character table for the I point group is

| **I** | E | $12C_5$ | $12C_5^2$ | $20C_3$ | $15C_2$ |
|---|---|---|---|---|---|
| A | 1 | 1 | 1 | 1 | 1 |
| $T_1$ | 3 | $\tau$ | $1-\tau$ | 0 | $-1$ |
| $T_2$ | 3 | $1-\tau$ | $\tau$ | 0 | $-1$ |
| G | 4 | $-1$ | $-1$ | 1 | 0 |
| H | 5 | 0 | 0 | $-1$ | 1 |

with $\tau$, again, the *golden ratio*. The difference between structures of I and $I_h$ symmetries rests on the absence of inversion symmetry in the lower point group. This observation means that

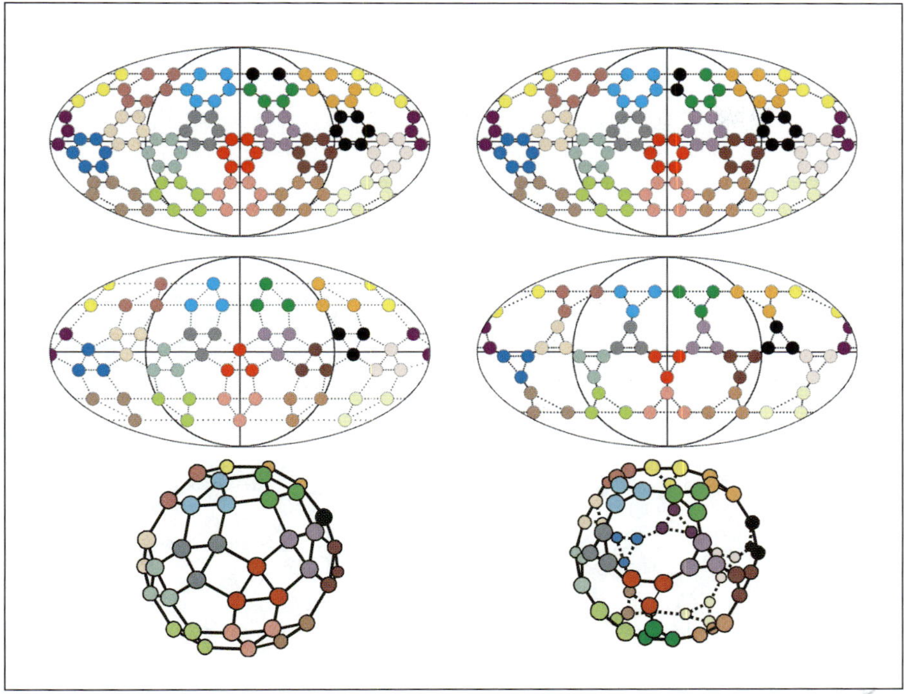

**Figure 2.22** Pairwise coalescence contractions of the regular orbit cage of $I_h$ on the 6-membered local sets about the poles of the three-fold axes, to give (first column) a further copy of the small rhombicosidodecahedron and (second column) the 3-valent $O_{60}$ orbit isomer of the fullerene cage of Figure 2.21, the truncated dodecahedron.

we can apply the procedure of Figure 2.14 to identify the chiral polyhedron with 60 vertices corresponding to the regular orbit of the I point group.

Thus, in Figure 2.24 the vertices of the regular orbit of the polyhedron of $I_h$ point symmetry are divided into two sets of sixty vertices, chosen to eliminate the inversion symmetry present in the *great* rhombicosidodecahedron, Figure 2.4l, but retaining all proper rotational symmetries. In the first column, this division is shown on the elliptical projection of the vertices of the regular orbit of $I_h$ and then the 60-vertex regular orbit cage of the I symmetry group is drawn in projection and 3d perspective. The orbit polyhedron is seen to be the Archimedean *snub* dodecahedron, Figure 2.4m.

This polyhedron is chiral and can be drawn as either enantiomer by appropriate choice of the 60 vertices, either red or blue, in Figure 2.24. All the lower orbit structures, $O_{12}$, $O_{20}$ and $O_{30}$ shown in the second column of projections in Figure 2.24 are achiral and identical to those found by coalescing local sets of 10, 6 and 4 vertices in full $I_h$ point symmetry.

## 2.8 Orbits in Space Group Theory

Space group theory is developed by the 'decoration' of the fourteen possible ways of arranging points in regular 3D arrays, in space, the Bravais lattices of Crystallography. In Crystallography, the decoration about a lattice point is called the 'basis' or, less commonly,

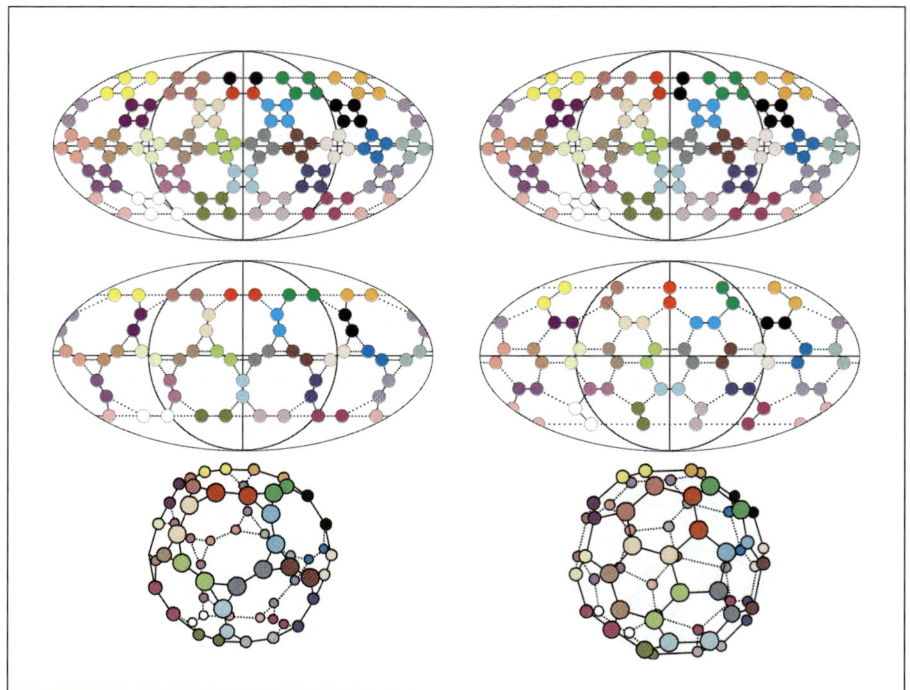

**Figure 2.23** Pairwise coalescence contractions of the regular orbit cage of $I_h$ on the four-membered local sets about the poles of the two-fold axes, to return in the first column a copy of the 3-valent $O_{60}$ orbit isomer of the fullerene cage of Figure 2.21, the truncated dodecahedron of geometry, while the small rhombicosidodecahedron results, in the second column, for the alternative sequence of pairwise contractions, which maintain $I_h$ point symmetry.

the 'lattice complex' of the structure. The effect of a particular decoration at each lattice point is to alter the site groups of the vertices, which can be grouped in orbits dictated by the lattice point symmetry. Some 73 *symmorphic* space groups arise directly in this manner, characterized by the property that all the symmetry operations, rotations, translations or combinations of these can be described entirely with respect to a single point of reference, with the effect that these groups are described completely as products of the point groups with the translational groups[7]. With the added device of periodic boundary conditions, so that for a large [*infinitely large*] number of associated translations of the lattice it is assumed that the origin of the lattice is reached again, the extra compound symmetry operations of *glide planes* and *screw rotations* can be defined. These 'extra' symmetry operations can arise from the presence of the point and translational operations. However, space groups can be constructed, too, with glide planes and screw rotations based on non-primitive fractional translations in the lattice. There are 157 *non-symmorphic* space groups of this kind and they are distinguished readily, in practice, from the *symmorphic* groups by the absence of any positions in the crystals exhibiting the full point symmetry of the *Crystallographic* point group.

[7]Note, that since five-fold rotational symmetry cannot be propagated on a lattice, there are only 32 *Crystallographic* point groups, since the icosahedral groups are excluded.

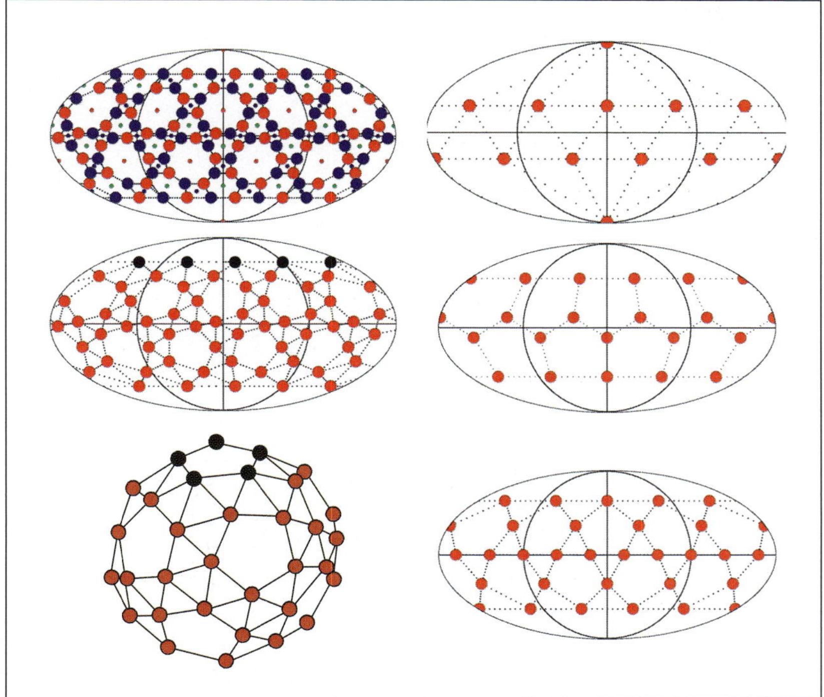

**Figure 2.24** Identification of one of two chiral 60-vertex cages, which correspond to the regular orbit of I symmetry and reduction of this regular orbit to find the lower structure orbits of the group. The set of five vertices about the topmost pole of a $C_5$ axis of the regular orbit of I are coloured black in the elliptical projection and the perspective drawing of the *snub* dodecahedron. The other structural orbits of the I group follow by the usual sequence of contractions of the local sets of 5, 3 and 2 onto the poles of the rotational axes.

For applications in Crystallography and Physics, it is common to find the 32 *Crystallographic* point groups identified as two-dimensional stereographic projections of their regular orbits as shown in Figure 2.25. In such stereograms, each point in the 'Northern' hemisphere is projected onto the equatorial plane using straight line projection through the 'South' pole and marked by a cross. Each point on the 'Southern' hemisphere of the inscribing unit sphere is projected similarly toward the 'North' pole [the +Z axis normal to the plane of the paper] and is marked by a circle. In the mapping, points in equivalent positions in the two hemispheres map onto a circle in the plane of the paper, but their centres do not map onto each other.

The stereograms include symbols to identify the locations of the symmetry elements of the structures with respect to the regular orbit points on the unit sphere. These are shown normally as filled polygons for proper rotational axes and empty polygons for improper rotations, which give rise to actions across the hemispherical plane, while binary rotations are shown as ellipses, either filled or empty, but mirror planes, the improper axes of binary rotation, are distinguished on the stereograms as solid lines.

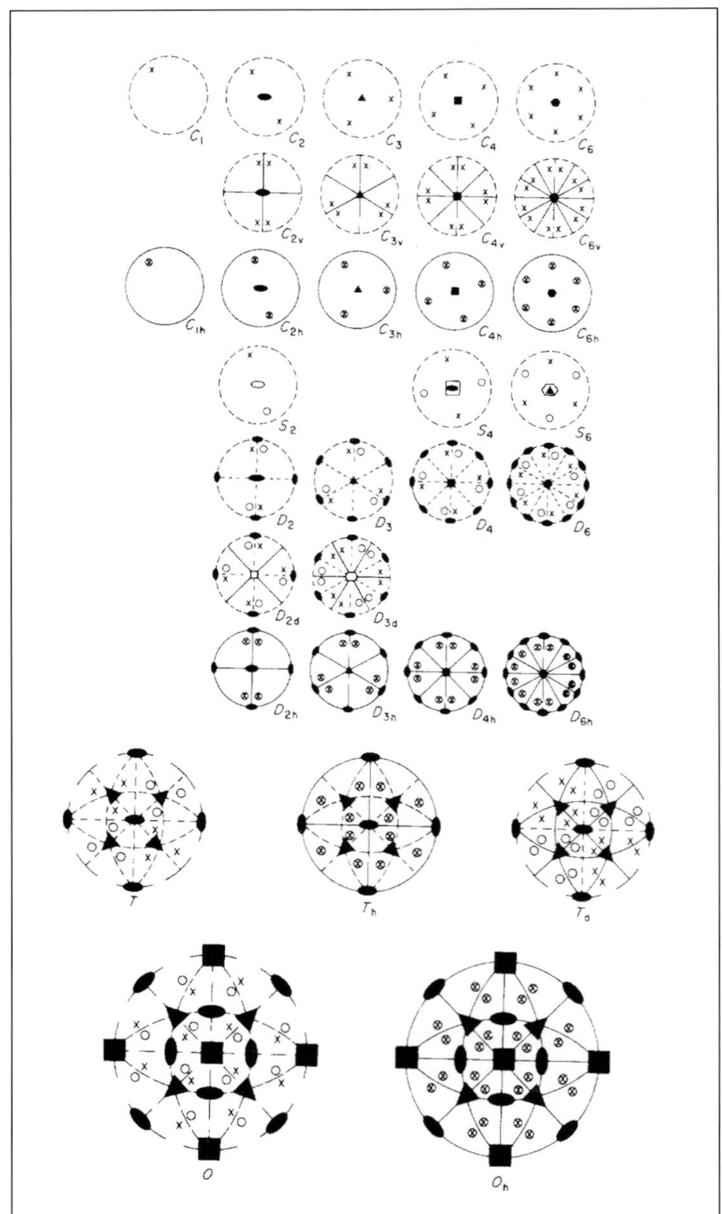


**Figure 2.25** Stereographic projections of the regular orbits of the Crystallographic point groups onto an inscribed sphere, showing vertices of the orbits (open-circles in the Northern hemisphere and crosses in the Southern) and the locations of symmetry elements as described in the text. [*From Symmetry in Physics, J.P. Elliot and P.G. Dawber, Macmillan, London, 1979.*]

**Table 2.3 Wyckoff Positions (WP) for symmorphic space group 187 (P-6m2/ $D^1_{3h}$) with the point group $D_{3h}$ as the factor group G/T, with T the group of translations. Note, especially that the symbols x, y and z in the tables are the magnitudes along the a, b and c edges of the hexagonal unit cell. The entries in the column SG identify the site groups for the different sets of equivalent positions in the unit cells distinguished by the different WP.**

| # | WP | SG | Coordinates | | | |
|---|---|---|---|---|---|---|
| 12 | o | 1 | (x, y, z) | (−y, x − y, z) | (−x + y, −x, z) | (x, y, −z) |
| | | | (−y, x − y, −z) | (−x + y, −x, −z) | (−y, −x, z) | (−x + y, y, z) |
| | | | (x, x − y, z) | (−y, −x, −z) | (−x + y, y, −z) | (x, x − y, −z) |
| 6 | n | m. | (x, −x, z) | (x, 2x, z) | (−2x, −x, z) | (x, −x, −z) |
| | | | (x, 2x, −z) | (−2x, −x, −z) | | |
| 6 | m | m.. | (x, y, 1/2) | (−y, x − y, 1/2) | (−x + y, −x, 1/2) | (−y, −x, 1/2) |
| | | | (−x + y, y, 1/2) | (x, x − y, 1/2) | | |
| 6 | l | m.. | (x, y, 0) | (−y, x − y, 0) | (−x + y, −x, 0) | (−y, −x, 0) |
| | | | (−x + y, y, 0) | (x, x − y, 0) | | |
| 3 | k | mm2 | (x, −x, 1/2) | (x, 2x, 1/2) | (−2x, −x, 1/2) | |
| 3 | j | mm2 | (x, −x, 0) | (x, 2x, 0) | (−2x, −x, 0) | |
| 3 | i | 3m. | (2/3, 1/3, z) | (2/3, 1/3, −z) | | |
| 2 | h | 3m. | (1/3, 2/3, z) | (1/3, 2/3, −z) | | |
| 2 | g | 3m. | (0, 0, z) | (0, 0, −z) | | |
| 1 | f | −6m2 | (2/3, 1/3, 1/2) | | | |
| 1 | e | −6m2 | (2/3, 1/3, 0) | | | |
| 1 | d | −6m2 | (1/3, 2/3, 1/2) | | | |
| 1 | c | −6m2 | (1/3, 2/3, 0) | | | |
| 1 | b | −6m2 | (0, 0, 1/2) | | | |
| 1 | a | −6m2 | (0, 0, 0) | | | |

Individual orbits, as sets of coordinate points, which are decorations of the lattice points by the 'basis' elements of structure are listed using Wyckoff numbers for individual unit cells or conveniently grouped sets of unit cells in the International Tables for Crystallography published. Two examples of these listings are reproduced in Tables 2.3 and 2.4.

Table 2.3 is the listing for Space Group 187, which group is realized on a hexagonal lattice by decorations exhibiting $D_{3h}$ point symmetry.

The sets of coordinates in the table identify the possible orbits of the point group sketched in Figure 2.2. However, because of the interactions of the translational and point operations, there are, 6 sites in the unit cell exhibiting $D_{3h}$ site symmetry, which correspond to the occurrences of the $O_1$ orbit, 3 occurrences of the $O_2$ orbit, 2 occurrences of the $O_3$ orbit, 2 occurrences of the $O_{6h}$ orbit, 1 occurrence of the $O_{6v}$ orbit and one regular orbit, $O_{12}$.

Table 2.4 shows the listing for the related *non-symmorphic* space group 188 (P-62m). The 'l' positions in the table do not identify the regular orbit of the point group $D_{3h}$ in this example since a non-primitive translation is required to render the 12 positions equivalent. Thus, the set of 12 positions divide into 2 sets of 6 exhibiting three-fold rotation and mirror-plane reflection, but they are not interchangeable under any point group operation of $D_{3h}$. However, there is a one-to-one correspondence of the factor group of the space group, G/T, with respect to the group of translations, T, and the elements of the crystallographic point

**Table 2.4 Wyckoff Positions (WP) for symmorphic space group 188 (P-6c2/ $D^2_{3h}$) with the point group $D_{3h}$ as the factor group G/T, with T the group of translations. Again, note, especially that the symbols x, y and z in the tables are the magnitudes along the a, b and c edges of the hexagonal unit cell. The entries in the column SG identify the site groups for the different sets of positions in the unit cells distinguished by the different WP, which are rendered equivalent under the operations of the factor group G/T as explained in the text.**

| # | WP | SG | Coordinates: | | | |
|---|---|---|---|---|---|---|
| 12 | o | 1 | (x, y, z) | (−y, x − y, z) | (−x + y, −x, z) | (x, x − y, −z) |
| | | | (−y, −x, −z) | (−x + y, −x, −z) | (x, x − y, z + 1/2) | (−x + y, y, z + 1/2) |
| | | | (−y, −x, z + 1/2) | (−y, x − y, −z + 1/2) | (−x + y, −x, −z + 1/2) | (−x + y, y, z + 1/2) |
| 6 | n | m. | (x, −x, z) | (x, 2x, z) | (−2x, −x, z) | (x, −x, −z) |
| | | | (x, 2x, −z) | (−2x, −x, −z) | | |
| 6 | m | m.. | (x, y, 1/2) | (−y, x − y, 1/2) | (−x + y, −x, 1/2) | (−y, −x, 1/2) |
| | | | (−x + y, y, 1/2) | (x, x − y, 1/2) | | |
| 6 | l | m.. | (x, y, 0) | (−y, x − y, 0) | (−x + y, −x, 0) | (−y, −x, 0) |
| | | | (−x + y, y, 0) | (x, x − y, 0) | | |
| 3 | k | mm2 | (x, −x, 1/2) | (x, 2x, 1/2) | (−2x, −x, 1/2) | |
| 3 | j | mm2 | (x, −x, 0) | (x, 2x, 0) | (−2x, −x, 0) | |
| 3 | i | 3m. | (2/3, 1/3, z) | (2/3, 1/3, −z) | | |
| 2 | h | 3m. | (1/3, 2/3, z) | (1/3, 2/3, −z) | | |
| 2 | g | 3m. | (0, 0, z) | (0, 0, −z) | | |
| 1 | f | −6m2 | (2/3, 1/3, 1/2) | | | |
| 1 | e | −6m2 | (2/3, 1/3, 0) | | | |
| 1 | d | −6m2 | (1/3, 2/3, 1/2) | | | |
| 1 | c | −6m2 | (1/3, 2/3, 0) | | | |
| 1 | b | −6m2 | (0, 0, 1/2) | | | |
| 1 | a | −6m2 | (0, 0, 0) | | | |

group $D_{3h}$. The elements of G/T are $\{R|\tau_i\}T$, with $\tau_i$ the non-primitive translation, either glide or screw, and correspond to the elements $\{R_i|0\}$ of the crystallographic point group and so the equivalence of the 12 positions of maximum multiplicity in structures exhibiting $D^2_{3h}$ symmetry is ensured.

## 2.9 Crystals as 'Point' Structures

### 2.9.1 Cubium

Within the context of an orbit-by-orbit analysis of structure, it is attractive to consider the growth of extended structures by 'decoration' of a point following the restrictions imposed by particular choices of overall symmetry.

As a first example, consider the growth of a cubic lattice about the origin. For the simple primitive lattice[8], most familiar as the structure of the model crystal 'cubium', the cubic array

[8] The $\alpha$-phase of polonium is the only known real example of a simple cubic structure.

**Table 2.5 The factor group[9], G/T, of the space group of the diamond lattice (Fd3m $O_h^7$). The operations are identified in the form {R(x,y,z)/$\tau$} with R a rotation about the axis (x,y,z) and $\tau$ a translation. The translational components of the factor group operations are listed in fractional unit cell coordinates, $\tau$ equal to 1/2,1/2,1/2 in, for example, {$C_2$(0,1,1)|1/2,1/2,1/2}. Note, that the first 24 operations listed identify the point group $T_d$.**

| {E\|000} | {$S_4$(0,0,1)\|000} | {$C_2$(1,1,0)\|1/2,1/2,1/2} | {i\|1/2,1/2,1/2} |
|---|---|---|---|
| {$C_3$(1,1,1)\|000} | {$S_4$(0,1,0)\|000} | {$C_2$(1,0,1)\|1/2,1/2,1/2} | {$S_6$(1,1,1)\|1/2,1/2,1/2} |
| {$C_3$(1,−1,−1)\|000} | {$S_4$(1,0,0)\|000} | {$C_2$(0,1,1)\|1/2,1/2,1/2} | {$S_6$(1,−1,−1)\|1/2,1/2,1/2} |
| {$C_3$(−1,1,−1)\|000} | {$S_4^2$(0,0,1)\|000} | {$C_2$(0,1,−1)\|1/2,1/2,1/2} | {$S_6$(−1,1,−1)\|1/2,1/2,1/2} |
| {$C_3$(−1,−1,1)\|000} | {$S_4^2$(0,1,0)\|000} | {$C_2$(1,0,−1)\|1/2,1/2,1/2} | {$S_6$(−1,−1,1)\|1/2,1/2,1/2} |
| {$C_3^2$(1,1,1)\|000} | {$S_4^2$(1,0,0)\|000} | {$C_2$(1,−1,0)\|1/2,1/2,1/2} | {$S_6^5$(1,1,1)\|1/2,1/2,1/2} |
| {$C_3^2$(1,−1,−1)\|000} | {$\sigma_d$(1,1,0)\|000} | {$C_4$(0,0,1)\|1/2,1/2,1/2} | {$S_6^5$(1,−1,−1)\|1/2,1/2,1/2} |
| {$C_3^2$(−1,1,−1)\|000} | {$\sigma_d$(1,0,1)\|000} | {$C_4$(0,1,0)\|1/2,1/2,1/2} | {$S_6^5$(−1,1,−1)\|1/2,1/2,1/2} |
| {$C_3^2$(−1,−1,1)\|000} | {$\sigma_d$(0,1,1)\|000} | {$C_4$(1,0,0)\|1/2,1/2,1/2} | {$S_6^5$(−1,−1,1)\|1/2,1/2,1/2} |
| {$C_2$(0,0,1)\|000} | {$\sigma_d$(1,−1,0)\|000} | {$C_4^3$(0,0,1)\|1/2,1/2,1/2} | {$\sigma_h$(0,0,1)\|1/2,1/2,1/2} |
| {$C_2$(0,1,0)\|000} | {$\sigma_d$(1,0,−1)\|000} | {$C_4^3$(0,1,0)\|1/2,1/2,1/2} | {$\sigma_h$(0,1,0)\|1/2,1/2,1/2} |
| {$C_2$(1,0,0)\|000} | {$\sigma_d$(0,1,−1)\|000} | {$C_4^3$(1,0,0)\|1/2,1/2,1/2} | {$\sigma_h$(1,0,0)\|1/2,1/2,1/2} |

of hydrogen atoms, all lattice points are determined by the lattice vector, $R_{mnp}$

$$R_{mnp} = m\mathbf{a} + n\mathbf{b} + p\mathbf{c}$$

with m, n and p integers and including zero values. For cubium, the lattice exhibits a distinct point group $O_h$ and space group 221 (Pm3m $O_h^1$) describes the crystal structure.

Actual cubic arrays are found in many simple metal structures, in which a cubic unit cell is maintained by allowing body-centring or face-centring of the primitive cubic unit cell, with single metal atoms at each lattice point, thereby simplifying the trigonometry required to calculate the important interplanar spacing parameter of Bragg's Law. For example, the space group 229 (Im3m $O_h^9$) describes the BCC crystal structure of tungsten, while space group 225 (Fm3m $O_h^5$) distinguishes FCC crystal structure of aluminium. To identify all the lattice points in these cubic arrays, it is necessary to relax the condition that all m, n and p coefficients be integer. For the FCC lattice, the rule is that half-integer values are allowed subject to the restriction that the sum $m+n+p$ be integer. For BCC lattice, again half-integer values are allowed, but the restriction is that the sums $m + n$, $n + p$ and $m + p$ must all be integer.

These observations provide a convenient method to classify the electronic densities of states in finite clusters for metals exhibiting cubic crystal structures. Each {m, n, p} set determining

[9] O.V. Kovalev, *Irreducible Representations of the Space Groups* [Gordon and Breach, New York, 1965]; O. V. Kovalev, *Representations of the Crystallographic Space Groups: Edition 2* [Gordon and Breach Science Publishers, Switzerland, 1993].

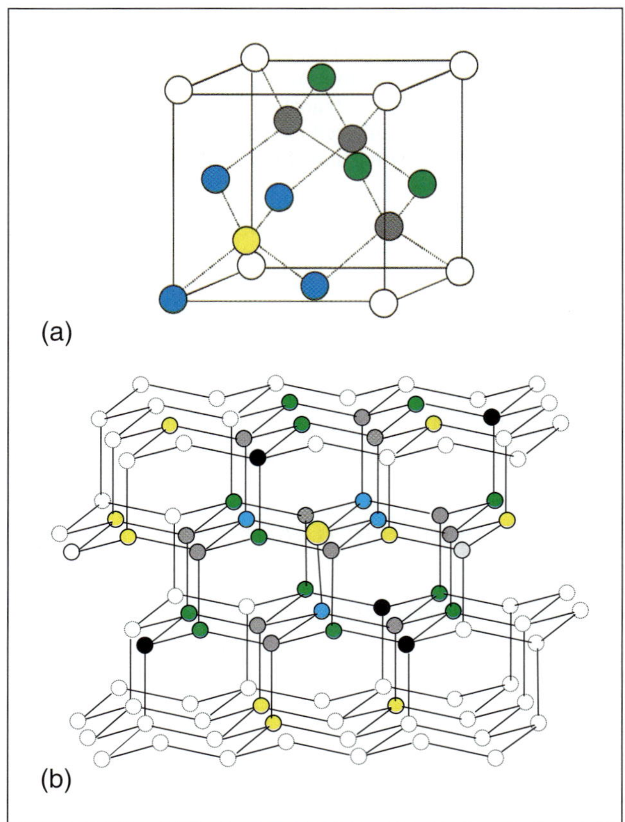


**Figure 2.26** Decoration of vertices in the diamond lattice to illustrate the possibility of growing the lattice by decoration about a single point using the orbits of the point group $T_d$: (a) within the conventional unit cell (b) about the larger circle, ◯, in the extended lattice of interlocked 'cyclohexane chairs'.

the lattice vector identifies a particular orbit of the point group of the structure and so the numbers of different kinds of irreducible symmetries possible [e.g. LCAO-MOS] are known from such analyses. Moreover, because in the extended crystal structure it follows that the regular orbit dominates increasingly as the radial distance is increased, this criterion can be applied to assess the modelling of the extended structure properties by the properties of a finite cluster.

For the idealized cubic arrays found as the crystal structures of simple metals, the lattice vector is determined solely as the square root of the sums of squares of the allowed m, n and p coefficient assuming that $|\mathbf{a}| = |\mathbf{b}| = |\mathbf{c}| = 1$. So the problem to find the distribution of the $O_h$ orbits as a function of cluster radius is reduced to the '3-squares' problem in mathematics[10].

[10] Charles M. Quinn, Densities of states in finite metal clusters, a group theory analysis, *Surface Science*, **156** (1985) 410. Charles. M. Quinn, Densities of states in particles and clusters: characterization of bulk and surface states, *Phil. Trans. Roy. Soc.* (London) **A318** (1980) 127.

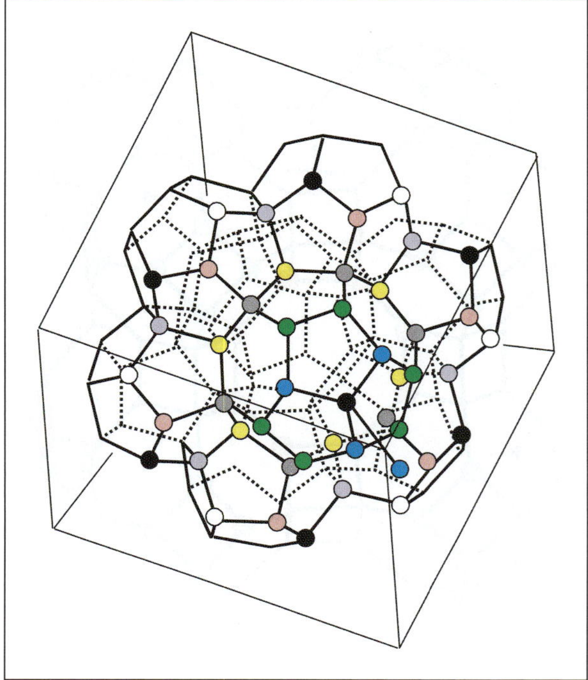

**Figure 2.27** The growth of the zeolite lattice MTN by decoration of a single vertex with the orbits of the point group $T_d$ for the structure exhibiting Fd3m symmetry as before, but now different because of the choice of the alternative geometry [Figure 2.12] for the first $O_{12}$ orbit of the group about the 'nearest neighbour' set of four vertices defining the basic tetrahedron.

### 2.9.2 Diamond

As a second example, consider the growth of the diamond lattice about a central point. Space group 227 (Fd3m $O_h^7$) describes the diamond crystal structure and we see from this information that the structure is non-symmorphic and that while the factor group is of order 48, not all of these symmetry operations are common to the point symmetry group $O_h$. Table 2.5 lists the 'point' symmetry operations of $O_h^7$, in which the non-primitive translation $(\frac{1}{2}, \frac{1}{2}, \frac{1}{2})$ accompanying some rotations is required to mimic some of the operations of the $O_h$ point group.

However, inspection of the list of symmetry operations of the factor group of space group 227 in Table 2.5 reveals that the 24 point symmetry operations of the $T_d$ point group are present amongst the 48 operations required to generate the isomorphic group to $O_h$. Thus, it is possible to construct the lattice by decorating a single point with appropriate orbits of the $T_d$ point group. For atoms within the conventional unit cell of the diamond lattice this is illustrated in Figure 2.26a and related to the extended array of 'cyclohexane chairs', present in the diamond lattice, Figure 2.26b.

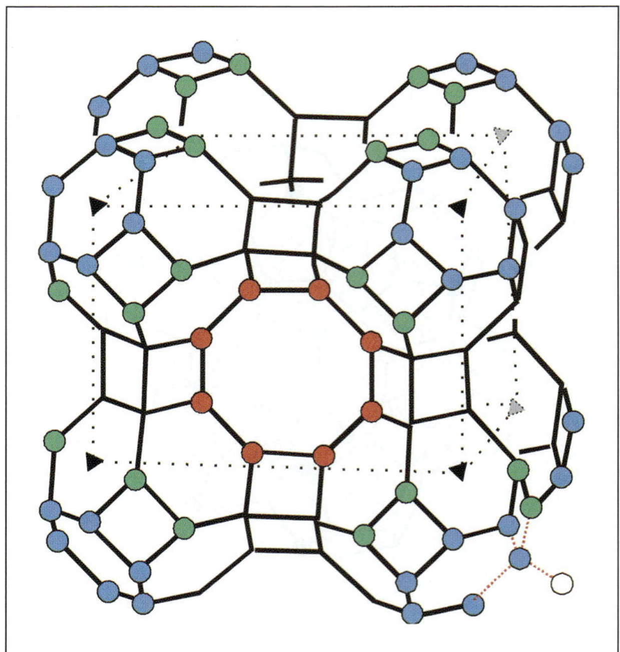

**Figure 2.28** The Zeolite A structure considered as an orbit-by-orbit building exercise: about a central position and a regular orbit [Figure 2.5], the addition of increasingly larger cubes decorated by regular orbits leads to the Pm-3m primitive cubic geometry of the framework. In the bottom right-hand corner of the diagram, the single tetrahedral unit is distinguished by the dotted red 'bonds'.

### 2.9.3 Silicates and zeolites

The growth of the diamond lattice as an orbit-by-orbit building sequence, Figure 2.26b, results from the choice that the arrangement, Figure 2.12, of the first $O_{12}$ orbit about the central position be 'staggered' with respect to the vertices of the underlying tetrahedron. A glimpse at the complexity of crystal structures is revealed if we make the alternative choice that the first $O_{12}$ orbit is 'eclipsed' in orientation with respect to these vertices. This leads to the alternate structure of vertex-shared tetrahedra if the mid-point of each bond are considered to be, for example, the oxygen positions in fused silicate anions, $SiO_4^{-2}$, as found in the zeolite structures. This space filling sequence resulting from the decoration of a single vertex with orbits of the $T_d$ group is shown in Figure 2.27. Characteristic *dodecahedral holes* form bounded by '$\beta$-cages' of truncated octahedra, which can link through the square or hexagonal faces. For hexagonal cross-section 'tunnels' the overall structure, again of Fd3m symmetry, corresponds to that found in the zeolite MTN.

Myriads of structures of isopoly and heteropolyacid anion cages are found as minerals with many different metal cations completing the chemical structures. The basic tetrahedral or near tetrahedral local geometry of, for example, component $SiO_4^{2-}$ units can be formed between neighbouring shells of decorations of cubic vertices. Three examples, which emphasize the orbit-by-orbit perspective are shown in Figures 2.28 to 2.30.

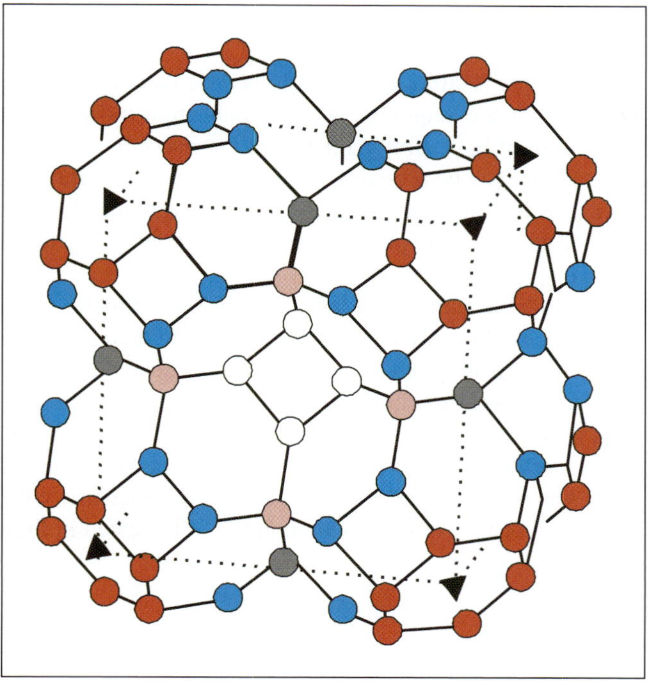

**Figure 2.29** The Sodalite/Ultramarine framework, which results by sharing of the square faces of the $O_{24h}$ orbit truncated octahedron over the extended lattice again, shown as a decoration of vertices of the cube.

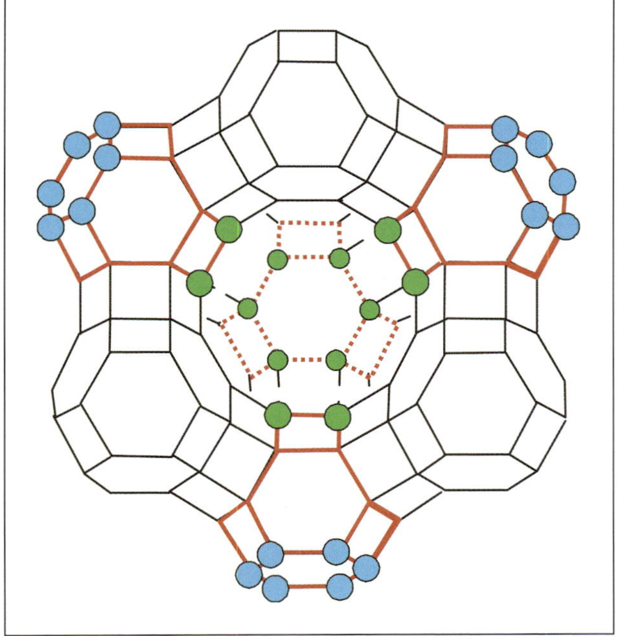

**Figure 2.30** The Faujasite structure of Fd3m space group symmetry with sets of $O_{24h}$ orbits identified about the tetrahedral vertices of a cube positioned at the centres of the truncated octahedra coloured in red.

The primitive cubic Zeolite A structure normally is described, in the literature, as a result of fulfilling the local tetrahedral requirement of 4-connection by making the square faces of the truncated octahedron [$O_{24h}$] orbit geometry of $O_h$ into cubic connecting channels by which the extended structure can be propagated. An alternative description of the structure is emphasized in Figure 2.28. The structure is considered to arise by the decoration of concentric cubes of increasing cube edges with regular orbits of the $O_h$ point group leading to the Pm-3m primitive lattice.

The ultramarine and sodalite framework lattice is shown in Figure 2.29. This framework results when the square faces of the truncated octahedron are shared. Again, the perspective in the figure emphasizes the possibility of constructing the extended lattice of P4-3m symmetry.

In contrast to these frameworks exhibiting symmorphic space groups, the extension of the truncated octahedron structure by fulfilling the 4-connection requirement in hexagonal channels about a single $O_{24h}$ orbit leads to the Faujasite framework exhibiting Fd3m space group symmetry. Thus, only sets of tetrahedral vertices of any one concentric cube are occupied in the initial stages of growth of the lattice about a point as is indicated in Figure 2.30.

# 3

# Decorations of orbits using local functions: reducible characters for s, p, d, ... local functions; central polynomial functions as basis sets for the irreducible representations of the point groups; the construction of group orbitals

Suppose that G is the group of symmetry operations of a polyhedron or polygon, with vertices corresponding to the atomic positions in a particular molecular structure. The division of the structure into orbits, as sets of vertices equivalent under the actions of the group symmetry operations and the calculation of associated permutation representations/characters were described in Chapter 2. In this chapter, the identity between the permutation representation/character on the labels of the vertices of an orbit and the $\sigma$ representation/character on sets of local s-orbitals or $\sigma$-oriented local functions is exploited to construct the characters of the representations that follow from the transformation properties of higher order local functions.

In this chapter you will learn:

1. how to decorate structure orbits with sets of local functions sited at the vertices of the orbit polyhedra or polygons;
2. how the characters generated by the symmetry operations of the group, on such sets of functions, follow from the permutation characters of the underlying structure orbits, since these characters are generated by the transformation properties of sets of local $\sigma$-functions [s-, p$\sigma$- and d$\sigma$-like], and all the characters for sets of higher order local functions [e.g. p$\pi$ or d$\pi$-like, d$\delta$-like, etc.] are found from recursion relationships;
3. how the list of such orbits is the only input required to determine all the group theoretical properties of a molecule, because these depend on how a molecular structure divides into its unique set of component orbits;
4. how convenient it is to have to hand sets of central functions, which provide distinct bases for the regular orbit characters as these provide bases for all possible cases because these can be applied to construct group orbitals and identify normal modes of vibration.

## 3.1 $\sigma$ Characters: Local $\sigma$, $\pi$ and $\delta$, ... Harmonic Functions

The observation that the permutation character on a set of vertices and the $\sigma$-character for a set of ns orbitals localized on these vertices are the same is illustrated in Figure 3.1, using as in Chapter 2, the example of the ammonia molecule, but, in this case, considering the permutations of the hydrogen 1s atomic orbitals about the equilateral triangular geometry of an $O_3$ orbit decorated in this manner.

Since the actions of the symmetry operators on local s-atomic orbitals are to leave the orbitals invariant, it is clear that the resultant matrices in Figure 3.1 are the same as those

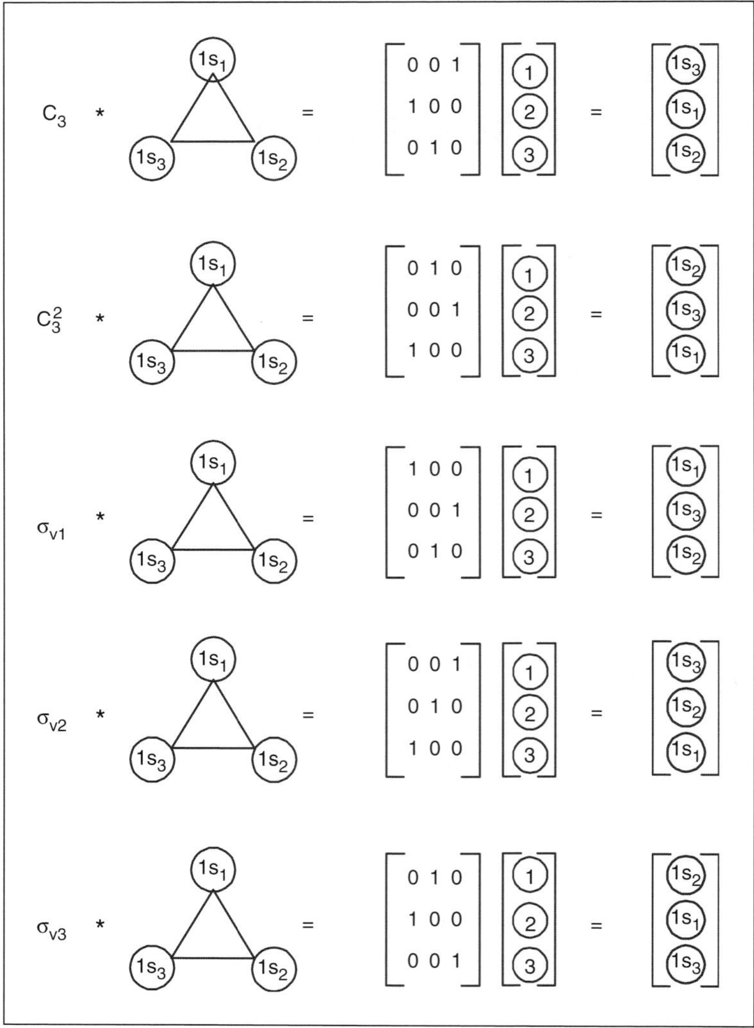


**Figure 3.1** Permutation matrices describing the actions of the symmetry operators of the point group $C_{3v}$ on 3 hydrogen 1s atomic orbitals localized on the vertices of the equilateral triangular base of the pyramidal geometry of $NH_3$.

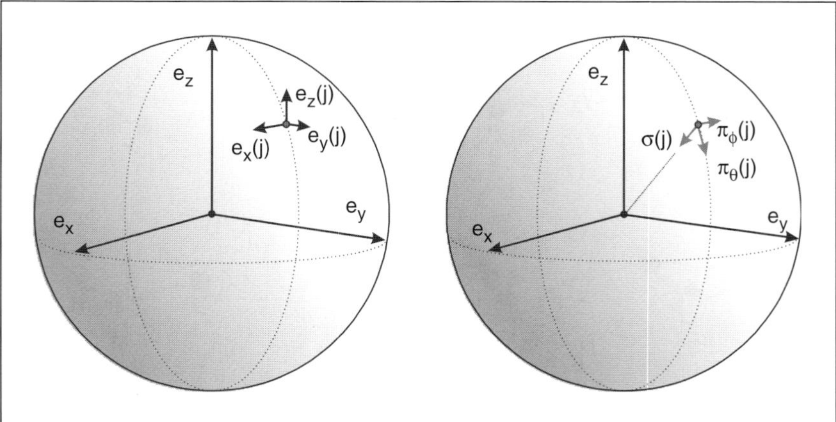


**Figure 3.2** Comparison of the local Cartesian and transformed coordinate systems at a general point on the unit sphere; the standard spherical harmonic functions as the angular parts of the appropriate atomic orbitals are defined with reference to the local Cartesian set $e_x(j)$, $e_y(j)$ and $e_z(j)$ for each atomic position $(j)$ with radius vector $R_j$ on the unit sphere. Then the transformation of equation 3.1 is applied to construct the new local coordinate system $\sigma(j)$, $\pi_\theta(j)$ and $\pi_\phi(j)$.

for the permutation of the hydrogen position labels displayed in Figure 2.1, because the only effects of the symmetry operations are to permute the arguments of the local functions, i.e. either to shift or not shift a local s function. Furthermore, if suitable linear combinations of local p- and d-atomic orbitals are formed so that the resultant functions, are rendered invariant under the actions of the rotations of the structure symmetry group, then these, also, give rise to the permutation character for labels of the vertices on that structure. Such linear combinations are known individually as p$\sigma$ and d$\sigma$ functions and so the permutation characters, which result from the actions of the symmetry operators, are called $\sigma$ characters.

Complementary to these radial linear combinations, we can construct pairs of local functions of $\pi$ and $\delta$ types and so on, with angular momentum quantum numbers ($\lambda$) about the radius vectors of $\pm 1$ for $\pi$, $\pm 2$ for $\delta$, $\pm 3$ for $\phi$, ... upon which the higher order representations and characters can be constructed.

Suppose that a vertex, j, carries a set of local unit axes $e_x(j)$, $e_y(j)$ and $e_z(j)$ that run parallel to the global coordinate axes at the origin. $\sigma(j)$, $\pi_\theta(j)$ and $\pi_\phi(j)$ are defined at the vertex j as shown in Figure 3.2. The $\sigma$ vector points from j along the radial directions towards the centre of the unit sphere, $\pi^\theta(j)$ points in the tangential direction of increasing $\theta$ and $\pi^\phi(j)$ points in the tangential direction of increasing $\phi$.

Elementary trigonometry gives the transformation relating the sets of local axes

$$\begin{pmatrix} \pi_\theta\ (j) \\ \pi_\phi\ (j) \\ \sigma\ (j) \end{pmatrix} = \begin{pmatrix} \cos\theta\cos\phi & \cos\theta\sin\phi & -\sin\theta \\ -\sin\phi & \cos\phi & 0 \\ -\sin\theta\cos\phi & -\sin\theta\sin\phi & -\cos\theta \end{pmatrix} \begin{pmatrix} e_x\ (j) \\ e_y\ (j) \\ e_z\ (j) \end{pmatrix} \qquad 3.1$$

This transformation matrix is the key to the construction of the correctly oriented local $\sigma$, $\pi$ and $\delta$ functions, which under the symmetry operations of the group of the molecular structure

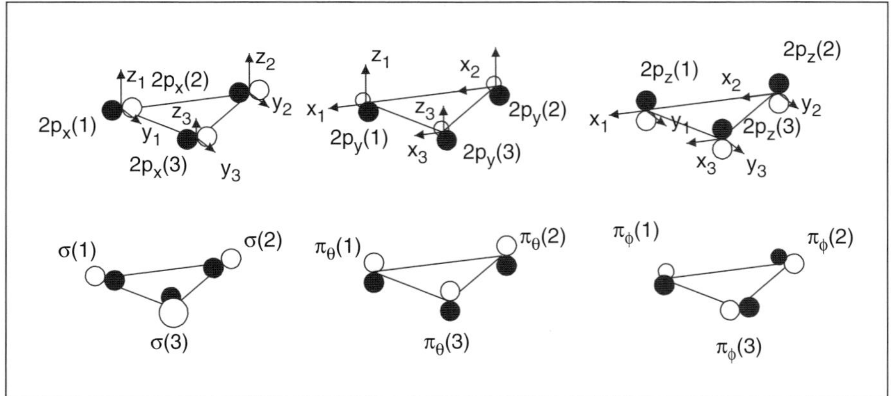


**Figure 3.3** The 2p orbitals of three atoms, arranged in equilateral triangular geometry, lying along standard sets of local Cartesian axes in the first row of diagrams. In the second row, appropriate linear combinations of these functions are taken to construct the p$\sigma$ and p$\pi$ local orbitals.

generate a reducible representation. The $\sigma$, $\pi$ and $\delta$ group orbitals are basis functions for the irreducible components of this representation. Local p orbitals are first-order polynomials in x, y and z, which requires one application of equation 3.1 to effect the transformation to the new coordinate system at each vertex; local d orbitals are second order, and so equation 3.1 is applied twice and so on.

For example, for the case of three fluorine atoms in the equilateral triangular geometry of $BF_3$, the local p$\sigma$, p$\pi_\theta$ and p$\pi_\phi$ orbitals are sketched in Figure 3.3. Similarly, for the same geometrical arrangement of three iron atoms in $Fe_3(CO)_{12}$, which molecule exhibits $D_{3h}$ point symmetry, Figure 3.4, it is possible to construct local d$\sigma$, d$\pi$ and d$\delta$ orbitals as sketched in Figure 3.5.

## 3.2 The Characters of the Representations Generated by Local Functions

We can build on the analysis for $\sigma$ characters in the following general manner. The reducible character, $\rho$, of the group G generated by local $\sigma$ functions on orbit vertices is the permutation character composed of the individual traces, $\rho(g)$, for the elements, g, of G, with $\rho(g)$ simply the number of vertices, $P_1 \ldots P_n$, unchanged by a representative symmetry operation in the class. Let $\Gamma_\varepsilon$ be the irreducible 1D character of the group with matrix traces +1 on the set of proper rotations of G and −1 on the set of improper rotations of G for the groups in which such symmetry operations are present. In turn, let $\gamma^0_{\text{origin}}$, $\gamma^1_{\text{origin}}$, $\gamma^2_{\text{origin}}$, etc. identify the characters of the group for which the s($\ell = 0$), p($\ell = 1$), d($\ell = 2$), etc. spherical harmonics, $[Y_{lm}(\theta,\phi)]$, provide bases, about the central origin in the global coordinate system and let $\Gamma_0, \Gamma_1, \Gamma_2, \ldots, \Gamma_\lambda$ [i.e. $\Gamma_\sigma$, $\Gamma_\pi$, $\Gamma_\delta$ and so on] be the components of the characters for the local functions. These are identified in Figure 3.4 for the example of the valence atomic orbitals of the fluorine triangle in $NF_3$ and in Figure 3.5 for the $Fe_3$ triangle 3d orbitals that can be transformed into $\sigma$, $\pi$ and $\delta$ sets. Let $\gamma^\ell_{\text{ligands}}$ be the character generated by transformation properties of a set

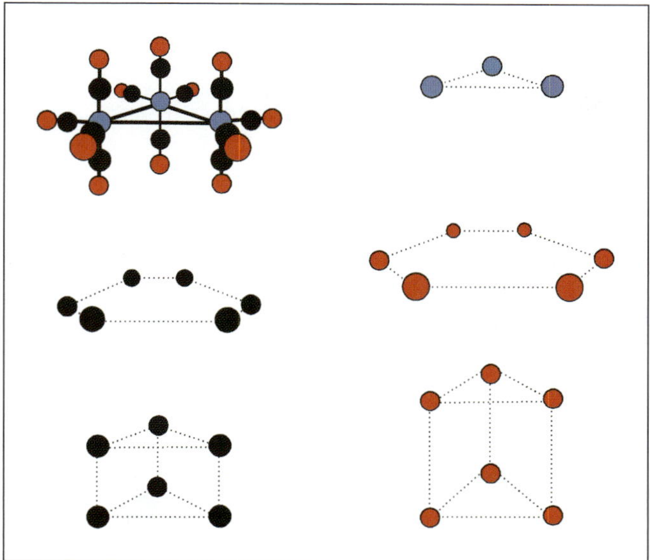

**Figure 3.4** Sketch of the molecular structure of $Fe_3(CO)_{12}$, which has $D_{3h}$ point symmetry, showing the central $O_3$ orbit of iron atoms, the planar $O_{6h}$ orbits of carbon and oxygen atoms and the non-planar (trigonal prismatic) $O_{6v}$ orbits of carbon and oxygen atoms.

of $Y_{lm}(\theta, \phi)$ sited on the vertices of an orbit of structure, for example, the angular parts of the iron atom valence atomic orbitals in $Fe_3(CO)_{12}$, Figure 3.5.

The characters defined in the last paragraph are related by the equation

$$\gamma^{\ell}_{\text{ligands}} = \rho \times \gamma^{\ell}_{\text{origin}} \qquad 3.2$$

which is a consequence of the fact that the space of the level $\ell$ harmonics, at the vertices of the structure polyhedron or polygon, which identifies the group G, is the tensor product of the permutation character space and the space of the level $\ell$ central harmonics.

Decomposition of the tensor product space of equation 3.2, which is the foundation for many of the group theory calculations that can be performed using the files on the CD-ROM, leads to the identity

$$\gamma^{\ell}_{\text{ligands}} = \Gamma_0 + \Gamma_1 + \Gamma_2 + \cdots + \Gamma_{\ell} \qquad 3.3$$

For example, for $f(\mathbf{r}-\mathbf{P_j})$ the local 3d orbitals of the iron atoms in the triangular geometrical orbit of $Fe_3(CO)_{12}$, we know that there can be $\sigma$ $[\Gamma_0]$, $\pi$ $[\Gamma_1]$ and $\delta$-type orientations, $[\Gamma_2]$ of the orbitals in the transformed local coordinate system at each Fe atom position and it follows directly from equation 3.3 that

$$\Gamma_{\ell} = \rho \times (\gamma^{\ell}_{\text{origin}} - \gamma^{\ell-1}_{\text{origin}}) \qquad 3.4$$

which can be applied to determine higher-order characters.

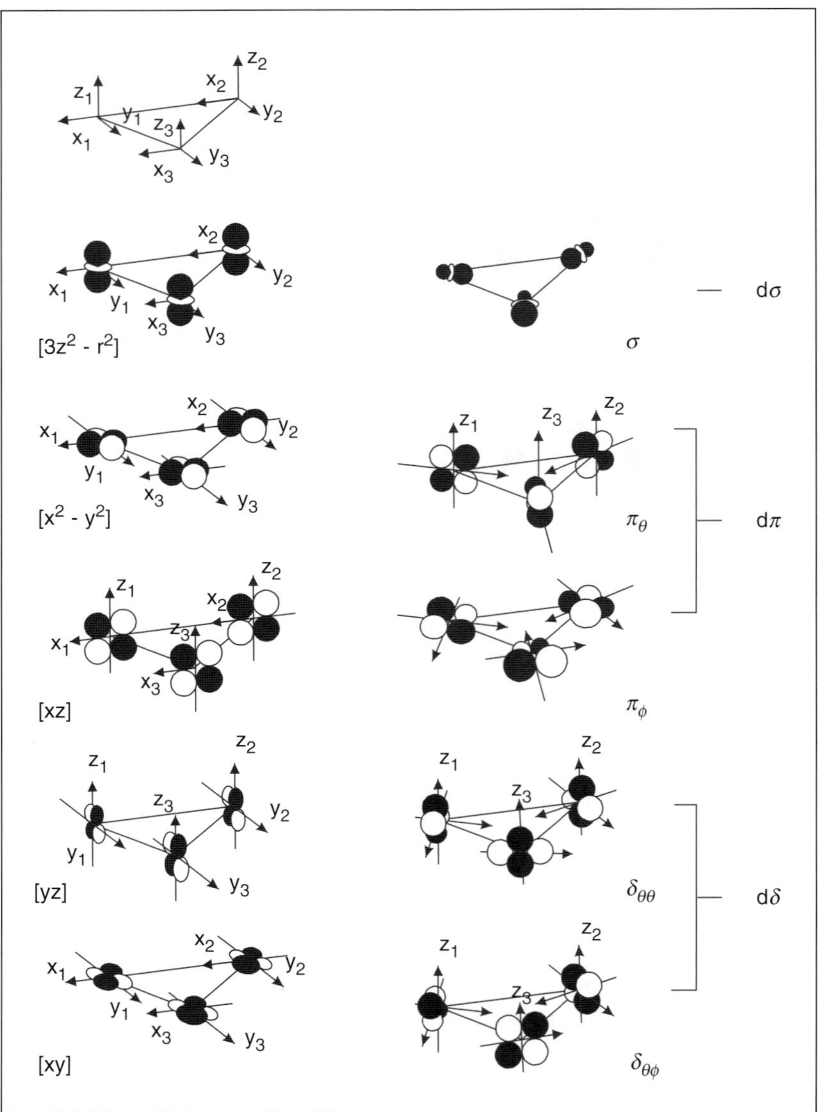


**Figure 3.5** Construction of local d$\sigma$, d$\pi$ and d$\delta$ local functions as linear combinations of the local 3d atomic orbitals in the $Fe_3$ triangle of $Fe_3(CO)_{12}$. The radially oriented d$\sigma$ local functions are sketched in the first diagram of the second column. Then, in the remaining diagrams of the second column, the complementary pairs of d$\pi$ and d$\delta$ group local functions are shown. For simplicity in these diagrams the transformed local functions are not identified individually, rather they are distinguished by type as $\sigma$; $\pi_\theta$ and $\pi_\phi$; $\delta_{\theta\theta}$ and $\delta_{\theta\phi}$ for later reference.

For the $Fe_3(CO)_{12}$ example, a full analysis requires the determination of $\pi$ and $\delta$ characters of the $Fe_3$ orbit and the four sets of $\pi$ characters arising from the 2p atomic orbitals of the two sets of $O_{6v}$ and $O_{6h}$ orbits of the $D_{3h}$ point symmetry molecular structure, shown in Figure 3.4, in addition to the individual permutation characters over the vertices of the orbits.

To determine the $\pi$ and $\delta$ characters, it is not necessary to evaluate the component traces separately for the individual classes of symmetry operations of the group G, which would often involve laborious trigonometry using standard methods. The application of equation 3.4 depends only on knowledge of the permutation character $\rho$ for each orbit and the readily determined transformation properties of the central harmonics under G.

The most practical procedure is to write out the implied recursive relationship between the characters generated by the local functions at the vertices giving rise to the permutation character of the orbit. Thus, for $\Gamma_{xyz}$ used to identify $\gamma^1_{origin}$, the character generated by central $p_x$, $p_y$ and $p_z$-like functions, $\Gamma_0$ the symmetric character of the group, with traces +1 for all symmetry operations, and $\Gamma_\varepsilon$ the antisymmetric character of the group, with traces +1 for all proper rotations and traces −1 for all improper rotations[1] we have from equation 3.4, for any group orbit with permutation character $\Gamma_\sigma$

$$\Gamma_\pi = \Gamma_\sigma \times \Gamma_{xyz} - \Gamma_\sigma \qquad 3.5$$

$$\Gamma_\delta = \Gamma_\pi \times \Gamma_{xyz} - \Gamma_\pi - \Gamma_\sigma \times (\Gamma_0 + \Gamma_\varepsilon) \qquad 3.6$$

and for higher order interactions $[\ell \geq 2]$,

$$\Gamma_{\ell+1} = \Gamma_\ell \times \Gamma_{xyz} - \Gamma_\ell - \Gamma_{\ell-1} \qquad 3.7$$

These results follow on application of the multiplication rules for the characters of the spherical group, $R_3$,

$$\gamma^\ell_{origin} \times \gamma^1_{origin} = \gamma^{\ell+1}_{origin} + \gamma^\ell_{origin} + \gamma^{\ell-1}_{origin} \qquad 3.8$$

to decompose the tensor product space corresponding to the character

$$\gamma^\ell_{ligands} \times \gamma^1_{origin} = \rho \times \left(\gamma^\ell_{origin} - \gamma^{\ell-1}_{origin}\right) \times \gamma^1_{origin} \qquad 3.9$$

We have

$$\begin{aligned}\gamma^\ell_{ligands} + \gamma^1_{origin} &= \rho \times \left(\gamma^{\ell+1}_{origin} + \gamma^\ell_{origin} + \gamma^{\ell-1}_{origin} - \gamma^\ell_{origin} - \gamma^{\ell-1}_{origin} - \gamma^{\ell-2}_{origin}\right)\\ &= \rho \times \left(\gamma^{\ell+1}_{origin} - \gamma^{\ell-2}_{origin}\right)\\ &= \rho \times \left(\gamma^{\ell+1}_{origin} - \gamma^\ell_{origin}\right) + \rho \times \left(\gamma^\ell_{origin} - \gamma^{\ell-1}_{origin}\right) + \rho \times \left(\gamma^{\ell-1}_{origin} - \gamma^{\ell-2}_{origin}\right)\end{aligned} \qquad 3.10$$

[1] Proper rotations are identified in character tables by the symbols $C_n$, improper rotations, $S_n$, are all other operations in the group, the special improper rotations corresponding to inversion and reflection, being identified separately by the symbols i, $\sigma_v$, $\sigma_d$, $\sigma_h$.

which is the general recursive form of equation 3.5 and 3.6, i.e. 3.7, the factor in $\Gamma_\varepsilon$ in equation 3.6 arising from the antisymmetric terms in the square

$$\gamma^1_{\text{origin}} \times \Gamma_{\text{xyz}} = (\gamma^1_{\text{origin}})^2 \qquad 3.11$$

i.e. the bracketed terms in the general expression for the square

$$\gamma^\ell_{\text{origin}} \times \gamma^\ell_{\text{origin}} = \gamma^{2\ell}_{\text{origin}} + \gamma^{2\ell-2}_{\text{origin}} + \cdots + \gamma^0_{\text{origin}} + \left[\gamma^{2\ell-1} + \cdots + \gamma^1\right] \qquad 3.12$$

In electronic structure problems, we would normally be interested in bonding interactions of s, p and d-atomic orbitals from atoms sited on the vertices of a molecular structure and hence only $\Gamma_\sigma$, $\Gamma_\pi$ and $\Gamma_\delta$. In vibrational problems, the mechanical representation is $\Gamma_{\text{coordinates}} = \Gamma_\sigma \times \Gamma_{\text{xyz}} = \Gamma_\sigma + \Gamma_\pi$ for an empty cluster and $\Gamma_\sigma + \Gamma_\pi + \Gamma_{\text{xyz}}$ for a cluster with a centrally placed atom. For the cases of interest in molecular problems, equations 3.5 to 3.7, which follow from a knowledge of the permutation characters alone, can be used to generate, once and for all, the full set of reducible characters for the orbits of the molecular point groups.

This analysis is applied to generate the group character results in the files on the CDROM and the complete set of $\Gamma_\pi$ and $\Gamma_\delta$ characters for all the orbits of the molecular point groups are set out in Tables 3.1 to 3.3. Tables of orbits of the useful point groups date from the work of Brester[2] and Jahn and Teller[3], while lists of site groups (column 3 data in the tables) were compiled by Rytter[4].

The data in column 4 of the tables, the m values, represent the numbers of times orbits occur in a particular molecular structure. Thus, $m_0$ is the number of atomic positions on all elements of symmetry of the structure point group; $m_2$, $m_3$, $m_4$, ..., $m_v$, $m_h$, $m_2'$ $m_{2x}$, ..., $m_{yz}$ are numbers of sets associated with symmetry elements $C_2$, $C_3$, $C_4$, ..., $\sigma_v$, $\sigma_d$, $\sigma_h$, $C_2'$, $C_{2x}$, ..., $\sigma_{yz}$, while the symbol, m, without subscripting identifies the number of sets of atoms on a general position, i.e. on no element of symmetry. A general formula for, for example, the vibrational character of a molecule[1] can be written in terms of the m numbers. Thus, we find for a $C_{3v}$ molecule

$$\Gamma_\sigma = (m_o + m_v + m)A_1 + mA_2 + (m_v + 2m)E$$

and

$$\Gamma_\pi = (m_v + 2m)A_1 + (m_v + 2m)A_2 + (m_0 + 2m_v + 4m)E$$

with $\Gamma_{\text{xyz}} = A_1 + E$ and $\Gamma_{\text{rotations}} = A_2 + E$. Thus, $\Gamma_{\text{vibrations}}$ is

$$\Gamma_{\text{vibrations}} = (m_0 + 2m_v + 3m - 1)A_1 + (m_v + 3m - 1)A_2 + (m_0 + 3m_v + 6m - 2)E$$

[2] *See* G. Herzberg, *Infrared and Raman Spectra of Polyatomic Molecules*, Chapter 2 [van Nostrand, New York 1945].

[3] H.A. Jahn and E. Teller, *Proc. Roy. Soc.*, **A161** (1937) 220.

[4] E. Rytter, *Chemical Physics*, **12** (1976) 355.

**Table 3.1** $\sigma, \pi, \delta$ **representations on the distinct orbits of the molecular point groups $C_1$, $C_s$, $C_i$, $C_n$, $C_{nh}$ and $C_{nv}$ up to n = 6 : $O_i$ is the orbit of order i and m is the number of times it occurs in a particular molecule as explained in the text. For each group the regular representation direct sum is given explicitly as the $\sigma$ representation for the largest orbit, but otherwise is abbreviated as $\Gamma_{regular}$; * in column 1 identifies a group in which a central $O_1$ orbit can be present. For this orbit the $\sigma$, $\pi$ and $\delta$ analysis does not apply.**

| Point Group | Orbit | Site Group | | $\Gamma_\sigma$ | $\Gamma_\pi$ | $\Gamma_\delta$ |
|---|---|---|---|---|---|---|
| $C_1$ | $O_1$ | $C_1$ | m | A | $2\Gamma_{regular}$ | $2\Gamma_{regular}$ |
| $C_s^*$ | $O_2$ | $C_1$ | m | $A' + A''$ | $2\Gamma_{regular}$ | $2\Gamma_{regular}$ |
| $C_i^*$ | $O_2$ | $C_1$ | m | $A_g + A_u$ | $2\Gamma_{regular}$ | $2\Gamma_{regular}$ |
| $C_2$ | $O_1$ | $C_2$ | $m_0$ | A | 2B | 2A |
| | $O_2$ | $C_1$ | m | A + B | $2\Gamma_{regular}$ | $2\Gamma_{regular}$ |
| $C_3$ | $O_1$ | $C_4$ | $m_0$ | A | E | E |
| | $O_3$ | $C_1$ | m | A + E | $2\Gamma_{regular}$ | $2\Gamma_{regular}$ |
| $C_4$ | $O_1$ | $C_5$ | $m_0$ | A | E | 2B |
| | $O_4$ | $C_1$ | m | A + B + E | $2\Gamma_{regular}$ | $2\Gamma_{regular}$ |
| $C_5$ | $O_1$ | $C_5$ | $m_0$ | A | $E_1$ | $E_2$ |
| | $O_5$ | $C_1$ | m | $A + E_1 + E_2$ | $2\Gamma_{regular}$ | $2\Gamma_{regular}$ |
| $C_6$ | $O_1$ | $C_6$ | $m_0$ | A | $E_1$ | $E_2$ |
| | $O_6$ | $C_1$ | m | $A + B + E_1 + E_2$ | $2\Gamma_{regular}$ | $2\Gamma_{regular}$ |
| $C_{2v}$ | $O_1$ | $C_{2v}$ | $m_0$ | $A_1$ | $B_1 + B_2$ | $A_1 + A_2$ |
| | $O_{2[xz]}$ | $C_s$ | $m_{[xz]}$ | $A_1 + B_1$ | $\Gamma_{regular}$ | $\Gamma_{regular}$ |
| | $O_{2[yz]}$ | $C_s$ | $m_{[yz]}$ | $A_1 + B_2$ | $\Gamma_{regular}$ | $\Gamma_{regular}$ |
| | $O_4$ | $C_1$ | m | $A_1 + A_2 + B_1 + B_2$ | $2\Gamma_{regular}$ | $2\Gamma_{regular}$ |
| $C_{3v}$ | $O_1$ | $C_{3v}$ | $m_0$ | $A_1$ | E | E |
| | $O_3$ | $C_s$ | | $A_1 + E$ | $\Gamma_{regular}$ | $\Gamma_{regular}$ |
| | $O_6$ | $C_1$ | | $A_1 + A_2 + 2E$ | $2\Gamma_{regular}$ | $2\Gamma_{regular}$ |

| Point Group | Orbit | Site Group | | $\Gamma_\sigma$ | $\Gamma_\pi$ | $\Gamma_\delta$ |
|---|---|---|---|---|---|---|
| $C_{4v}$ | $O_1$ | $C_{4v}$ | $m_0$ | $A_1$ | E | $B_1 + B_2$ |
| | $O_{4d}$ | $C_s$ | $m_d$ | $A_1 + B_2 + E$ | $\Gamma_{regular}$ | $\Gamma_{regular}$ |
| | $O_{4v}$ | $C_s$ | $m_v$ | $A_1 + B_1 + E$ | $\Gamma_{regular}$ | $\Gamma_{regular}$ |
| | $O_8$ | $C_1$ | m | $A_1 + A_2 + B_1 + B_2 + 2E$ | $2\Gamma_{regular}$ | $2\Gamma_{regular}$ |
| $C_{5v}$ | $O_1$ | $C_{5v}$ | $m_0$ | $A_1$ | $E_1$ | $E_2$ |
| | $O_5$ | $C_s$ | $m_v$ | $A_1 + E_1 + E_2$ | $\Gamma_{regular}$ | $\Gamma_{regular}$ |
| | $O_{10}$ | $C_1$ | m | $A_1 + A_2 + 2E_1\ 2E_2$ | $2\Gamma_{regular}$ | $2\Gamma_{regular}$ |
| $C_{6v}$ | $O_1$ | $C_{6v}$ | $m_0$ | $A_1$ | $E_1$ | $E_2$ |
| | $O_{6d}$ | $C_s$ | $m_d$ | $A_1 + B_2 + E_1 + E_2$ | $\Gamma_{regular}$ | $\Gamma_{regular}$ |
| | $O_{6v}$ | $C_s$ | $m_v$ | $A_1 + B_1 + E_1\ E_2$ | $\Gamma_{regular}$ | $\Gamma_{regular}$ |
| | $O_{12}$ | $C_1$ | m | $A_1 + A_2 + B_1 + B_2$ $+ 2E_1 + 2E_2$ | $2\Gamma_{regular}$ | $2\Gamma_{regular}$ |
| $C^*_{2h}$ | $O_2$ | $C_2$ | $m_2$ | $A_g + A_u$ | $2B_g + 2B_u$ | $2A_g + 2A_u$ |
| | $O_{2h}$ | $C_s$ | $m_h$ | $A_g + B_u$ | $\Gamma_{regular}$ | $\Gamma_{regular}$ |
| | $O_4$ | $C_1$ | m | $A_g + A_u + B_g + B_u$ | $2\Gamma_{regular}$ | $2\Gamma_{regular}$ |
| $C^*_{3h}$ | $O_2$ | $C_3$ | $m_3$ | $A' + A''$ | $E' + E''$ | $E' + E''$ |
| | $O_3$ | $C_s$ | $m_h$ | $A' + E'$ | $\Gamma_{regular}$ | $\Gamma_{regular}$ |
| | $O_6$ | $C_1$ | m | $A' + A'' + E' + E''$ | $2\Gamma_{regular}$ | $2\Gamma_{regular}$ |
| $C^*_{4h}$ | $O_2$ | $C_4$ | $m_4$ | $A_g + A_u$ | $E_g + E_u$ | $2B_g + 2B_u$ |
| | $O_4$ | $C_s$ | $m_h$ | $A_g + B_g + E_u$ | $\Gamma_{regular}$ | $\Gamma_{regular}$ |
| | $O_8$ | $C_1$ | m | $A_g + A_u + B_g + B_u\ E_g + E_u$ | $2\Gamma_{regular}$ | $2\Gamma_{regular}$ |
| $C^*_{5h}$ | $O_2$ | $C_5$ | $m_5$ | $A' + A''$ | $E'_1 + E''_1$ | $E'_2 + E''_2$ |
| | $O_5$ | $C_s$ | $m_h$ | $A' + E'_1 + E'_2$ | $\Gamma_{regular}$ | $\Gamma_{regular}$ |
| | $O_{10}$ | $C_1$ | m | $A' + A'' + E'_1 + E'_2 + E''_1 + E''_2$ | $2\Gamma_{regular}$ | $2\Gamma_{regular}$ |
| $C^*_{6h}$ | $O_2$ | $C_6$ | $m_6$ | $A_g + A_u$ | $E_{1g} + E_{1u}$ | $E_{2g} + E_{2u}$ |
| | $O_6$ | $C_s$ | $m_h$ | $A_g + B_u + E_{2g} + E_{1u}$ | $\Gamma_{regular}$ | $\Gamma_{regular}$ |
| | $O_{12}$ | $C_1$ | m | $A_g + A_u + B_g + B_u$ $E_{1g} + E_{1u} + E_{2g} + E_{2u}$ | $2\Gamma_{regular}$ | $2\Gamma_{regular}$ |

**Table 3.2** **$\sigma, \pi, \delta$ representations on the distinct orbits of the molecular point groups $D_n$, $D_{nh}$, $D_{nd}$, $S_n$, up to $n = 6$ : $O_i$ is the orbit of order i and m is the number of times it occurs in a particular molecule as explained in the text. For each group the regular representation direct sum is given explicitly as the $\sigma$ representation for the largest orbit, but otherwise is abbreviated as $\Gamma_{regular}$; * in column 1 identifies a group in which a central $O_1$ orbit can be present. For this orbit the $\sigma$, $\pi$ and $\delta$ analysis does not apply.**

| Point Group | Orbit | Site Group | | $\Gamma_\sigma$ | $\Gamma_\pi$ | $\Gamma_\delta$ |
|---|---|---|---|---|---|---|
| $D_2^*$ | $O_{2[x]}$ | $C_2$ | $m_{2[x]}$ | $A + B_3$ | $2B_1 + 2B_2$ | $2A + 2B_3$ |
| | $O_{2[y]}$ | $C_2$ | $m_{2[y]}$ | $A + B_2$ | $2B_1 + 2B_3$ | $2A + 2B_2$ |
| | $O_{2[z]}$ | $C_2$ | $m_{2[z]}$ | $A + B_1$ | $2B_2 + 2B_3$ | $2A + 2B_1$ |
| | $O_4$ | $C_1$ | m | $A + B_1 + B_2 + B_3$ | $2\Gamma_{regular}$ | $2\Gamma_{regular}$ |
| $D_3^*$ | $O_2$ | $C_3$ | $m_3$ | $A_1 + A_2$ | 2E | 2E |
| | $O_3$ | $C_2$ | $m_2$ | $A_1 + E$ | $2A_2 + 2E$ | $2A_1 + 2E$ |
| | $O_6$ | $C_1$ | m | $A + A_2 + 2E$ | $2\Gamma_{regular}$ | $2\Gamma_{regular}$ |
| $D_4^*$ | $O_2$ | $C_4$ | $m_4$ | $A_1 + A_2$ | 2E | $2B_1 + 2B_2$ |
| | $O_{4'}$ | $C_2$ | $m_{2'}$ | $A_1 + B_2 + E$ | $2A_2 + 2B_1 + 2E$ | $2A_1 + 2B_2 + 2E$ |
| | $O_4$ | $C_2$ | $m_2$ | $A_1 + B_1 + E$ | $2A_2 + 2B_2 + 2E$ | $2A_1 + 2B_1 + 2E$ |
| | $O_8$ | $C_1$ | m | $A_1 + A_2 + B_1 + B_2$ $+ 2E$ | $2\Gamma_{regular}$ | $2\Gamma_{regular}$ |
| $D_5^*$ | $O_2$ | $C_5$ | $m_5$ | $A_1 + A_2$ | $2E_1$ | $2E_2$ |
| | $O_5$ | $C_2$ | $m_2$ | $A_1 + E_1 + E_2$ | $2A_2 + 2E_1 + 2E_2$ | $2A_1 + 2E_1 + 2E_2$ |
| | $O_{10}$ | $C_1$ | m | $A_1 + A_2 + 2E_1 + 2E_2$ | $2\Gamma_{regular}$ | $2\Gamma_{regular}$ |
| $D_6^*$ | $O_2$ | $C_6$ | $m_6$ | $A_1 + A_2$ | $2E_1$ | $2E_2$ |
| | $O_{6'}$ | $C_2$ | $m_{2'}$ | $A_1 + B_2 + E_1 + E_2$ | $2A_2 + 2B_1 + 2E_1 + 2E_2$ | $2A_1 + 2B_2 + 2E_1$ $+ 2E_2$ |
| | $O_6$ | $C_2$ | $m_2$ | $A_1 + B_1 + E_1 + E_2$ | $2A_2 + 2B_2 + 2E_1 + 2E_2$ | $2A_1 + 2B_1 + 2E_1$ $+ 2E_2$ |
| | $O_{12}$ | $C_1$ | m | $A_1 + A_2 + B_1 + B_2$ $+ 2E_1 + 2E_2$ | $2\Gamma_{regular}$ | $2\Gamma_{regular}$ |
| $D_{2h}^*$ | $O_{2[z]}$ | $C_{2v}$ | $m_{2[z]}$ | $A_g + B_{1u}$ | $B_{2g} + B_{2u} + B_{3g} + B_{3u}$ | $A_g + A_u + B_{1g} + B_{1u}$ |
| | $O_{2[y]}$ | $C_{2v}$ | $m_{2[y]}$ | $A_g + B_{2u}$ | $B_{1g} + B_{1u} + B_{3g} + B_{3u}$ | $A_g + A_u + B_{2g} + B_{2u}$ |
| | $O_{2[x]}$ | $C_{2v}$ | $m_{2[x]}$ | $A_g + B_{3u}$ | $B_{1g} + B_{1u} + B_{2g} + B_{2u}$ | $A_g + A_u + B_{3g} + B_{3u}$ |
| | $O_{4[yz]}$ | $C_s$ | $m_{[yz]}$ | $A_g + B_{3g} + B_{1u} + B_{2u}$ | $\Gamma_{regular}$ | $\Gamma_{regular}$ |
| | $O_{4[xz]}$ | $C_s$ | $m_{[xz]}$ | $A_g + B_{2g} + B_{1u} + B_{3u}$ | $\Gamma_{regular}$ | $\Gamma_{regular}$ |
| | $O_{4[xy]}$ | $C_s$ | $m_{[xy]}$ | $A_g + B_{1g} + B_{2u} + B_{3u}$ | $\Gamma_{regular}$ | $\Gamma_{regular}$ |
| | $O_8$ | $C_1$ | m | $A_g + A_u + B_{1g} + B_{1u}$ $B_{2g} + B_{2u} + B_{3g} + B_{3u}$ | $2\Gamma_{regular}$ | $2\Gamma_{regular}$ |

| Point Group | Orbit | Site Group | | $\Gamma_\sigma$ | $\Gamma_\pi$ | $\Gamma_\delta$ |
|---|---|---|---|---|---|---|
| $D^*_{3h}$ | $O_2$ | | $m_3$ | $A'_1 + A''_2$ | $E' + E''$ | $E' + E''$ |
| | $O_3$ | | $m_2$ | $A'_1 + E'$ | $A'_1 + A''_2 + E' + E''$ | $A'_1 + A''_1 + E' + E''$ |
| | $O_{6h}$ | | $m_h$ | $A'_1 + A'_2 + 2E'$ | $\Gamma_{regular}$ | $\Gamma_{regular}$ |
| | $O_{6v}$ | | $m_v$ | $A'_1 + A''_2 + E' + E''$ | $\Gamma_{regular}$ | $\Gamma_{regular}$ |
| | $O_{12}$ | | m | $A'_1 + A''_1 + A'_2 + A''_2$ $+ 2E' + 2E''$ | $2\Gamma_{regular}$ | $2\Gamma_{regular}$ |
| $D^*_{4h}$ | $O_2$ | $C_{3v}$ | $m_4$ | $A_{1g} + A_{2u}$ | $E_g + E_u$ | $B_{1g} + B_{1u} + B_{2g}$ $+ B_{2u}$ |
| | $O_{4'}$ | $C_{2v}$ | $m_{2'}$ | $A_{1g} + B_{2g} + E_u$ | $A_{2g} + A_{2u} + B_{1g} +$ $B_{1u} + E_g + E_u$ | $A_{1g} + A_{1u} + B_{2g}$ $+ B_{2u} + E_g + E_u$ |
| | $O_4$ | $C_{2v}$ | $m_2$ | $A_{1g} + B_{1g} + E_u$ | $A_{2g} + A_{2u} + B_{2g} +$ $B_{2u}\ E_g + E_u$ | $A_{1g} + A_{1u} + B_{1g}$ $+ B_{1u} + E_g + E_u$ |
| | $O_{8h}$ | $C_s$ | $m_h$ | $A_{1g} + A_{2g} + B_{1g} + B_{2g}\ 2E_u$ | $\Gamma_{regular}$ | $\Gamma_{regular}$ |
| | $O_{8d}$ | $C_s$ | $m_d$ | $A_{1g} + A_{2u} + B_{1u} + B_{2g}$ $E_g + E_u$ | $\Gamma_{regular}$ | $\Gamma_{regular}$ |
| | $O_{8v}$ | $C_s$ | $m_v$ | $A_{1g} + A_{2u} + B_{1g} + B_{2u}\ E_g + E_u$ | $\Gamma_{regular}$ | $\Gamma_{regular}$ |
| | $O_{16}$ | $C_1$ | m | $A_{1g} + A_{1u} + A_{2g} + A_{2u} + B_{1g} +$ $B_{1u} + B_{2g} + B_{2u} + 2E_g + 2E_u$ | $2\Gamma_{regular}$ | $2\Gamma_{regular}$ |
| $D^*_{5h}$ | $O_2$ | $C_{5v}$ | $m_5$ | $A'_1 + A''_2$ | $E'_1 + E''_1$ | $E'_2 + E''_2$ |
| | $O_5$ | $C_{2v}$ | $m_2$ | $A'_1 + E'_1 + E'_2$ | $A'_2 + A''_2 + E'_1 + E''_1$ $+ E'_2 + E''_2$ | $A'_1 + A''_1 + E'_1 + E''_1$ $+ E'_2 + E''_2$ |
| | $O_{10h}$ | $C_s$ | $m_h$ | $A'_1 + A'_2 + 2E'_1 + 2E'_2$ | $\Gamma_{regular}$ | $\Gamma_{regular}$ |
| | $O_{10v}$ | $C_s$ | $m_v$ | $A'_1 + A''_2 + E'_1 + E''_1 + E'_2 + E''_2$ | $\Gamma_{regular}$ | $\Gamma_{regular}$ |
| | $O_{20}$ | $C_1$ | m | $A'_1 + A''_1 + A'_2 + A''_2 + 2E'_1$ $+ 2E''_1 + 2E'_2 + 2E''_2$ | $2\Gamma_{regular}$ | $2\Gamma_{regular}$ |
| $D^*_{6h}$ | $O_2$ | $C_{6v}$ | $m_6$ | $A_{1g} + A_{2u}$ | $E_{1g} + E_{1u}$ | $E_{2g} + E_{2u}$ |
| | $O_{6'}$ | $C_{2v}$ | $m_{2'}$ | $A_{1g} + B_{2u} + E_{1u} + E_{2g}$ | $A_{2g} + A_{2u} + B_{1g}$ $+ B_{1u} + E_{1g} + E_{1u}$ $+ E_{2g} + E_{2u}$ | $A_{1g} + A_{1u} + B_{2g}$ $+ B_{2u} + E_{1g} + E_{1u}$ $+ E_{2g} + E_{2u}$ |
| | $O_6$ | $C_{2v}$ | $m_2$ | $A_{1g} + B_{1u} + E_{1u} + E_{2g}$ | $A_{2g} + A_{2u} + B_{2g}$ $+ B_{2u} + E_{1g} + E_{1u}$ $+ E_{2g} + E_{2u}$ | $A_{1g} + A_{1u} + B_{1g}$ $+ B_{1u} + E_{1g} + E_{1u}$ $+ E_{2g} + E_{2u}$ |
| | $O_{12h}$ | $C_s$ | $m_h$ | $A_{1g} + A_{2g} + B_{1u} + B_{2u}$ $+ 2E_{1u} + 2E_{2g}$ | $\Gamma_{regular}$ | $\Gamma_{regular}$ |
| | $O_{12d}$ | $C_s$ | $m_d$ | $A_{1g} + A_{2g} + B_{1u} + B_{2u}$ $+ E_{1g} + E_{1u} + E_{2g} + E_{2u}$ | $\Gamma_{regular}$ | $\Gamma_{regular}$ |
| | $O_{12v}$ | $C_s$ | $m_v$ | $A_{1g} + A_{2u} + B_{1u} + B_{2g}$ $+ E_{1g} + E_{1u} + E_{2g} + E_{2u}$ | $\Gamma_{regular}$ | $\Gamma_{regular}$ |

| Point Group | Orbit | Site Group | | $\Gamma_\sigma$ | $\Gamma_\pi$ | $\Gamma_\delta$ |
|---|---|---|---|---|---|---|
| | $O_{24}$ | $C_1$ | m | $A_{1g} + A_{1u} + A_{2g} + A_{2u} + B_{1g} + B_{1u} + B_{2g} + B_{2u} + 2E_{1g} + 2E_{1u} + 2E_{2g} + 2E_{2u}$ | $2\Gamma_{regular}$ | $2\Gamma_{regular}$ |
| $D^*_{2d}$ | $O_2$ | $C_{2v}$ | $m_4$ | $A_1 + B_2$ | $2E$ | $A_1 + A_2 + B_1 + B_2$ |
| | $O_4$ | $C_2$ | $m_2$ | $A_1 + B_1 + E$ | $2A_2 + 2B_2 + 2E$ | $2A_1 + 2B_1 + 2E$ |
| | $O_{4d}$ | $C_s$ | $m_d$ | $A_1 + B_2 + E$ | $\Gamma_{regular}$ | $\Gamma_{regular}$ |
| | $O_8$ | $C_1$ | m | $A_1 + A_2 + B_1 + B_2 + 2E$ | $2\Gamma_{regular}$ | $2\Gamma_{regular}$ |
| $D^*_{3d}$ | $O_2$ | $C_{3v}$ | $m_6$ | $A_{1g} + A_{2u}$ | $E_g + E_u$ | $E_g + E_u$ |
| | $O_6$ | $C_2$ | $m_2$ | $A_{1g} + A_{1u} + E_g + E_u$ | $2A_{2g} + 2A_{2u} + 2E_g + 2E_u$ | $2A_{1g} + 2A_{1u} + 2E_g + 2E_u$ |
| | $O_{6d}$ | $C_s$ | $m_d$ | $A_{1g} + A_{2u} + E_g + E_u$ | $\Gamma_{regular}$ | $\Gamma_{regular}$ |
| | $O_{12}$ | $C_1$ | m | $A_{1g} + A_{1u} + A_{2g} + A_{2u} + 2E_g + 2E_u$ | $2\Gamma_{regular}$ | $2\Gamma_{regular}$ |
| $D^*_{4d}$ | $O_2$ | $C_{4v}$ | $m_8$ | $A_1 + B_2$ | $E_1 + E_3$ | $2E_2$ |
| | $O_8$ | $C_2$ | $m_2$ | $A_1 + B_1 + E_1 + E_2 + E_3$ | $2A_2 + 2B_2 + 2E_1 + 2E_2 + 2E_3$ | $2A_1 + 2B_1 + 2E_1 + 2E_2 + 2E_3$ |
| | $O_{8d}$ | $C_s$ | $m_d$ | $A_1 + B_2 + E_1 + E_2 + E_3$ | $\Gamma_{regular}$ | $\Gamma_{regular}$ |
| | $O_{16}$ | $C_1$ | m | $A_1 + A_2 + B_1 + B_2 + 2E_1 + 2E_2 + 2E_3$ | $2\Gamma_{regular}$ | $2\Gamma_{regular}$ |
| $D^*_{5d}$ | $O_2$ | $C_{5v}$ | $m_{10}$ | $A_{1g} + A_{2u}$ | $E_{1g} + E_{1u}$ | $E_{2g} + E_{2u}$ |
| | $O_{10}$ | $C_2$ | $m_2$ | $A_{1g} + A_{1u} + E_{1g} + E_{1u} + E_{2g} + E_{2u}$ | $2A_{2g} + 2A_{2u} + 2E_{1g} + 2E_{1u} + 2E_{2g} + 2E_{2u}$ | $2A_{1g} + 2A_{1u} + 2E_{1g} + 2E_{1u} + 2E_{2g} + 2E_{2u}$ |
| | $O_{10d}$ | $C_s$ | $m_d$ | $A_{1g} + A_{2u} + E_{1g} + E_{1u} + E_{2g} + E_{2u}$ | $\Gamma_{regular}$ | $\Gamma_{regular}$ |
| | $O_{20}$ | $C_1$ | m | $A_{1g} + A_{1u} + A_{2g} + A_{2u} + 2E_{1g} + 2E_{1u} + 2E_{2g} + 2E_{2u}$ | $2\Gamma_{regular}$ | $2\Gamma_{regular}$ |
| $D^*_{6d}$ | $O_2$ | $C_{6v}$ | $m_{12}$ | $A_1 + B_2$ | $E_1 + E_5$ | $E_2 + E_4$ |
| | $O_{12}$ | $C_2$ | $m_2$ | $A_1 + B_1 + E_1 + E_2 + E_3 + E_4 + E_5$ | $2A_2 + 2B_2 + 2E_1 + 2E_2 + 2E_3 + 2E_4 + 2E_5$ | $2A_1 + 2B_1 + 2E_1 + 2E_2 + 2E_3 + 2E_4 + 2E_5$ |
| | $O_{12d}$ | $C_s$ | $m_d$ | $A_1 + B_2 + E_1 + E_2 + E_3 + E_4 + E_5$ | $\Gamma_{regular}$ | $\Gamma_{regular}$ |
| | $O_{24}$ | $C_1$ | m | $A_1 + A_2 + B_1 + B_2 + 2E_1 + 2E_2 + 2E_3 + 2E_4 + 2E_5$ | $2\Gamma_{regular}$ | $2\Gamma_{regular}$ |
| $S^*_4$ | $O_2$ | $C_2$ | $m_2$ | $A + B$ | $2E$ | $2A + 2B$ |
| | $O_4$ | $C_1$ | m | $A + B + E$ | $2\Gamma_{regular}$ | $2\Gamma_{regular}$ |
| $S^*_6$ | $O_2$ | $C_3$ | $m_3$ | $A_g + A_u$ | $E_g + E_u$ | $E_g + E_u$ |
| | $O_6$ | $C_1$ | m | $A_g + A_u + E_g + E_u$ | $2\Gamma_{regular}$ | $2\Gamma_{regular}$ |

**Table 3.3** **$\sigma$, $\pi$, $\delta$ representations on the distinct orbits of the molecular point groups T, $T_d$, $T_h$, O and $O_h$: $O_i$ is the orbit of order i and m is the number of times it occurs in a particular molecule as explained in the text. For each group the regular representation direct sum is given explicitly as the $\sigma$ representation for the largest orbit, but otherwise is abbreviated as $\Gamma_{regular}$; * in column 1 identifies a group in which a central $O_1$ orbit can be present. For this orbit the $\sigma$, $\pi$ and $\delta$ analysis does not apply.**

| Point Group | Orbit | Site Group | | $\Gamma_\sigma$ | $\Gamma_\pi$ | $\Gamma_\delta$ |
|---|---|---|---|---|---|---|
| T* | $O_4$ | $C_3$ | $m_3$ | A + T | E + 2T | E + 2T |
| | $O_6$ | $C_2$ | $m_2$ | A + E + T | 4T | 2A + 2E + 2T |
| | $O_{12}$ | $C_1$ | m | A + E + 3T | $2\Gamma_{regular}$ | $2\Gamma_{regular}$ |
| $T_d^*$ | $O_4$ | $C_{3v}$ | $m_3$ | $A_1 + T_2$ | $E + T_1 + T_2$ | $E + T_1 + T_2$ |
| | $O_6$ | $C_{2v}$ | $m_2$ | $A_1 + E + T_2$ | $2T_1 + 2T_2$ | $A_1 + A_2 + 2E$ $+T_1 + T_2$ |
| | $O_{12}$ | $C_s$ | $m_d$ | $A_1 + E + T_1 + 2T_2$ | $\Gamma_{regular}$ | $\Gamma_{regular}$ |
| | $O_{24}$ | $C_1$ | m | $A_1 + A_2 + 2E$ $+3T_1 + 3T_2$ | $2\Gamma_{regular}$ | $2\Gamma_{regular}$ |
| $T_h^*$ | $O_6$ | $C_{2v}$ | $m_2$ | $A_g + E_g + T_u$ | $2T_g + 2T_u$ | $A_g + A_u + E_g +$ $E_u + T_g + T_u$ |
| | $O_8$ | $C_3$ | $m_3$ | $A_g + A_u + T_g + T_u$ | $E_g + E_u + 2T_g + 2T_u$ | $E_g + E_u + 2T_g + 2T_u$ |
| | $O_{12}$ | $C_s$ | $m_h$ | $A_g + E_g + T_g + 2T_u$ | $\Gamma_{regular}$ | $\Gamma_{regular}$ |
| | $O_{24}$ | $C_1$ | m | $A_g + A_u + E_g + E_u$ $+ 3T_g + 3T_u$ | $2\Gamma_{regular}$ | $2\Gamma_{regular}$ |
| O* | $O_6$ | $C_4$ | $m_4$ | $A_1 + E + T_1$ | $2T_1 + 2T_2$ | $2A_2 + 2E + 2T_2$ |
| | $O_8$ | $C_3$ | $m_3$ | $A_1 + A_2 + T_1 + T_2$ | $2E + 2T_1 + 2T_2$ | $2E + 2T_1 + 2T_2$ |
| | $O_{12}$ | $C_s$ | $m_2$ | $A_1 + E + T_1 + 2T_2$ | $2A_2 + 2E + 4T_1 + 2T_2$ | $2A_1 + 2E + 2T_1 + 4T_2$ |
| | $O_{24}$ | $C_1$ | m | $A_1 + A_2 + 2E + 3T_1 + 3T_2$ | $2\Gamma_{regular}$ | $2\Gamma_{regular}$ |
| $O_h^*$ | $O_6$ | $C_{4v}$ | $m_4$ | $A_{1g} + E_g + T_{1u}$ | $T_{1g} + T_{1u} + T_{2g} + T_{2u}$ | $A_{2g} + A_{2u} + E_g + E_u$ $+ T_{2g} + T_{2u}$ |
| | $O_8$ | $C_{3v}$ | $m_3$ | $A_{1g} + A_{2u} + T_{2g} + T_{1u}$ | $E_g + E_u + T_{1g} + T_{1u} +$ $T_{2g} + T_{2u}$ | $E_g + E_u + T_{1g} + T_{1u}$ $+ T_{2g} + T_{2u}$ |
| | $O_{12}$ | $C_{2v}$ | $m_2$ | $A_{1g} + E_g + T_{1u}$ $+ T_{2g} + T_{2u}$ | $A_{2g} + A_{2u} + E_g + E_u$ $+ 2T_{1g} + 2T_{1u}$ $+ 2T_{2g} + T_{2u}$ | $A_{1g} + A_{1u} + E_g + E_u$ $+ T_{1g} + T_{1u}$ $+ 2T_{2g} + T_{2u}$ |
| | $O_{24d}$ | $C_s$ | $m_d$ | $A_{1g} + A_{2u} + E_g + E_u$ $+ T_{1g} + 2T_{1u} + 2T_{2g} + T_{2u}$ | $\Gamma_{regular}$ | $\Gamma_{regular}$ |
| | $O_{24h}$ | $C_s$ | $m_h$ | $A_{1g} + A_{2g} + 2E_g + T_{1g}$ $+ 2T_{1u} + 2T_{2g} + 2T_{2u}$ | $\Gamma_{regular}$ | $\Gamma_{regular}$ |
| | $O_{48}$ | $C_1$ | m | $A_{1g} + A_{1u} + A_{2g} + A_{2u}$ $+ 2E_g + 2E_u + 3T_{1g}$ $+ 3T_{1u} + 3T_{2g} + 3T_{2u}$ | $2\Gamma_{regular}$ | $2\Gamma_{regular}$ |

**Table 3.4** $\sigma, \pi, \delta$ **representations on the distinct orbits of the molecular point groups I and $I_h$: $O_i$ is the orbit of order i and m is the number of times it occurs in a particular molecule as explained in the text. For each group the regular representation direct sum is given explicitly as the $\sigma$ representation for the largest orbit, but otherwise is abbreviated as $\Gamma_{regular}$; * in column 1 identifies a group in which a central $O_1$ orbit can be present. For this orbit the $\sigma$, $\pi$ and $\delta$ analysis does not apply.**

| Point Group | Orbit | Site Group | | $\Gamma_\sigma$ | $\Gamma_\pi$ | $\Gamma_\delta$ |
|---|---|---|---|---|---|---|
| I* | $O_{12}$ | $C_5$ | $m_5$ | $A + T_1 + T_2 + H$ | $2T_1 + 2G + 2H$ | $2T_2 + 2G + 2H$ |
| | $O_{20}$ | $C_3$ | $m_3$ | $A + T_1 + T_2 + 2G + H$ | $2T_1 + 2T_2 + 2G + 4H$ | $2T_1 + 2T_2 + 2G + 4H$ |
| | $O_{30}$ | $C_2$ | $m_2$ | $A + T_1 + T_2 + 2G + 3H$ | $4T_1 + 4T_2 + 4G + 4H$ | $2A + 2T_1 + 2T_2 + 4G + 6H$ |
| | $O_{60}$ | $C_1$ | m | $A + 3T_1 + 3T_2 + 4G + 5H$ | $2\Gamma_{regular}$ | $2\Gamma_{regular}$ |
| $I_h^*$ | $O_{12}$ | $C_{5v}$ | $m_5$ | $A_g + T_{1u} + T_{2u} + H_g$ | $T_{1g} + T_{1u} + G_g + G_u + H_g + H_u$ | $T_{2g} + T_{2u} + G_g + G_u + H_g + H_u$ |
| | $O_{20}$ | $C_{3v}$ | $m_3$ | $A_g + T_{1u} + T_{2u} + G_g + G_u + H_g$ | $T_{1g} + T_{1u} + T_{2g} + T_{2u} + G_g + G_u + 2H_g + 2H_u$ | $T_{1g} + T_{1u} + T_{2g} + T_{2u} + G_g + G_u + 2H_g + 2H_u$ |
| | $O_{30}$ | $C_{2v}$ | $m_2$ | $A_g + T_{1u} + T_{2u} + G_g + G_u + 2H_g + H_u$ | $2T_{1g} + 2T_{1u} + 2T_{2g} + 2T_{2u} + 2G_g + 2G_u + 2H_g + 2H_u$ | $A_g + A_u + T_{1g} + T_{1u} + T_{2g} + T_{2u} + 2G_g + 2G_u + 3H_g + 3H_u$ |
| | $O_{60}$ | $C_s$ | $m_d$ | $A_g + T_{1g} + 2T_{1u} + T_{2g} + 2T_{2u} + 2G_g + 2G_u + 3H_g + 2H_u$ | $\Gamma_{regular}$ | $\Gamma_{regular}$ |
| | $O_{120}$ | $C_1$ | m | $A_g + A_u + 3T_{1g} + 3T_{1u} + 3T_{2g} + 3T_{2u} + 4G_g + 4G_u + 5H_g + 5H_u$ | $2\Gamma_{regular}$ | $2\Gamma_{regular}$ |

## 3.3 The General, *Kubic* and Icosahedral Harmonics

The theorems described in section 3.2 provide for the ready calculation of the $\Gamma_\sigma$, $\Gamma_\pi$ and $\Gamma_\delta$ reducible characters, generated by the transformation properties of s, p and d-atomic orbitals distributed over the vertices of the structure orbits of the various point groups, which decompose into the direct sums of irreducible components listed in Tables 3.1 to 3.4. Application of the theorems requires the identification of sufficient numbers of central harmonics to act as basis functions for the irreducible components of the regular orbits of these molecular point groups.

The general spherical harmonics are familiar, in low order, as the mutually orthonormal angular components of valence atomic orbitals. Now, the sufficient number of these functions to provide basis functions for the regular representations of the molecular point groups, in

**Table 3.5 Classification of the general spherical harmonics up to angular momentum level 4 by descent in symmetry, into their irreducible components for the molecular point groups $C_i$, $C_s$, $C_2$, $C_3$, $C_4$, $C_5$, $C_6$, $C_7$ and $C_8$.**

<table>
<tr><th></th><th></th><th>$C_i$</th><th>$C_s$</th><th>$C_2$</th><th>$C_3$</th><th>$C_4$</th><th>$C_5$</th><th>$C_6$</th><th>$C_7$</th><th>$C_8$</th></tr>
<tr><td>1s</td><td>(000)</td><td>$1A_g$</td><td>$1A'$</td><td>1A</td><td>1A</td><td>1A</td><td>1A</td><td>1A</td><td>1A</td><td>1A</td></tr>
<tr><td>2p</td><td>(001)</td><td>$1A_u$</td><td>$1A''$</td><td>2A</td><td>2A</td><td>2A</td><td>2A</td><td>2A</td><td>2A</td><td>2A</td></tr>
<tr><td>2p</td><td>(100)</td><td>$2A_u$</td><td>$2A'$</td><td>1B</td><td rowspan="2">1E</td><td rowspan="2">1E</td><td rowspan="2">$1E_1$</td><td rowspan="2">$1E_1$</td><td rowspan="2">$1E_1$</td><td rowspan="2">$1E_1$</td></tr>
<tr><td>2p</td><td>(010)</td><td>$3A_u$</td><td>$3A'$</td><td>2B</td></tr>
<tr><td>3d</td><td>3(002)−(000)</td><td>$2A_g$</td><td>$4A'$</td><td>3A</td><td>3A</td><td>3A</td><td>3A</td><td>3A</td><td>3A</td><td>3A</td></tr>
<tr><td>3d</td><td>(101)</td><td>$3A_g$</td><td>$2A''$</td><td>3B</td><td rowspan="2">2E</td><td rowspan="2">2E</td><td rowspan="2">$2E_1$</td><td rowspan="2">$2E_1$</td><td rowspan="2">$2E_1$</td><td rowspan="2">$2E_1$</td></tr>
<tr><td>3d</td><td>(011)</td><td>$4A_g$</td><td>$3A''$</td><td>4B</td></tr>
<tr><td>3d</td><td>(200)−(020)</td><td>$5A_g$</td><td>$5A'$</td><td>4A</td><td rowspan="2">3E</td><td>1B</td><td rowspan="2">$1E_2$</td><td rowspan="2">$1E_2$</td><td rowspan="2">$1E_2$</td><td rowspan="2">$1E_2$</td></tr>
<tr><td>3d</td><td>(110)</td><td>$6A_g$</td><td>$6A'$</td><td>5A</td><td>2B</td></tr>
<tr><td>4f</td><td>5(003)−3(001)</td><td>$4A_u$</td><td>$4A''$</td><td>6A</td><td>4A</td><td>4A</td><td>4A</td><td>4A</td><td>4A</td><td>4A</td></tr>
<tr><td>4f</td><td>5(102)−(100)</td><td>$5A_u$</td><td>$7A'$</td><td>5B</td><td rowspan="2">4E</td><td rowspan="2">3E</td><td rowspan="2">$3E_1$</td><td rowspan="2">$3E_1$</td><td rowspan="2">$3E_1$</td><td rowspan="2">$3E_1$</td></tr>
<tr><td>4f</td><td>5(012)−(010)</td><td>$6A_u$</td><td>$8A'$</td><td>6B</td></tr>
<tr><td>4f</td><td>(201)−(021)</td><td>$7A_u$</td><td>$5A''$</td><td>7A</td><td rowspan="2">5E</td><td>3B</td><td rowspan="2">$2E_2$</td><td rowspan="2">$2E_2$</td><td rowspan="2">$2E_2$</td><td rowspan="2">$2E_2$</td></tr>
<tr><td>4f</td><td>(111)</td><td>$8A_u$</td><td>$6A''$</td><td>8A</td><td>4B</td></tr>
<tr><td>4f</td><td>(300)−3(120)</td><td>$9A_u$</td><td>$9A'$</td><td>7B</td><td>5A</td><td rowspan="2">4E</td><td rowspan="2">$3E_2$</td><td>1B</td><td rowspan="2">$1E_3$</td><td rowspan="2">$1E_3$</td></tr>
<tr><td>4f</td><td>3(210)−(030)</td><td>$10A_u$</td><td>$10A'$</td><td>8B</td><td>6A</td><td>2B</td></tr>
<tr><td>5g</td><td>35(004)−30(002)+3(000)</td><td>$7A_g$</td><td>$11A'$</td><td>9A</td><td>7A</td><td>5A</td><td>5A</td><td>5A</td><td>7A</td><td>5A</td></tr>
<tr><td>5g</td><td>7(103)−3(101)</td><td>$8A_g$</td><td>$7A''$</td><td>9B</td><td rowspan="2">6E</td><td rowspan="2">5E</td><td rowspan="2">$4E_1$</td><td rowspan="2">$4E_1$</td><td rowspan="2">$4E_1$</td><td rowspan="2">$4E_1$</td></tr>
<tr><td>5g</td><td>7(013)−3(011)</td><td>$9A_g$</td><td>$8A''$</td><td>10B</td></tr>
<tr><td>5g</td><td>7(202)−7(022)−(200)+(020)</td><td>$10A_g$</td><td>$12A'$</td><td>10A</td><td rowspan="2">7E</td><td>5B</td><td rowspan="2">$4E_2$</td><td rowspan="2">$3E_2$</td><td rowspan="2">$3E_2$</td><td rowspan="2">$3E_2$</td></tr>
<tr><td>5g</td><td>7(112)−(110)</td><td>$11A_g$</td><td>$13A'$</td><td>11A</td><td>6B</td></tr>
<tr><td>5g</td><td>(301)−3(121)</td><td>$12A_g$</td><td>$9A''$</td><td>11B</td><td>8A</td><td rowspan="2">6E</td><td rowspan="2">$5E_2$</td><td>3B</td><td rowspan="2">$2E_3$</td><td rowspan="2">$2E_3$</td></tr>
<tr><td>5g</td><td>3(211)−(031)</td><td>$13A_g$</td><td>$10A''$</td><td>12B</td><td>9A</td><td>4B</td></tr>
<tr><td>5g</td><td>(400)+(040)−6(220)</td><td>$14A_g$</td><td>$14A'$</td><td>12A</td><td rowspan="2">8E</td><td>6A</td><td rowspan="2">$5E_1$</td><td rowspan="2">$4E_2$</td><td rowspan="2">$3E_3$</td><td>1B</td></tr>
<tr><td>5g</td><td>(310)−(130)</td><td>$15A_g$</td><td>$15A'$</td><td>13A</td><td>7A</td><td>2B</td></tr>
</table>

which an axis of principal rotation is present, are listed in Tables 3.5 to 3.8 using Elert's notation [Chapter 1, page 24].

Extra considerations are required to construct suitable sets of polynomials, which provide basis functions for the irreducible subspaces of the cubic and icosahedral point groups. Clearly, such a set of central functions is invariant under the point group G. For such a function, f, then $\langle f_{g1}, f_{g2}, \ldots, f_{gn} \rangle$ is a subspace of the central functions invariant under G. But it is not, in general, an irreducible subspace, i.e. it may contain further subspaces that transform according to different irreducible representations.

**Table 3.6 Classification of the general spherical harmonics up to angular momentum level 4 by descent in symmetry, into their irreducible components for the molecular point groups $C_{2v}$, $C_{3v}$, $C_{4v}$, $C_{5v}$, $C_{6v}$, $C_{2h}$, $C_{3h}$, $C_{4h}$, $C_{5h}$ and $C_{6h}$.**

| | | $C_{2v}$ | $C_{3v}$ | $C_{4v}$ | $C_{5v}$ | $C_{6v}$ | $C_{2h}$ | $C_{3h}$ | $C_{4h}$ | $C_{5h}$ | $C_{6h}$ |
|---|---|---|---|---|---|---|---|---|---|---|---|
| 1s | (000) | $1A_1$ | $1A_1$ | $1A_1$ | $1A_1$ | $1A_1$ | $1A_g$ | $1A'$ | $1A_g$ | $1A'$ | $1A_g$ |
| 2p | (001) | $2A_1$ | $2A_1$ | $2A_1$ | $2A_1$ | $2A_1$ | $1A_u$ | $1A''$ | $1A_u$ | $1A''$ | $1A_u$ |
| 2p | (100) | $1B_1$ | $1E$ | $1E$ | $1E_1$ | $1E_1$ | $1B_u$ | $1E''$ | $1E_u$ | $1E'_1$ | $1E_{1u}$ |
| 2p | (010) | $1B_2$ | | | | | $2B_u$ | | | | |
| 3d | 3(002)−(000) | $3A_1$ | $3A_1$ | $3A_1$ | $3A_1$ | $3A_1$ | $2A_g$ | $2A'$ | $2A_g$ | $2A'$ | $2A_g$ |
| 3d | (101) | $2B_1$ | $2E$ | $2E$ | $2E_1$ | $2E_1$ | $1B_g$ | $1E''$ | $1E_g$ | $1E''_1$ | $1E_{1g}$ |
| 3d | (011) | $2B_2$ | | | | | $2B_g$ | | | | |
| 3d | (200)−(020) | $4A_1$ | $3E$ | $1B_1$ | $1E_2$ | $1E_2$ | $3A_g$ | $2E'$ | $1B_g$ | $1E'_2$ | $1E_{2g}$ |
| 3d | (110) | $1A_2$ | | $1B_2$ | | | $4A_g$ | | $2B_g$ | | |
| 4f | 5(003)−3(001) | $5A_1$ | $4A_1$ | $4A_1$ | $4A_1$ | $4A_1$ | $2A_u$ | $2A''$ | $2A_u$ | $2A''$ | $2A_u$ |
| 4f | 5(102)−(100) | $3B_1$ | $4E$ | $3E$ | $3E_1$ | $3E_1$ | $2B_u$ | $3E'$ | $2E_u$ | $2E'_1$ | $2E_{1u}$ |
| 4f | 5(012)−(010) | $3B_2$ | | | | | $3B_u$ | | | | |
| 4f | (201)−(021) | $6A_1$ | $5E$ | $2B_1$ | $2E_2$ | $2E_2$ | $3A_u$ | $2E''$ | $1B_u$ | $1E''_2$ | $1E_{2u}$ |
| 4f | (111) | $2A_2$ | | $3B_1$ | | | $4A_u$ | | $2B_u$ | | |
| 4f | (300)−3(120) | $4B_1$ | $5A_1$ | $4E$ | $3E_2$ | $1B_1$ | $4B_u$ | $3A'$ | $3E_u$ | $2E'_2$ | $1B_u$ |
| 4f | 3(210)−(030) | $4B_2$ | $1A_2$ | | | $1B_2$ | $5B_u$ | $4A'$ | | $2E'_2$ | $2B_u$ |
| 5g | 35(004)−30(002)+3(000) | $7A_1$ | $6A_1$ | $5A_1$ | $5A_1$ | $5A_1$ | $5A_g$ | $5A'$ | $3A_g$ | $3A'$ | $3A_g$ |
| 5g | 7(103)−3(101) | $5B_1$ | $6E$ | $5E$ | $4E_1$ | $4E_1$ | $3B_g$ | $3E''$ | $2E_g$ | $2E''_1$ | $2E_{1g}$ |
| 5g | 7(013)−3(011) | $5B_2$ | | | | | $4B_g$ | | | | |
| 5g | 7(202)−7(022)−(200)+(020) | $8A_1$ | $7E$ | $3B_1$ | $4E_2$ | $3E_2$ | $6A_g$ | $4E'$ | $3B_g$ | $3E'_2$ | $2E_{2g}$ |
| 5g | 7(112)−(110) | $3A_2$ | | $3B_2$ | | | $7A_g$ | | $4B_g$ | | |
| 5g | (301)−3(121) | $5B_1$ | $8A_1$ | $6E$ | $5E_2$ | $2B_1$ | $5B_g$ | $3A''$ | $3E_g$ | $2E''_2$ | $1B_g$ |
| 5g | 3(211)−(031) | $5B_2$ | $2A_2$ | | | $2B_2$ | $6B_g$ | $4A''$ | | | $2B_g$ |
| 5g | (400)+(040)−6(220) | $9A_1$ | $8E$ | $6A_1$ | $5E_1$ | $4E_2$ | $8A_g$ | $5E'$ | $4A_g$ | $3E'_1$ | $2E_{2g}$ |
| 5g | (310)−(130) | $4A_2$ | | $1A_2$ | | | $9A_g$ | | $5A_g$ | | |

To generate an irreducible G subspace, for particular cases, f needs to be chosen with care. In the case of the *kubic* harmonics, first defined by Bethe in 1929[5], suitable functions are the mononomials $x^m y^n z^p$, which we identify in Elert's notation as (mnp). The *kubic* harmonics up to level 4 and their maps onto the irreducible representations of the cubic groups are listed in Table 3.9.

[5] H. Bethe, *Ann. der Physik*, **3** (1929) 133.

**Table 3.7 Classification of the general spherical harmonics up to angular momentum level 4 by descent in symmetry, into their irreducible components for the molecular point groups $D_2$, $D_3$, $D_4$, $D_5$, $D_6$, $D_{2h}$, $D_{3h}$, $D_{4h}$, $D_{5h}$ and $D_{6h}$.**

<table>
<tr><th></th><th></th><th>$D_2$</th><th>$D_3$</th><th>$D_4$</th><th>$D_5$</th><th>$D_6$</th><th>$D_{2h}$</th><th>$D_{3h}$</th><th>$D_{4h}$</th><th>$D_{5h}$</th><th>$D_{6h}$</th></tr>
<tr><td>1s</td><td>(000)</td><td>1A</td><td>$1A_1$</td><td>$1A_1$</td><td>$1A_1$</td><td>$1A_1$</td><td>$1A_g$</td><td>$1A_1'$</td><td>$1A_{1g}$</td><td>$1A_1'$</td><td>$1A_{1g}$</td></tr>
<tr><td>2p</td><td>(001)</td><td>$1B_1$</td><td>$1A_2$</td><td>$1A_2$</td><td>$1A_2$</td><td>$1A_2$</td><td>$1B_{1u}$</td><td>$1A_2''$</td><td>$1A_{2u}$</td><td>$1A_2''$</td><td>$1A_{2u}$</td></tr>
<tr><td>2p</td><td>(100)</td><td>$1B_3$</td><td rowspan="2">1E</td><td rowspan="2">1E</td><td rowspan="2">$1E_1$</td><td rowspan="2">$1E_1$</td><td>$1B_{3u}$</td><td rowspan="2">$1E'$</td><td rowspan="2">$1E_u$</td><td rowspan="2">$1E_1'$</td><td rowspan="2">$1E_{1u}$</td></tr>
<tr><td>2p</td><td>(010)</td><td>$1B_2$</td><td>$1B_{2u}$</td></tr>
<tr><td>3d</td><td>3(002)−(000)</td><td>2A</td><td>$2A_1$</td><td>$2A_1$</td><td>$2A_1$</td><td>$2A_1$</td><td>$2A_g$</td><td>$2A_1'$</td><td>$2A_{1g}$</td><td>$2A_1'$</td><td>$2A_{1g}$</td></tr>
<tr><td>3d</td><td>(101)</td><td>$2B_2$</td><td rowspan="2">2E</td><td rowspan="2">2E</td><td rowspan="2">2E</td><td rowspan="2">$2E_1$</td><td>$1B_{2g}$</td><td rowspan="2">$1E''$</td><td rowspan="2">$1E_g$</td><td rowspan="2">$1E_1''$</td><td rowspan="2">$1E_{1g}$</td></tr>
<tr><td>3d</td><td>(011)</td><td>$2B_3$</td><td>$1B_{3g}$</td></tr>
<tr><td>3d</td><td>(200)−(020)</td><td>3A</td><td rowspan="2">3E</td><td>$1B_1$</td><td rowspan="2">$1E_2$</td><td rowspan="2">$1E_2$</td><td>$3A_g$</td><td rowspan="2">$2E'$</td><td>$1B_{1g}$</td><td rowspan="2">$1E_2'$</td><td rowspan="2">$1E_{2g}$</td></tr>
<tr><td>3d</td><td>(110)</td><td>$2B_1$</td><td>$1B_2$</td><td>$1B_{1g}$</td><td>$1B_{2g}$</td></tr>
<tr><td>4f</td><td>5(003)−3(001)</td><td>$3B_1$</td><td>$2A_2$</td><td>$2A_2$</td><td>$2A_2$</td><td>$2A_2$</td><td>$2B_{1u}$</td><td>$2A_2''$</td><td>$2A_{2u}$</td><td>$2A_2''$</td><td>$2A_{2u}$</td></tr>
<tr><td>4f</td><td>5(102)−(100)</td><td>$3B_3$</td><td rowspan="2">4E</td><td>3E</td><td rowspan="2">$3E_1$</td><td rowspan="2">$3E_1$</td><td>$2B_{3u}$</td><td rowspan="2">$3E'$</td><td rowspan="2">$2E_u$</td><td rowspan="2">$2E_1'$</td><td rowspan="2">$2E_{1u}$</td></tr>
<tr><td>4f</td><td>5(012)−(010)</td><td>$3B_2$</td><td>3E</td><td>$3B_{2u}$</td></tr>
<tr><td>4f</td><td>(201)−(021)</td><td>$4B_1$</td><td rowspan="2">5E</td><td>$2B_2$</td><td rowspan="2">$2E_2$</td><td rowspan="2">$2E_2$</td><td>$3B_{1u}$</td><td rowspan="2">$2E''$</td><td>$1B_{2u}$</td><td rowspan="2">$2E_2''$</td><td rowspan="2">$1E_{2u}$</td></tr>
<tr><td>4f</td><td>(111)</td><td>3A</td><td>$2B_1$</td><td>$1A_u$</td><td>$1B_{1u}$</td></tr>
<tr><td>4f</td><td>(300)−3(120)</td><td>$3B_3$</td><td>$3A_1$</td><td rowspan="2">4E</td><td rowspan="2">$3E_2$</td><td>$1B_1$</td><td>$3B_{3u}$</td><td>$3A_1'$</td><td rowspan="2">$3E_u$</td><td>$2E_2'$</td><td>$1B_{2u}$</td></tr>
<tr><td>4f</td><td>3(210)−(030)</td><td>$4B_2$</td><td>$3A_2$</td><td>$1B_2$</td><td>$3B_{2u}$</td><td>$1A_2'$</td><td>$2E_2'$</td><td>$1B_{1u}$</td></tr>
<tr><td>5g</td><td>35(004)−30(002)+3(000)</td><td>4A</td><td>$4A_1$</td><td>$3A_1$</td><td>$3A_1$</td><td>$3A_1$</td><td>$4A_g$</td><td>$4A_1'$</td><td>$3A_{1g}$</td><td>$3A_1'$</td><td>$3A_{1g}$</td></tr>
<tr><td>5g</td><td>7(103)−3(101)</td><td>$5B_2$</td><td rowspan="2">6E</td><td rowspan="2">5E</td><td rowspan="2">$4E_1$</td><td rowspan="2">$4E_1$</td><td>$2B_{2g}$</td><td rowspan="2">$3E''$</td><td rowspan="2">$2E_g$</td><td rowspan="2">$2E_1''$</td><td rowspan="2">$2E_{1g}$</td></tr>
<tr><td>5g</td><td>7(013)−3(011)</td><td>$4B_3$</td><td>$2B_{3g}$</td></tr>
<tr><td>5g</td><td>7(202)−7(022)−(200)+(020)</td><td>5A</td><td rowspan="2">7E</td><td>$3B_1$</td><td rowspan="2">$4E_2$</td><td rowspan="2">$3E_2$</td><td>$4A_g$</td><td rowspan="2">$4E'$</td><td>$2B_{1g}$</td><td rowspan="2">$3E_2'$</td><td rowspan="2">$2E_{2g}$</td></tr>
<tr><td>5g</td><td>7(112)−(110)</td><td>$5B_1$</td><td>$3B_2$</td><td>$2B_{1g}$</td><td>$2B_{2g}$</td></tr>
<tr><td>5g</td><td>(301)−3(121)</td><td>$6B_2$</td><td>$4A_2$</td><td rowspan="2">6E</td><td rowspan="2">$5E_2$</td><td>$2B_2$</td><td>$3B_{2g}$</td><td>$3A_2''$</td><td rowspan="2">$3E_g$</td><td rowspan="2">$2E_2''$</td><td>$1B_{1g}$</td></tr>
<tr><td>5g</td><td>3(211)−(031)</td><td>$5B_3$</td><td>$5A_1$</td><td>$2B_1$</td><td>$3B_{3g}$</td><td>$1A_1''$</td><td>$1B_{2g}$</td></tr>
<tr><td>5g</td><td>(400)+(040)−6(220)</td><td>6A</td><td rowspan="2">7E</td><td>$4A_1$</td><td rowspan="2">$5E_1$</td><td rowspan="2">$4E_2$</td><td>$5A_g$</td><td rowspan="2">$5E'$</td><td>$4A_{1g}$</td><td rowspan="2">$3E_1'$</td><td rowspan="2">$3E_{2g}$</td></tr>
<tr><td>5g</td><td>(310)−(130)</td><td>$6B_1$</td><td>$3A_2$</td><td>$3B_{1g}$</td><td>$1A_{2g}$</td></tr>
</table>

It is necessary to identify combinations of the spherical harmonics

$$Y_{\ell m} = P_\ell^{|m|}(\cos\theta)\, e^{im\phi} \qquad -\ell \le m \le \ell$$

or their real and imaginary forms

$$\begin{matrix} P_\ell^{|m|}(\cos\theta)\cos(m\phi) \\ P_\ell^{|m|}(\cos\theta)\sin(m\phi) \end{matrix} \qquad 0 \le m \le \ell$$

**Table 3.8 Classification of the general spherical harmonics up to angular momentum level 4 by descent in symmetry, into their irreducible components for the molecular point groups $D_{2d}$, $D_{3d}$, $D_{4d}$, $D_{5d}$, $D_{6d}$, $S_4$, $S_6$, $S_8$.**

| | | $D_{2d}$ | $D_{3d}$ | $D_{4d}$ | $D_{5d}$ | $D_{6d}$ | $S_4$ | $S_6$ | $S_8$ |
|---|---|---|---|---|---|---|---|---|---|
| 1s | (000) | $1A_1$ | $1A_{1g}$ | $1A_1$ | $1A_{1g}$ | $1A_1$ | 1A | $1A_g$ | 1A |
| 2p | (001) | $1A_2$ | $1A_{2u}$ | $1B_2$ | $1A_{2u}$ | $1B_2$ | 1B | $1A_u$ | 1B |
| 2p | (100) | 1E | $1E_u$ | $1E_1$ | $1E_{1u}$ | $1E_1$ | 1E | $1E_u$ | $1E_1$ |
| 2p | (010) | | | | | | | | |
| 3d | 3(002)−(000) | $2A_1$ | $2A_{1g}$ | $2A_1$ | $2A_{1g}$ | $2A_1$ | 2A | $2A_g$ | 2A |
| 3d | (101) | 2E | $2E_g$ | $1E_3$ | $1E_{1g}$ | $1E_5$ | 2E | $1E_g$ | $1E_3$ |
| 3d | (011) | | | | | | | | |
| 3d | (200)−(020) | $1B_1$ | $3E_g$ | $1E_2$ | $1E_{2g}$ | $1E_2$ | 2B | $2E_g$ | $1E_2$ |
| 3d | (110) | $1B_2$ | | | | | 3B | | |
| 4f | 5(003)−3(001) | $2A_1$ | $2A_{2u}$ | $2B_2$ | $2A_{2u}$ | $2B_2$ | 4B | $2A_u$ | 2B |
| 4f | 5(102)−(100) | 3E | $2E_u$ | $2E_1$ | $2E_{1u}$ | $2E_1$ | 3E | $2E_u$ | $2E_1$ |
| 4f | 5(012)−(010) | | | | | | | | |
| 4f | (201)−(021) | $2B_2$ | $3E_u$ | $2E_2$ | $1E_{2u}$ | $1E_4$ | 3A | $3E_u$ | $2E_2$ |
| 4f | (111) | $2B_1$ | | | | | 4A | | |
| 4f | (300)−3(120) | 4E | $1A_{1u}$ | $2E_3$ | $2E_{2u}$ | $1E_3$ | 4E | $3A_u$ | $2E_3$ |
| 4f | 3(210)−(030) | | $3A_{2u}$ | | | | | $4A_u$ | |
| 5g | 35(004)−30(002)+3(000) | $3A_1$ | $3A_{1g}$ | $3A_1$ | $3A_{1g}$ | $3A_1$ | 5A | $3A_g$ | 3A |
| 5g | 7(103)−3(101) | 5E | $3E_g$ | $2E_3$ | $3E_{1u}$ | $2E_5$ | 5E | $3E_g$ | $3E_3$ |
| 5g | 7(013)−3(011) | | | | | | | | |
| 5g | 7(202)−7(022)−(200)+(020) | $3B_1$ | $4E_g$ | $3E_2$ | $2E_{2g}$ | $2E_2$ | 5B | $4E_g$ | $3E_2$ |
| 5g | 7(112)−(110) | $3B_2$ | | | | | 6B | | |
| 5g | (301)−3(121) | 6E | $1A_{2g}$ | $3E_1$ | $3E_{2g}$ | $2E_3$ | 6E | $4A_g$ | $3E_1$ |
| 5g | 3(211)−(031) | | $4A_{1g}$ | | | | | $5A_g$ | |
| 5g | (400)+(040)−6(220) | $4A_1$ | $5E_g$ | $3B_1$ | $2E_{1g}$ | $2E_4$ | 6A | $5E_g$ | 3B |
| 5g | (310)−(130) | $3A_2$ | | $3B_2$ | | | 7A | | 4B |

where

$$P_\ell^{|m|}(z) = \sqrt{\frac{(\ell - m)!}{(\ell + m)!}} \left(1 - z^2\right)^{m/2} \frac{d^m}{dz^m} P_\ell(z)$$

with

$$P_\ell(z) = \frac{1}{2^\ell \ell!} \frac{d^\ell}{dz^\ell} \left( \left(z^2 - 1\right)^\ell \right)$$

**Table 3.9 Classification of the *kubic* spherical harmonics up to angular momentum level 4 by descent in symmetry, into their irreducible components for the molecular point groups $O_h$, O, $T_d$, $T_h$ and T.**

| | | $O_h$ | O | $T_d$ | $T_h$ | T |
|---|---|---|---|---|---|---|
| 1s | (000) | $1A_{1g}$ | $1A_1$ | $1A_1$ | $1A_g$ | 1A |
| 2p | (001) | $1T_{1u}$ | $1T_1$ | $1T_2$ | $1T_u$ | 1T |
| 2p | (100) | | | | | |
| 2p | (010) | | | | | |
| 3d | (110) | $1T_{2g}$ | $1T_2$ | $2T_2$ | $1T_g$ | 2T |
| 3d | (101) | | | | | |
| 3d | (011) | | | | | |
| 3d | 2(002)–(200)–(020) | $1E_g$ | 1E | 1E | $1E_g$ | 1E |
| 3d | (200)–(020) | | | | | |
| 4f | (111) | $1A_{2u}$ | $1A_2$ | $2A_1$ | $1A_u$ | 2A |
| 4f | 5(300)–3(100) | $2T_{1u}$ | $2T_1$ | $3T_2$ | $2T_u$ | 3T |
| 4f | 5(030)–3(010) | | | | | |
| 4f | (003)–3(001) | | | | | |
| 4f | (201)–(021) | $1T_{2u}$ | $2T_2$ | $1T_1$ | $3T_u$ | 4T |
| 4f | (102)–(120) | | | | | |
| 4f | (210)–(012) | | | | | |
| 5g | 35(004)–30(002)+3(000) | $2A_{1g}$ | $2A_1$ | $3A_1$ | $2A_g$ | 3A |
| 5g | 7(022)–7(202)–(020)+(200) | $2E_g$ | 2E | 2E | $2E_g$ | 2E |
| 5g | 14(220)–7(022)–7(202)–(200)–(020)+2(002) | | | | | |
| 5g | (013)–(031) | $1T_{1g}$ | $3T_1$ | $2T_1$ | $2T_g$ | 5T |
| 5g | (301)–(103) | | | | | |
| 5g | (130)–(310) | | | | | |
| 5g | 7(112)–(110) | $2T_{2g}$ | $3T_2$ | $4T_2$ | $3T_g$ | 6T |
| 5g | 7(211)–(011) | | | | | |
| 5g | 7(121)–(101) | | | | | |

and, as usual, $z = \mathrm{Cos}\,\theta$, $x = \mathrm{Sin}\,\theta \mathrm{Cos}\,\phi$ and $y = \mathrm{Sin}\,\theta \mathrm{Sin}\,\phi$, which will generate the appropriate irreducible subspaces.

So, for instance, (100) generates $\langle(100), (010), (001)\rangle$ of type $T_{1u}$ under the actions of the symmetry operators of the $O_h$ point group. But, if it is required that the second $T_{1u}$ is required to be orthogonal to the first one, with respect to integration over the unit sphere, then it is necessary to modify this second function with a Gram-Schmidt type transformation to obtain the distinct second set of $T_{1u}$ symmetry, $\langle 5(300)-3(100), 5(030)-3(010), 5(003)-3(001)\rangle$.

Other polynomials can be treated in a similar fashion with results depending on the parity/equality of the exponents. So (113) generates $\langle(113), (131), (311)\rangle$, which in turn

**Table 3.10 Classification of the icosahedral harmonics up to angular momentum level 4 by descent in symmetry, into their irreducible components for the molecular point groups $I_h$ and I.**

| 1s | (000) | $1A_g$ | 1A |
|---|---|---|---|
| 2p | (100) | | |
| 2p | (010) | $1T_{1u}$ | $1T_1$ |
| 2p | (001) | | |
| 3d | 2(002)−(200)−9(020) | | |
| 3d | (200)−(020) | | |
| 3d | (110) | $1H_g$ | 1H |
| 3d | (101) | | |
| 3d | (011) | | |
| 4f | 5(003)−3(001) | | |
| 4f | (300)−3(120)+3(201)−3(021) | $1T_{2u}$ | $1T_2$ |
| 4f | 3(210)−(030)−6(111) | | |
| 4f | 5(102)−(100) | | |
| 4f | 5(012)−(010) | $1G_u$ | 1G |
| 4f | (300)−3(120)−2(201)+2(021) | | |
| 4f | 3(210)−(030)+4(111) | | |
| 5g | 35(004)−30(002)+3(000) | | |
| 5g | 14(301)−42(121)+7(202)−7(022)−(200)+(020) | | |
| 5g | 21(211)−7(031)−7(112)+(110) | $2H_g$ | 2H |
| 5g | 7(400)−42(220)+7(040)−56(103)+24(101) | | |
| 5g | 7(310)−7(130)+14(013)−6(011) | | |
| 5g | (301)−3(121)−7(202)+7(022)+(200)−(020) | | |
| 5g | 3(211)−(031)+14(112)−2(110) | $1G_g$ | 2G |
| 5g | (400)−6(220)+(040)+7(103)−3(101) | | |
| 5g | 4(310)−4(130)−7(013)+3(011) | | |

decomposes into

$$\begin{aligned} &\langle (113) + (131) + (311) \rangle + \langle 2\,(113) - (131) - (311), (311) - (131) \rangle \\ &\cong A_{2u} + E_u \end{aligned}$$

In the case of the icosahedral point groups, $I_h$ and I, Table 3.10, the analysis is more complicated and there is a need to identify the combinations of the spherical harmonics, which will generate higher dimensional irreducibile subspaces. For example, at level 3, there are 7 harmonics, but the irreducible subspaces in icosahedral symmetry are four-fold [$G_u$] and three-fold [$T_{2u}$]. It is found that three of the original functions can be carried over to provide basis functions in icosahedral, symmetry but that four distinct linear combinations of the originals are required to complete the two sets, with the results

$$
T_{2u} = \begin{cases} P_3^0\,(\mathrm{Cos}\,\theta) = 5z^3 - 3z \\ P_3^3\,(\mathrm{Cos}\,\theta \mathrm{Cos}\,3\phi) + \sqrt{\tfrac{3}{2}} P_3^2\,(\mathrm{Cos}\,\theta\ \mathrm{Sin}\,2\phi) = \left(x^2 - 3xy^2\right) + 3\left(x^2 - y^2\right) z \\ P_3^3\,(\mathrm{Cos}\,\theta \mathrm{Cos}\,3\phi) - \sqrt{\tfrac{3}{2}} P_3^2\,(\mathrm{Cos}\,\theta\ \mathrm{Sin}\,2\phi) = \left(3xy^2 - y^3\right) + 3\,(2xy)\, z \end{cases}
$$

$$
G_u = \begin{cases} P_3^1\,(\mathrm{Cos}\,\theta\ \mathrm{Cos}\,\phi) = \left(5z^2 - 1\right) x \\ P_3^1\,(\mathrm{Cos}\,\theta\ \mathrm{Sin}\,\phi) = \left(5z^2 - 1\right) y \\ P_3^3\,(\mathrm{Cos}\,\theta \mathrm{Cos}\,3\phi) - \sqrt{\tfrac{2}{3}} P_3^2\,(\mathrm{Cos}2\theta) = \left(x^3 - 3xy^2\right) - 2\left(x^2 - y^2\right) z \\ P_3^3\,(\mathrm{Cos}\,\theta \mathrm{Cos}\,3\phi) + \sqrt{\tfrac{2}{3}} P_3^2\,(\mathrm{Cos}2\theta) = \left(3x^2y - y^3\right) + 2\,(2xy)\, z \end{cases}
$$

In addition, in the icosahedral groups five-fold degenerate irreducible characters are found, and so five distinct functions, mutually orthogonal on the unit sphere, are required.

It is important to emphasize, again, that the harmonic functions listed in the tables of this section[6] and in the CD files are mutually orthogonal only on the unit sphere. Projections onto the vertices of structure orbits preserve mutual orthogonality only between the central functions of different symmetries. So, for example, an A-type group orbital is a linear combination of the local functions on the vertices of the orbit, which is orthogonal to a B-type or E-type group orbital. However, in the case of the regular orbits of the point groups, the orthogonality of the two linear combinations of E-type symmetry, which result from application of the 1E and 2E central functions listed in the tables is not guaranteed, or indeed even likely.

## 3.4 Examples

It is instructive to include here several examples involving the use of the general, *kubic* and icosahedral harmonics and the theorems to establish the methodology of our group theory. These examples illustrate:

3.4.1 the construction of group orbitals, as basis functions for irreducible components of $\sigma$, $\pi$ and $\delta$ group characters, using the central polynomials;

3.4.2 how the group orbitals over the vertices of an orbit, $O_n$, found in environments of different symmetries reflect the irreducible characteristics of the total structure, leading, generally, to the requirement that several linear combinations be found and be rendered mutually orthogonal, when, more than one copy of a particular irreducible component is present in the reducible character over the orbit;

[6] Note that the lists of the central polynomial functions in the tables up to level 4 often are insufficient to identify suitable basis functions for the regular character irreducible components, because longer lists for all the groups would increase greatly the complexity of the tables. Complete sets of these functions for the components of the regular characters of the groups are given in the files on the CD. In a few cases, for example, see 3.4.2, it is necessary to identify certain components by calculation using Table 3.17.

3.4.3 how in multi-orbit structures, it is much simpler to construct group orbital components on each orbit and then combine these to make the group orbital over all the vertices of the molecule;

3.4.4 the ease with which this methodology can be applied to form group orbitals for structures for which the traditional methods of group theory require considerable trigonometric skills.

### 3.4.1 Group orbitals modelled on central polynomials as basis functions for the $\sigma$, $\pi$ and $\delta$ group orbitals that can be formed on different structure orbits of the point groups

The general problem to construct basis functions, as linear combinations of local orbitals, for irreducible representation, is managed most conveniently if the reducible characters on decorated orbits are built using the alternative local sets of harmonic functions sketched in Figure 3.2, Figure 3.3 and Figure 3.5.

Basis functions of $\sigma$, $\pi$ and $\delta$-types, for the irreducible components of the reducible characters of each kind, then, can be formed by the taking of the linear combinations generated by the actions of the structure point group symmetry operations on the sets $[\sigma]_j$, $[\pi_\theta, \pi_\phi]_j$ and $[\delta_{\theta\theta}, \delta_{\theta\phi}]_j$ about each vertex, $P_j$, of a structure orbit with the rendering of these functions to be mutually orthonormal, where this is necessary as an additional step.

With reference to Figure 3.1, we have defined at each position, $P_j$, a local coordinate system $\sigma(j)$, $\pi_\theta(j)$, $\pi_\phi(j)$: $\sigma(j)$ is the unit vector towards the central origin on the unit sphere; while $\pi_\theta(j)$ points south; and $\pi_\phi(j)$ points to the east[7]. This local coordinate system provides for the construction of local functions of $\sigma$, $\pi$ and $\delta$ orientations upon which group orbitals of these irreducible symmetries can be formed as linear combinations exhibiting angular momentum components $\lambda = 0, \pm 1$ and $\pm 2$ about the radial vectors to each vertex of the structure orbit.

These can be constructed with the help of a particular complex differential operator[8]. Let

$$D_j^\lambda = \left(\frac{\partial}{\partial \pi_\theta\,(j)} + i\frac{\partial}{\partial \pi_\phi\,(j)}\right)^\lambda \qquad 3.13$$

for the local coordinate system at the vertex point $P_j$ on the unit sphere inscribing the structure orbit cage and let

$$U_j^\lambda = \mathrm{Re}\left(D_j^\lambda\right) \qquad 3.14$$

$$U_j^\lambda = \mathrm{Im}\left(D_j^\lambda\right) \qquad 3.15$$

[7] It is convenient to adopt the following conventions at the North and South poles: at the North pole, choose $\pi_\theta$ to lie as the tangent to the Great Circle through $+e_z$, $+e_x$, $-e_z$, $-e_x$ and of positive sign toward $+e_x$; choose $\pi_\phi$ as the tangent to the Great Circle through $+e_z$, $+e_y$, $-e_z$, $-e_y$ and to be of positive sign toward $+e_y$: at the South pole, choose $\pi_\theta$ to lie as the tangent to the Great Circle through $-e_z$, $-e_x$, $+e_z$, $+e_x$ and of positive sign toward $-e_x$; choose $\pi_\phi$ as the tangent to the Great Circle through $-e_z$, $-ey$, $+e_z$, $+e_y$ and to be of positive sign toward $-e_y$.

[8] David B. Redmond, Charles M. Quinn and John G. McKiernan, *J. Chem. Soc. Faraday II*, **79** (1983) 1791, Anthony J. Stone, *Molecular Physics*, **41** (1980) 1339.

represent the real and imaginary parts of equation 3.16 taken with respect to complex conjugation of the operator.

Finally, let

$$\delta^{\lambda}\left(f\left(x, y, z\right)\right) = \sum_{j=1}^{n} \left| \left( U_j^{\lambda}\left(f\left(x, y, z\right)\right) u_j^{\lambda} + V_j^{\lambda}\left(f\left(x, y, z\right)\right) v_j^{\lambda} \right) \right|_{P_j} \qquad 3.16$$

for a central polynomial $f(x, y, z)$, with the $u_j^{\lambda}$ and $v_j^{\lambda}$ the transformed local functions of Figure 3.4.

Operating with $D^{\lambda}$ on the central functions generates the map $\delta^{\lambda}$ from central harmonics to harmonics on the point $P_j$, i.e.

$$\delta^{\lambda} : \text{central functions} \rightarrow \text{level } \ell \text{ harmonics on } P_j \qquad 3.17$$

$D^{\lambda}$ commutes with the operations of G and so preserves the symmetry of the projection operations upon $\sigma$, $\pi$, $\delta$ ... harmonics at $P_j$. Thus, the problem of finding bases for the irreducible subspaces of the components of $\Gamma_{\lambda}$ generated by the transformation properties of the local $\ell$-level harmonic functions at the vertices of any structure orbit of the group G is solved if bases for the same irreducible subspaces of the central harmonics or appropriate copies of these harmonics are found.

It is convenient to write out $D^{\lambda}$ in polar coordinates, since this is the common coordinate system used to present the spherical harmonics, $Y_{lm}(\theta, \phi)$, as the angular components of atomic orbitals. This change of variables from the $(\pi_{\theta}(j), \pi_{\phi}(j))$ set at each vertex of a structure orbit on the unit sphere leads to the transformation relationships

$$\frac{\partial}{\partial \pi_{\theta(j)}} = \frac{\cos\left(\theta_j\right) \sin\left(\phi_j - \phi\right)}{\sin\left(\theta\right)} \frac{\partial}{\partial \phi} + \frac{\cos\left(\theta_j\right) \cos\left(\phi_j - \phi\right)}{\cos\left(\theta\right)} \frac{\partial}{\partial \theta} \qquad 3.18$$

$$\frac{\partial}{\partial \pi_{\phi(j)}} = \frac{\cos\left(\phi - \phi_j\right)}{\sin\left(\theta\right)} \frac{\partial}{\partial \phi} + \frac{\sin\left(\phi - \phi_j\right)}{\cos\left(\theta\right)} \frac{\partial}{\partial \theta} \qquad 3.19$$

for a given $Y_{lm}(\theta, \phi)$ wherein $(\theta_j, \phi_j)$ are the angular coordinates of the vertex $P_j$ of the structure orbit on the unit sphere.

In summary, therefore, the following equations can be applied to describe the $\sigma$-, $\pi$- and $\delta$-type interactions of harmonic components on a structure orbit cage. For $\sigma$-type interactions, the case $\lambda = 0$,

$$\begin{aligned} U_j^0\left(Y_{\ell m}\left(\theta, \phi\right)\right) &= \left|Y_{\ell m}\left(\theta, \phi\right)\right|_j \\ V_j^0\left(Y_{\ell m}\left(\theta, \phi\right)\right) &= 0 \\ u_j^0 = 1, \; v_j^0 &= 0 \end{aligned} \qquad 3.20$$

For $\pi$-type interactions, the case $\lambda = |1|$, we have

$$\mathrm{U}_j^1\left(\mathrm{Y}_{\ell m}(\theta,\phi)\right) = \left|\frac{\partial}{\partial\theta}\left(\mathrm{Y}_{\ell m}(\theta,\phi)\right)\right|_j$$

$$\mathrm{V}_j^1\left(\mathrm{Y}_{\ell m}(\theta,\phi)\right) = \left|\frac{1}{\sin(\theta)}\frac{\partial}{\partial\phi}\left(\mathrm{Y}_{\ell m}(\theta,\phi)\right)\right|_j \qquad 3.21$$

$$\mathrm{u}_j^1 = \pi_{\theta_j},\ \mathrm{v}_j^1 = \pi_{\phi_j}$$

and the local functions at each structure orbit vertex are linear combinations of the local $\pi_\theta|_j$ and $\pi_\phi|_j$ functions. The $\mathrm{U}_{lm}^1$ and $\mathrm{V}_{lm}^1$ are known as Vector Surface Harmonics.

For $\delta$-type interactions, the case $\lambda = |2|$, we have

$$\mathrm{U}_j^2\left(\mathrm{Y}_{\ell m}(\theta,\phi)\right) = \left|\left(\frac{\partial^2}{\partial\pi_{\theta_j}^2} - \frac{\partial^2}{\partial\pi_{\phi_j}^2}\right)\left(\mathrm{Y}_{\ell m}(\theta,\phi)\right)\right|_j$$

$$= \left|\left(\frac{\partial^2}{\partial\theta^2} - \frac{1}{\sin^2(\theta)}\frac{\partial^2}{\partial\phi^2} - \cot(\theta)\frac{\partial}{\partial\phi}\right)\left(\mathrm{Y}_{\ell m}(\theta,\phi)\right)\right|_j$$

$$\mathrm{V}_j^2\left(\mathrm{Y}_{\ell m}(\theta,\phi)\right) = \left|2\left(\frac{\partial}{\partial\pi_{\theta_j}}\frac{\partial}{\partial\pi_{\phi_j}}\right)\mathrm{Y}_{\ell m}(\theta,\phi)\right|_j \qquad 3.22$$

$$= \left|\frac{\partial}{\partial\theta}\frac{1}{\sin(\theta)}\frac{\partial}{\partial\phi}\left(\mathrm{Y}_{\ell m}(\theta,\phi)\right)\right|_j$$

$$\mathrm{u}_j^2 = \delta_{\theta\theta_j},\ \mathrm{v}_j^2 = \delta_{\theta\phi_j}$$

and the local functions at each structure orbit vertex are linear combinations of the local $\delta_{\theta\theta}|_j$ and $\delta_{\theta\phi}|_j$ functions. The $\mathrm{U}_{lm}^2$ and $\mathrm{V}_{lm}^2$ are known as the Tensor Surface Harmonics.

For $\sigma$ group orbitals, therefore, it suffices to find the local magnitudes of the harmonic at the vertices. For $\pi$ we need the local gradients [as first derivatives] and for $\delta$ the local concavities [as second derivatives]. These determine the combination coefficients of the local resultant functions in the group orbitals, which local functions are found by making the appropriate linear combinations of $\sigma$; $\pi_\theta$ and $\pi_\phi$; $\delta_{\theta\theta}$ and $\delta_{\theta\phi}$ combinations of the local atomic orbitals decorating the orbit vertices.

Thus, for any central function, $\mathrm{Y}_{lm}(\theta,\phi)$, the $\sigma$ group orbital of the corresponding irreducible symmetry in the $\Gamma_\sigma^\lambda$ over a particular structure orbit of the point group G is the linear combination of the $\sigma$-oriented local functions with the coefficients modulated by the magnitude of the central function at the vertices of the orbit, equations 3.20.

For $\pi$-oriented local functions, equations 3.21 dictate the mixing of the $\pi_\theta|_j$ and $\pi_\phi|_j$ at each vertex of the structure orbit so that the local resultant lies in the tangent plane towards the direction of maximum gradient of the central function and modulated by the local magnitude of that gradient.

For $\delta$-oriented local functions, the local concavity of the central function is the property of interest, which is sampled by equations 3.22. Thus, the local $\delta$-function as a linear combination of the $\delta_{\theta\theta}|_j$ and $\delta_{\theta\phi}|_j$ is formed so that the positive lobes of the resultant local function lie along the direction of maximum and the negative lobes are, of necessity, in the direction of minimum concavity. The modulating coefficients of the local functions distributed over the vertices of the structure orbit are

$$\left|\left(\frac{\partial^2}{\partial\pi_{\theta_j}^2} - \frac{\partial^2}{\partial\pi_{\phi_j}^2}\right)\right|_j$$

corresponding, at each vertex, $P_j$, to the magnitude of the difference between the local maxima and minima of the central function concavity.

When, as is often the case, the vertices of structure orbits lie on axes of symmetry or anti-symmetry of the central functions, simple rules can be applied to complete the construction of all the group orbitals for a particular decoration of orbit vertices with valence atomic orbitals. For vertices lying on great circles of symmetry of the central functions, the gradient directions lie in the tangent planes, at these vertices, and so coincide with the orientation of the $u_j^1$ suitably oriented to match the central function phases. Moreover, for such $\pi$ group orbitals, exhibiting an irreducible symmetry distinguished by the character $\Gamma_\chi$, the complementary set of $\pi$ group orbitals, based on the transformation properties of the $v_j^1$ local orbit, exhibiting the irreducible symmetry $\Gamma_\chi \times \Gamma_\varepsilon$, follow by rotation of the $u_j^1$ locally by $\pi/2$ in the same sense at each vertex of the orbit. The orientations of $\delta$-group orbital components at vertices on the great circles of symmetry of central functions correspond to aligning the positive phase of the $u_j^2$ local functions to match with the local phases of the central function. For any such a group orbital of $\Gamma_\chi$ irreducible symmetry, a complementary group orbital based on the $v_j^2$ local functions and exhibiting the irreducible symmetry $\Gamma_\chi \Gamma_\varepsilon$, follow by concerted local rotations of the $u_j^2$ functions, by $\pi/4$ in the tangent planes at each vertex.

The local rotational property, which converts any $u_j^\lambda$ into the complementary function $v_j^\lambda$ is manifest in the transformations

$$\begin{aligned} U_{\pi/2}^1 &= -V^1 \\ V_{\pi/2}^1 &= -U^1 \\ U_{\pi/4}^2 &= -V^2 \\ V_{\pi/4}^2 &= -U^2 \quad \text{all at } P_j \end{aligned} \qquad 3.23$$

in the linear combinations of equation 3.16, which return the basis functions for the irreducible components of the $\Gamma_\pi$ and $\Gamma_\delta$ characters of the different structure orbits.

Table 3.11 lists the real forms for the $U_j^\lambda$ and $V_j^\lambda$ spherical ($\lambda = 0$) and vector ($\lambda = |1|$) harmonics for the $l = 0$ to 3 central harmonic functions. Group orbitals of the particular irreducible symmetries for which these central functions provide bases follow simply by making linear combinations of the $u_j^\lambda$ and $v_j^\lambda$ at each vertex point modulated by the values of $U_j^\lambda$ and $V_j^\lambda$ to form local resultants, which interconvert from vertex to vertex of the orbit under the actions of the symmetry operations of the point group.

Figure 3.6 displays the group orbitals over the valence orbital decoration of the triangular orbit, for example, in $NF_3$ or $BF_3$ as transformed using equation 3.1 and the data of Table 3.11

**Table 3.11 The spherical ($U^0_{lm}$, equations 3.20) and vector surface ($U^1_{lm}$, $V^1_{lm}$, equation 3.21) harmonics for central s-, p-, d- and f-atomic orbitals.**

| | $U^0_{lm} = Y_{lm}(\theta, \phi)$ | $U^1_{lm}$ | $V^1_{lm}$ |
|---|---|---|---|
| s | $+1$ | $0$ | $0$ |
| z | $+\cos\theta$ | $-\sin\theta$ | $0$ |
| y | $+\surd(1/2)\sin\theta\cos\phi$ | $+\surd(1/2)\cos\theta\sin\phi$ | $+\surd(1/2)\sin\theta\cos\phi$ |
| x | $+\surd(1/2)\sin\theta\sin\phi$ | $+\surd(1/2)\cos\theta\cos\phi$ | $-\surd(1/2)\sin\theta\sin\phi$ |
| $z^2$ | $+(1/2)(3\cos^2\phi - 1)$ | $-(3/2)\sin 2\theta$ | $0$ |
| yz | $+\surd(3/8)\sin\theta\cos\theta\sin\phi$ | $+\surd(3/2)\cos 2\theta\sin\phi$ | $+\surd(3/2)\cos\theta\cos\phi$ |
| xz | $+\surd(3/8)\sin(\theta)\cos(\phi)\cos(\phi)$ | $+\surd(3/2)\cos 2\theta\cos\phi$ | $-\surd(3/2)\cos\theta\sin\phi$ |
| $x^2 - y^2$ | $+\surd(3/8)\sin^2(\theta)\sin(2\phi)$ | $+\surd(3/8)\sin 2\theta\cos 2\phi$ | $-\surd(3/2)\sin\theta\sin 2\phi$ |
| xy | $+\surd(3/8)\sin(\theta)\cos(2\phi)$ | $+\surd(3/8)\sin 2\theta\sin 2\phi$ | $+\surd(3/2)\sin\theta\cos 2\phi$ |
| $z^3$ | $+(1/2)(5\cos^2(\theta) - 3\cos(\theta))$ | $-(3/2)\sin\theta(5\cos^2(\theta) - 1)$ | $0$ |
| $yz^2$ | $+\surd(3/16)\sin\theta(5\cos^2\theta - 1)\sin\phi$ | $+\surd(3/16)\cos\theta(15\cos^2\theta - 11)\sin\phi$ | $+\surd(3/16)(5\cos^2\theta - 1)\cos\phi$ |
| $xz^2$ | $+\surd(3/16)\sin\theta(5\cos^2\theta - 1)\cos\phi$ | $+\surd(3/16)\cos\theta(15\cos^2\theta - 11)\cos\phi$ | $-\surd(3/16)(5\cos^2\theta - 1)\sin\phi$ |
| $z(x^2 - y^2)$ | $+\surd(15/8)\sin^2\theta\cos\theta\cos 2\phi$ | $+\surd(15/8)\sin\theta(3\cos^2(\theta) - 1)\cos 2\phi$ | $-\surd(15/8)\sin 2\theta\sin 2\phi$ |
| z(xy) | $+\surd(15/8)\sin^2\theta\cos\theta\sin 2\phi$ | $+\surd(15/8)\sin\theta(3\cos^2(\theta) - 1)\sin 2\phi$ | $+\surd(15/8)\sin 2\theta\cos 2\phi$ |
| $y(3x^2 - y^2)$ | $+\surd(5/16)\sin^3\theta\sin 3\phi$ | $+(3/4)\surd 5\sin^2\theta\cos\theta\sin 3\phi$ | $+(3/4)\surd 5\sin^2\theta\cos 3\phi$ |
| $x(x^2 - 3y^2)$ | $+\surd(5/16)\sin^3\theta\cos 3\phi$ | $+(3/4)\surd 5\sin^2\theta\cos\theta\cos 3\phi$ | $-(3/4)\surd 5\sin^2\theta\sin 3\phi$ |

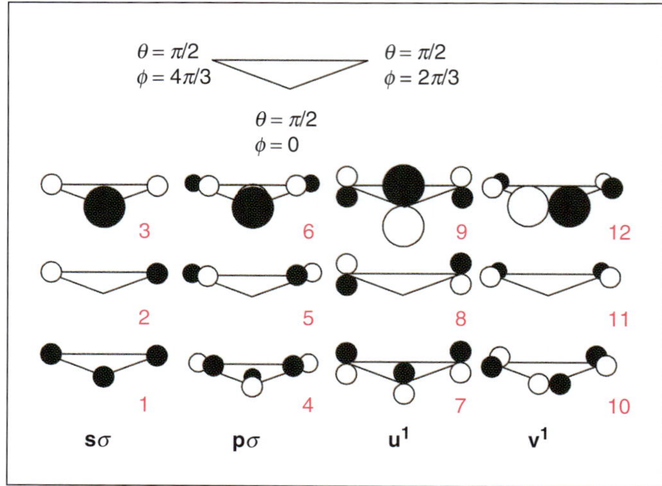


**Figure 3.6** The 12 linear combinations that can be formed from valence s- and p-atomic orbitals on an equilateral triangular orbit, assumed to be in a structure of overall $D_{3h}$ point symmetry.

assuming that the coordinate origin is at the triangle centre so that the intrinsic $D_{3h}$ symmetry of the orbit can be applied in both cases.

With reference to Table 3.11 and the linear combinations over local s-orbitals, sketched in diagrams, 1, 2 and 3 in Figure 3.6, there is a central s-like linear combination and the contributions from local s-orbitals are of equal weight, and so the icons for these orbitals are of equal size and shading in diagram 1. There is no z-like combination, since $\theta = \pi/2$ for all vertices of the triangular orbit. However, we see that the azimuthal angle sine and cosine values modulate the contributions of the local s-orbitals to the remaining pair of the three possible linear combinations in the form of equation 3.16, with the results sketched in diagrams 2 and 3 of Figure 3.6.

The diagrams in the second column of Figure 3.6 are sketches of the linear combinations over the local pσ functions at each of the vertices of the triangular structure orbit. With the convention that the normal orientation of the pσ functions is with the positive phase towards the origin, the next three linear combinations are formed using the same modulations imposed by the $Y_{lm}(\theta, \phi)$ of Table 3.11 and can be viewed also to have been constructed by substitution of the local pσ orbitals in the group orbitals formed as the combinations of the local s-orbitals.

The sketches in the 3rd and 4th columns of Figure 3.6 relate to the linear combinations formed using the vector surface harmonics of Table 3.11 as the modulating coefficients multiplying the local π-orbitals and these values are listed in Table 3.12. The first coefficients in equation 3.19 are $-\sin\theta$ for the z-like $U^1_{lm}$ surface harmonic and so lead to the equal weight linear combination of local $p_z$ orbitals sketched in diagram 7. Following the hierachical sequence of the surface harmonics listed in Table 3.5, we see that the next two non-zero $U^1_{lm}$ coefficients for the $u^1_j$ and $v^1_j$ local orbitals involve only the $v^1_j$ at each vertex, and are found for the xz and yz central functions. Thus we identify the linear combinations sketched

**Table 3.12 The modulating coefficients, reduced to their simplest ratios, for the linear combinations over the $u_j^1$[here $\pi_\theta$] and $v_j^1$ [here $\pi_\phi$] from equation 3.21 to form the group orbitals of the $F_3$ triangle displayed in the 3rd and 4th columns of Figure 3.6 using the vector surface harmonics of Table 3.11.**

| $\theta$ | $\pi/2$ | | $\pi/2$ | | $\pi/2$ | |
|---|---|---|---|---|---|---|
| $\phi$ | 0 | | $2\pi/3$ | | $4\pi/3$ | |
| $Y^0_{lm}(\theta,\phi)$ | $U^1_{lm}(\theta,\phi)$ | $V^1_{lm}(\theta,\phi)$ | $U^1_{lm}(\theta,\phi)$ | $V^1_{lm}(\theta,\phi)$ | $U^1_{lm}(\theta,\phi)$ | $V^1_{lm}(\theta,\phi)$ |
| s | | | | | | |
| z | −1 | 0 | −1 | 0 | −1 | 0 |
| y | 0 | 1 | 0 | −½ | 0 | −½ |
| x | 0 | 0 | 0 | −1 | 0 | 1 |
| $z^2$ | 0 | 0 | 0 | 0 | 0 | 0 |
| yz | 0 | 0 | 1 | 0 | −1 | 0 |
| xz | −1 | 0 | ½ | 0 | ½ | 0 |
| xy | 0 | 0 | 0 | 1 | 0 | −1 |
| $x^2 - y^2$ | 0 | 1 | 0 | −½ | 0 | −½ |
| $5z^3 - 3z$ | 1 | 0 | 1 | 0 | 1 | 0 |
| $yz^2$ | 0 | 0 | 0 | 0 | 0 | 0 |
| $xz^2$ | 0 | 0 | 0 | 0 | 0 | 0 |
| $z(x^2 - y^2)$ | −1 | 0 | ½ | 0 | ½ | 0 |
| z(xy) | 0 | 0 | 1 | 0 | −1 | 0 |
| $y(3x^2 - y^2)$ | 0 | 1 | 0 | 1 | 0 | 1 |
| $x(x^2 - 3y^2)$ | 0 | 0 | 0 | 0 | 0 | 0 |

in diagrams 11 and 12 of Figure 3.6. Two further linear combinations, again over the local $\pi_\theta$ [$-p_z$] orbitals at the vertices of the structure orbit, result by applying the yz and xz-Surface Harmonics in equation 3.16 and these give rise to diagrams 8 and 9 in Figure 3.6.

Diagram 10 shows the remaining linear combination, which is realized by local $\pi/2$ rotations of diagram 9 to give the group orbital $a_1' \times \Gamma_\varepsilon = a_1''$ ($a_2$ in $C_{3v}$).

Useful pictorial displays of these results can be constructed if the group orbitals for an $O_3$ orbit decorated with atomic orbitals, for example, as in Figure 3.6, are displayed on elliptical projections, Figure 3.7, of the generating central functions, with the orbit vertex decorations superimposed as familiar icons of different sizes, shaded, as necessary, to reflect the modulations from the different harmonic functions and in directions to reflect the directions of the local gradients and concavities for $\pi$ and $\delta$ group orbitals. Figure 3.8 displays the $\sigma$ group orbitals of Figure 3.6 in this manner. A similar construction is illustrated in Figure 3.9 for the same $O_3$ orbit, but in this case the decoration motifs are the tangential $\pi$-type components for which the modulations are determined by equations 3.21 and listed in Table 3.13.

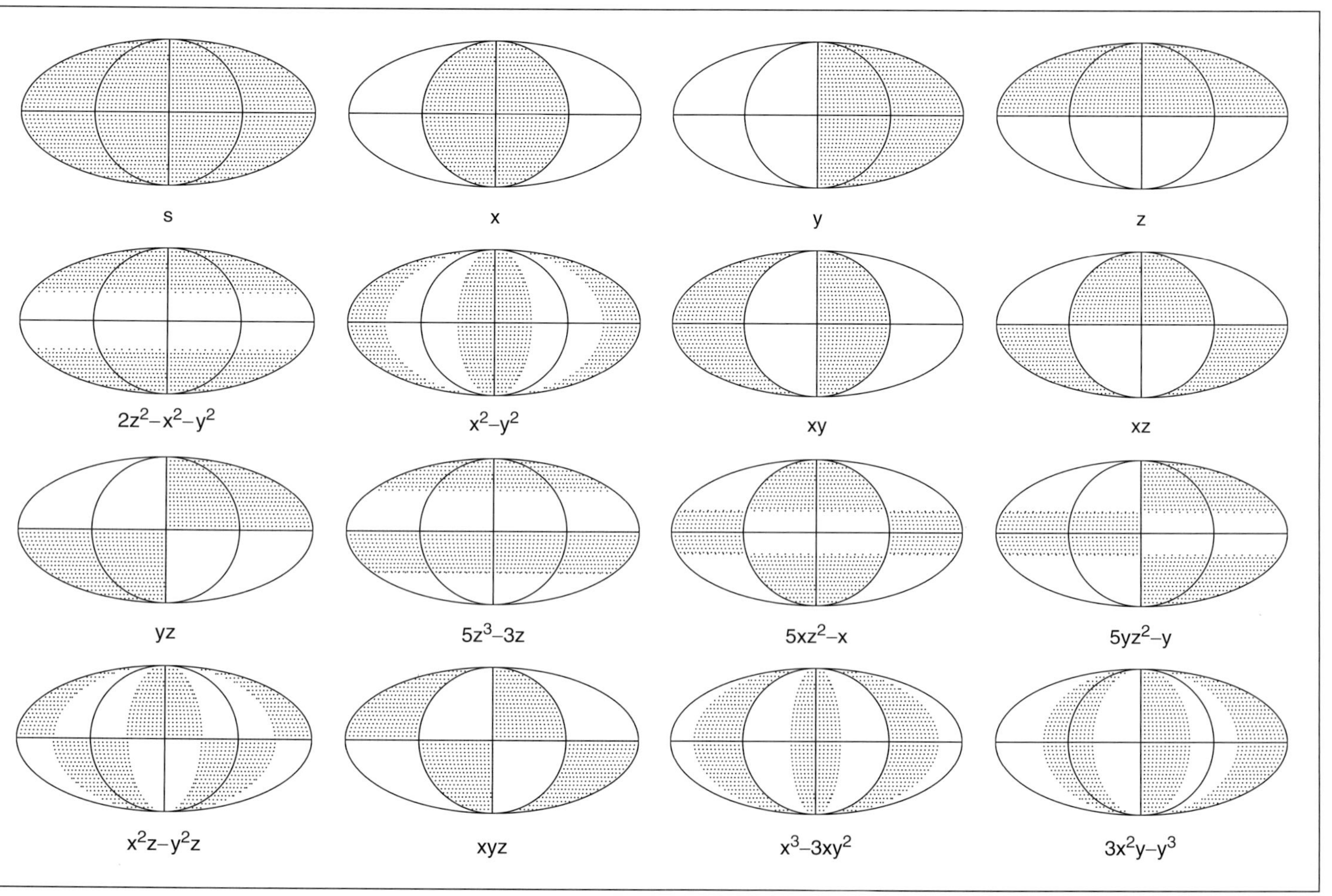


**Figure 3.7** The general spherical harmonics in the range $0 \leq \ell \leq 4$ displayed as elliptical projections on the unit sphere. The shadings in the diagrams reflect areas on the unit sphere of positive function amplitude.

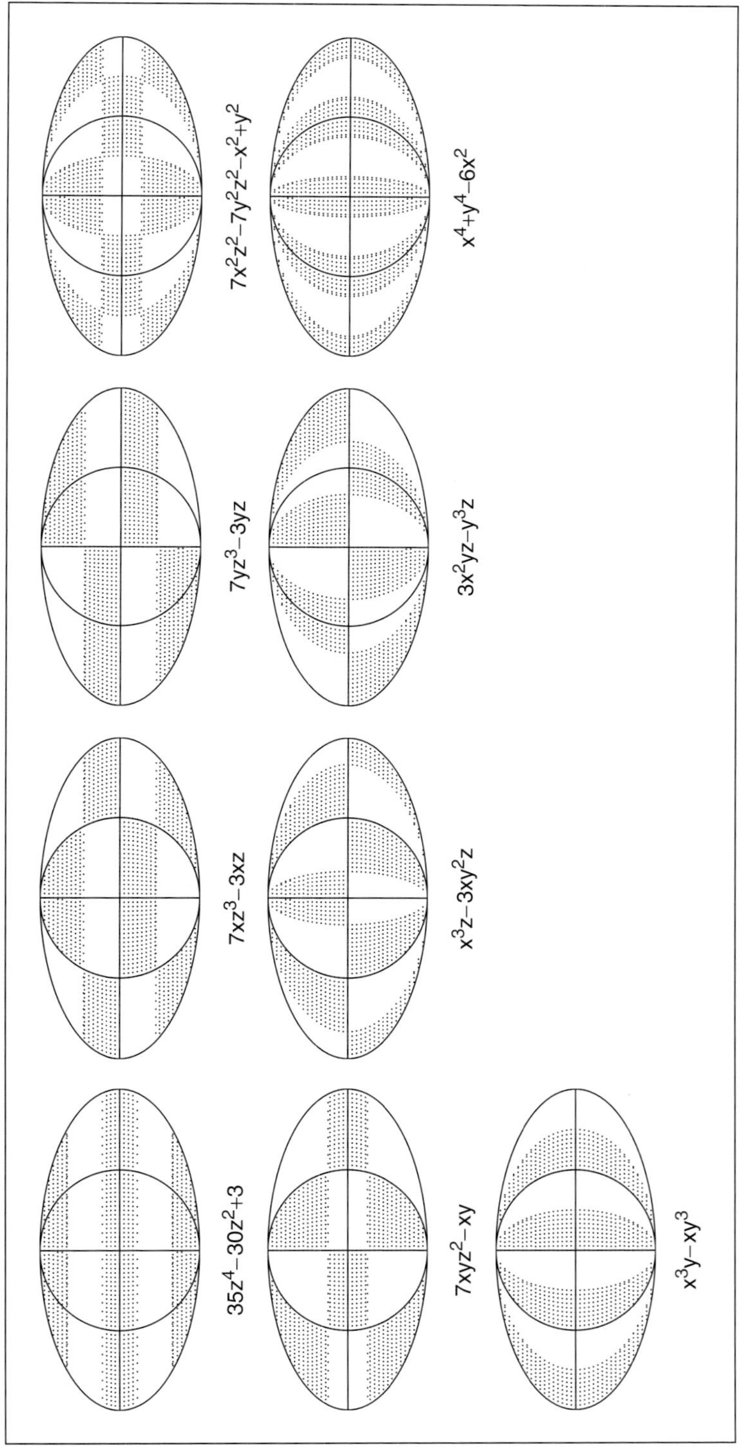


**Figure 3.7** Continued.

**Table 3.13 The spherical $Y_{lm}(\theta, \phi)$ and corresponding tensor surface ($U^2_{lm}$, $V^2_{lm}$, equations 3.22) harmonics for central s-, p-, d- and f-atomic orbitals.**

| | $U^0_{lm} = Y_{lm}(\theta, \phi)$ | $U^2_{lm} = Y_{lm}(\theta, \phi)$ | $V^2_{lm}$ |
|---|---|---|---|
| s | $+1$ | 0 | 0 |
| z | $+\cos\theta$ | 0 | 0 |
| y | $+\surd(1/2)\sin\theta\cos\phi$ | 0 | 0 |
| x | $+\surd(1/2)\sin\theta\sin\phi$ | 0 | 0 |
| $z^2$ | $+(1/2)(3\cos^2\phi - 1)$ | $+(3/2)\sin^2\theta$ | 0 |
| yz | $+\surd(3/8)\sin\theta\cos\theta\sin\phi$ | $-\surd(3/2)\sin\theta\cos\theta\sin\phi$ | $-\surd(3/2)\sin\theta\cos\phi$ |
| xz | $+\surd(3/8)\sin(\theta)\cos(\theta)\cos(\phi)$ | $-\surd(3/2)\sin\theta\cos\theta\cos\phi$ | $+\surd(3/2)\sin\theta\sin\phi$ |
| $x^2 - y^2$ | $+\surd(3/8)\sin^2(\theta)\sin(2\phi)$ | $+\surd(3/8)(1 + \cos^2\theta)\cos 2\phi$ | $-\surd(3/2)\cos\theta\sin 2\phi$ |
| xy | $+\surd(3/8)\sin(\theta)\cos(2\phi)$ | $+\surd(3/8)(1 + \cos^2\theta)\sin 2\phi$ | $+\surd(3/2)\cos\theta\cos 2\phi$ |
| $z^3$ | $+(1/2)(5\cos^2(\theta) - 3\cos(\theta))$ | $-(15/2)\cos\theta\sin^2\theta$ | 0 |
| $yz^2$ | $+\surd(3/16)\sin\theta(5\cos^2\theta - 1)\sin\phi$ | $-(5/4)\surd 3\sin\theta(3\cos^2\theta - 1)\sin\phi$ | $-(5/4)\surd 3\sin 2\theta\cos\phi$ |
| $xz^2$ | $+\surd(3/16)\sin\theta(5\cos^2\theta - 1)\cos\phi$ | $-(5/4)\surd 3\sin\theta(3\cos^2\theta - 1)\cos\phi$ | $+(5/4)\surd 3\sin 2\theta\sin\phi$ |
| $z(x^2 - y^2)$ | $+\surd(15/8)\sin^2\theta\cos\theta\cos 2\phi$ | $+\surd(15/8)\cos\theta(3\cos^2\theta - 1)\cos 2\phi$ | $-\surd(15/2)\cos 2\theta\sin 2\phi$ |
| z(xy) | $+\surd(15/8)\sin^2\theta\cos\theta\sin 2\phi$ | $+\surd(15/8)\cos\theta(3\cos^2\theta - 1)\sin 2\phi$ | $+\surd(15/2)\cos 2\theta\cos 2\phi$ |
| $y(3x^2 - y^2)$ | $+\surd(5/16)\sin^3\theta\sin 3\phi$ | $+(3/4)\surd 5\sin\theta(2\cos^2\theta + 1)\sin 3\phi$ | $+(3/4)\surd 5\sin 2\theta\cos 3\phi$ |
| $x(x^2 - 3y^2)$ | $+\surd(5/16)\sin^3\theta\cos 3\phi$ | $+(3/4)\surd 5\sin\theta(2\cos^2\theta + 1)\cos 3\phi$ | $-(3/4)\surd 5\sin 2\theta\sin 3\phi$ |

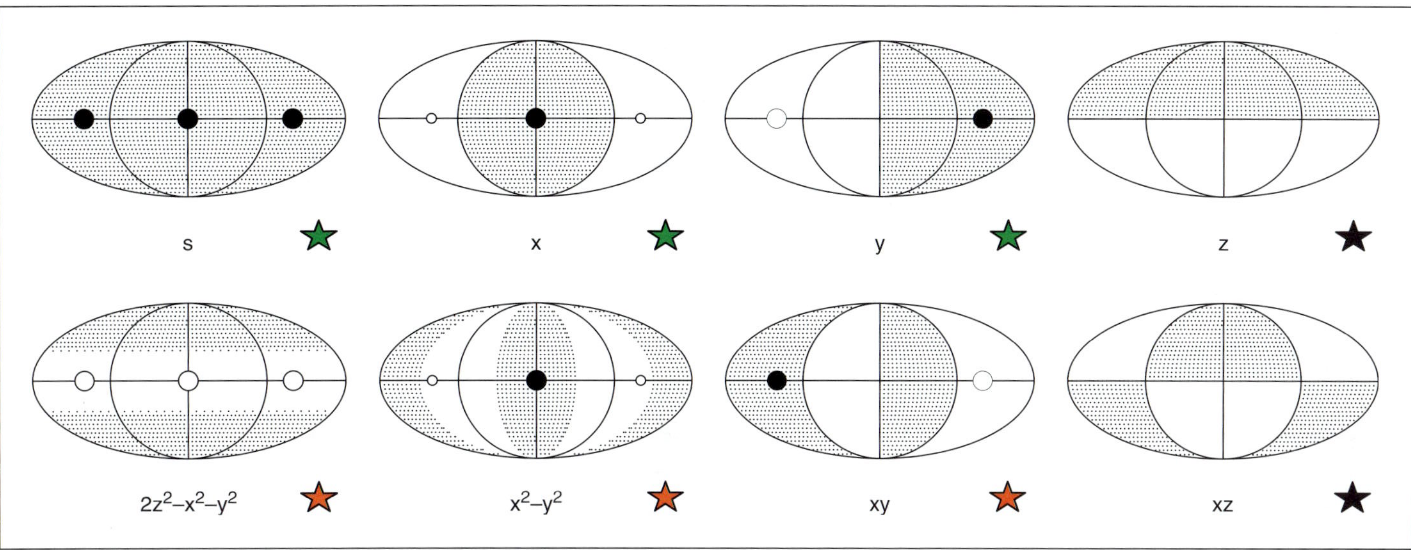


**Figure 3.8** The $\sigma$-type group orbitals on the vertices of an $O_3$ structure orbit exhibiting $D_{3h}$ point symmetry displayed on the elliptical projections of Figure 3.7. The circular icons, filled and open circles, identify $\sigma$-oriented orbital components at the vertices, sized to reflect the coefficients of the linear combinations, equation 3.20, for the spherical harmonics in Table 3.11. The icons ★, ★ and ★ identify the distinct group orbitals transforming as the irreducible components of the reducible character over the decorated orbit, unnecessary repetitions of these components and central functions for which no group orbital can be constructed owing to the locations of the decorated vertices of the orbit in the Cartesian coordinate system.

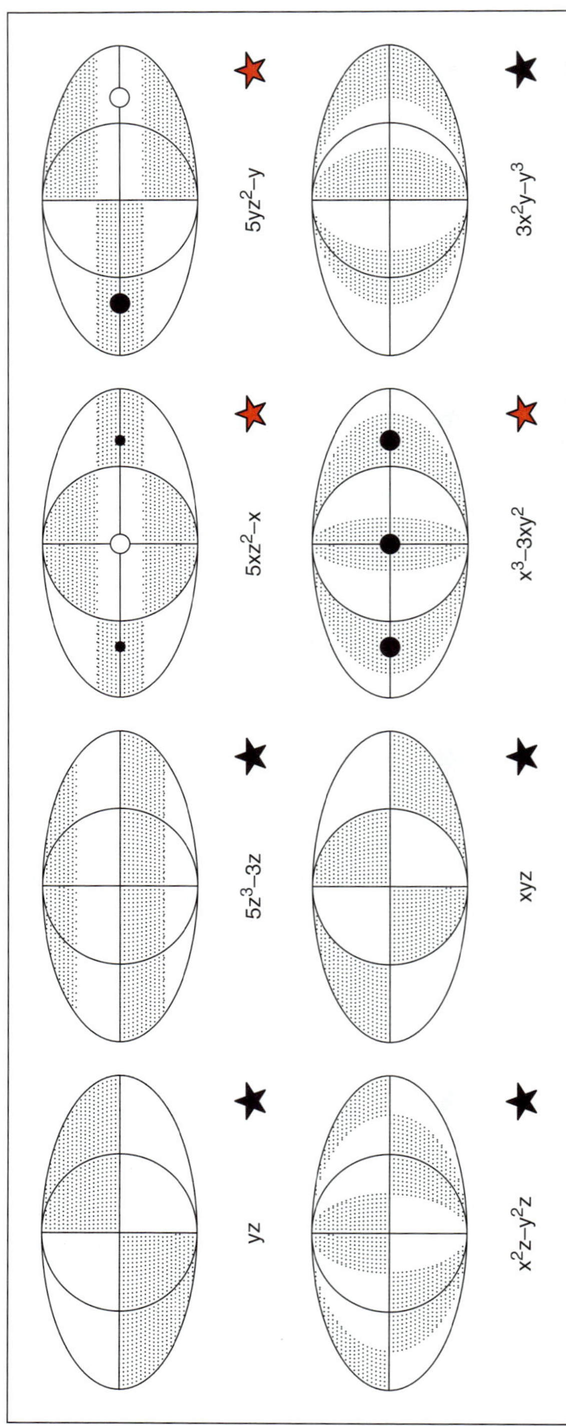


**Figure 3.8** Continued.

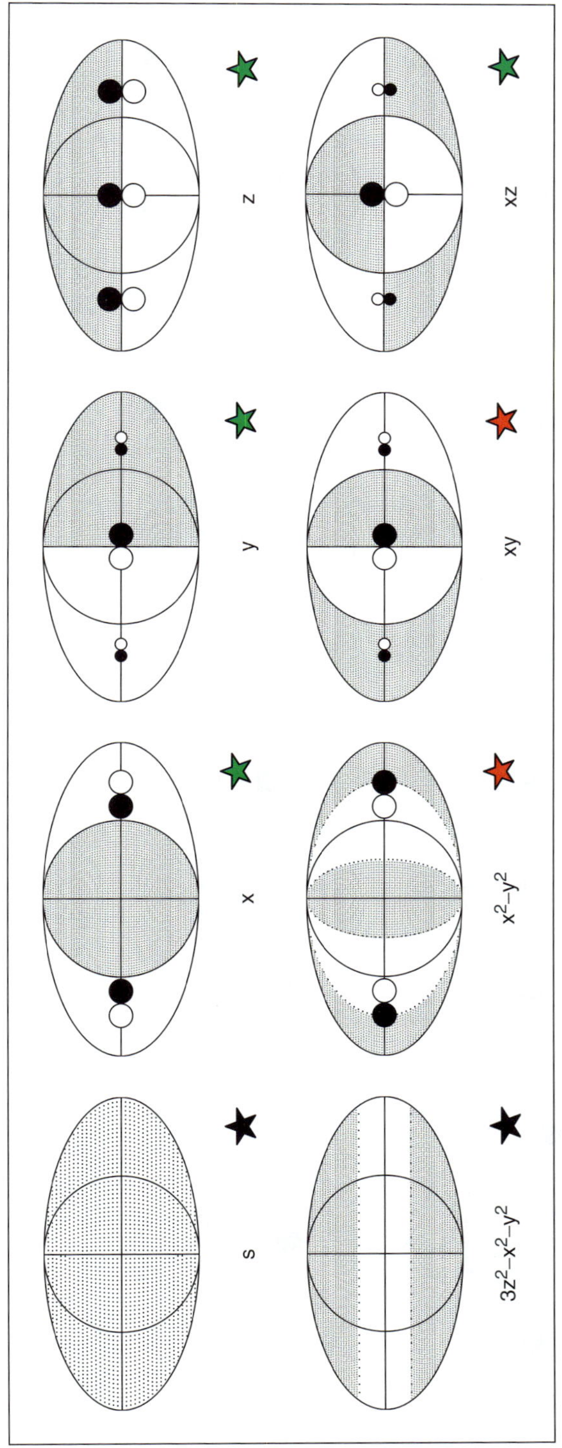


**Figure 3.9** The $\pi$-type group orbitals on the vertices of an $O_3$ structure orbit exhibiting $D_{3h}$ point symmetry, displayed superimposed on the elliptical projections of the generating functions of Table 3.11. The familiar 'dumbbells' identify $\pi$-oriented atomic orbital components at the vertices, equation 3.21 sized to reflect the values, Table 3.12, of the corresponding vector harmonics at the orbit vertices. The motifs ★, ★ and ★ are applied as in Figure 3.8.

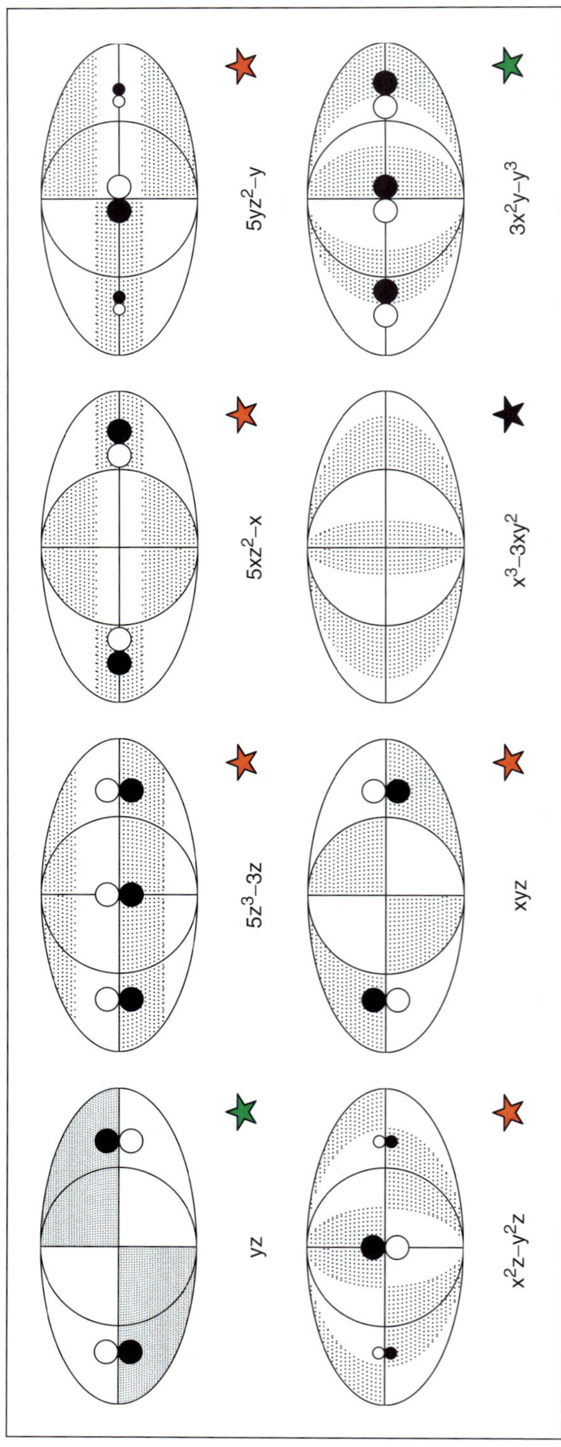


**Figure 3.9** Continued.

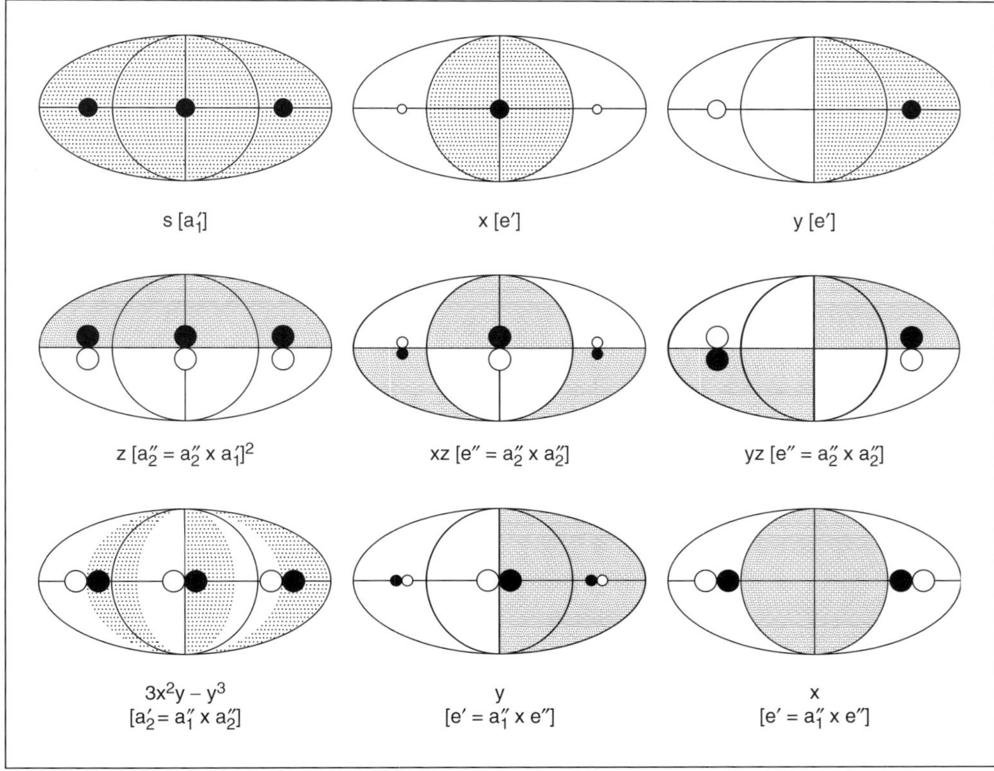


**Figure 3.10** The consolidation of Figure 3.8 and Figure 3.9 to summarize the superposition procedure for the construction of $\sigma$ and $\pi$ group orbitals on the vertices of the equilateral triangular orbit for the choice that the intrinsic symmetry, $D_{3h}$, of the orbit is preserved. Note with reference to the Character Table for $D_{3h}$ that $\Gamma_z$ is $a_2''$ and $\Gamma_\varepsilon$ is $a_1''$.

Figure 3.10 summarizes the distinct results displayed in Figure 3.8 and 3.9 and makes evident the origins of the diagrams in Figure 3.6. The first row displays the $\sigma$ group orbitals over the $O_3$ orbit of an object of $D_{3h}$ point symmetry, the result that the group orbitals are of central s-like, $p_x$-like and $p_y$-like symmetries. In the second row of diagrams in the figure, superimposition of the local $\pi_\theta$ atomic function[9] leads to the $\pi$-type group orbitals based on the local atomic functions oriented in the directions of the local gradients of the matching central functions. Thus we find a z-like linear combination, a second linear combination of xz-like character and the third function, which transforms as central yz. In the third row of diagram, these linear combinations are converted to the final three possible over the local functions, by local rotations of the $\pi_\theta$ atomic functions, through $\pi/2$ in the same sense about each radial line joining the vertices to the central origin. These final linear combinations are the $\pi_\phi$ $[v^1(j)]$ set over the $O_3$ orbit. The $\pi_\theta$ set of group orbitals span $\Gamma_z \times \Gamma_\sigma$ and the $\pi_\phi$ set span $\Gamma_\varepsilon \times \Gamma_z \times \Gamma_\sigma$ symmetries as marked in the different diagrams.

[9]In this simple example, in which the principal rotational axis is coincident with central z, the $u_1$ and $v_1$ of equation 3.16 are the local $\pi_\theta$ and $\pi_\phi$ basis functions of the coordinate system $(\sigma(j), \pi_\theta(j), \pi_\phi(j))$ at each vertex.

The summary highlights an important observation, which identifies a general procedure that is particularly easy to apply for the cases of molecules in which a principal rotational axes can be identified [the dihedral groups]. For an orbit of n vertices, it is necessary only to identify the first occurrences of n distinct $\sigma$ group orbitals, which occur in the hierarchical order s-, p-, d-, ... like, with respect to the central functions. Then the construction of the other group orbitals of the valence set decorations are formed by superimposition of the $u^1_{lm}$ [$\pi_\theta$ here] which operation leads to the $\Gamma_z \times \Gamma_\sigma$ symmetry group orbitals, with the remainder identified on concerted local rotation of the $u^1_{lm}$ into the $v^1_{lm}$ [$\pi_\phi$ here] and of symmetries $\Gamma_\varepsilon \times \Gamma_z \times \Gamma_\sigma$.

| $D_{3h}$ | E | $2C_3$ | $3C_2'$ | $\sigma_h$ | $2S_3$ | $\sigma_v$ | |
|---|---|---|---|---|---|---|---|
| $A_1'$ | 1 | 1 | 1 | 1 | 1 | 1 | (s) |
| $A_2'$ | 1 | 1 | −1 | 1 | 1 | −1 | $3x^2y - y^3$ |
| $E'$ | 2 | −1 | 0 | 2 | −1 | 0 | $(x, y)\ (x^2 - y^2, xy)$ |
| $A_1''$ $[\Gamma_\varepsilon]$ | 1 | 1 | 1 | −1 | −1 | −1 | $3x^2yz - y^3$ |
| $A_2''$ | 1 | 1 | −1 | −1 | −1 | 1 | z |
| $E''$ | 2 | −1 | 0 | −2 | 1 | 0 | (xz, yz) |

Figure 3.11 presents a similar summary for the arrangement where the decorated $O_3$ orbit lies below the coordinate origin, as, for example, in the case of $NF_3$, with the N atom sited at the origin, so that the overall molecule is of $C_{3v}$ point symmetry, for which the Character Table is

| $C_{3v}$ | E | $2C_3$ | $3\sigma_v$ | |
|---|---|---|---|---|
| $A_1$ | 1 | 1 | 1 | (s) |
| $A_2[\Gamma_\varepsilon]$ | 1 | 1 | −1 | |
| E | 2 | −1 | 0 | $(x, y), (x^2 - y^2, xy), (xz, yz)$ |

The linear combinations corresponding to the group orbitals displayed in Figure 3.11 do not differ greatly from those found for the analysis of the decorated $O_3$ orbit in the higher symmetry arrangement of Figure 3.10. The $\sigma$ group orbitals exhibit the same nodal features, of which observation is to be expected. However, because of the reduced point symmetry of the $O_3$ orbit, when it is taken to be a part of a $C_{3v}$ structure, the $u^1_{lm}/v^1_{lm}$ resultants are not exclusively either the $\pi_\theta$ or $\pi_\phi$ of the transformed coordinate systems of Figure 3.1. In most of the linear combinations, the $u^1_{lm}$ and $v^1_{lm}$ functions are rotated by different angles, with respect to the $\pi_\theta/\pi_\phi$ axes.

Moreover, there is no discrimination in the transformation properties of central functions of the form f(xyz) and z × f(xyz) and so, for example, the s-like group orbital and the z-like orbital are of the same symmetry, which reflects the mixing of s and $p_z$ in $C_{3v}$ structures and the absence of such mixing in $D_{3h}$ structures. Since, also, all the other group orbitals in Figure 3.11 are of the same symmetry, it is not surprising that there is some mixing of these linear combinations, for example, to return the x-type linear combination of Figure 3.11, which is substantially $\pi_\phi$ at $(\pi/2, 2\pi/3)$ and $(\pi/2, 4\pi/3)$, but, in which the small $\pi_\theta$ contribution is evident at $(\pi/2, 0)$.

The tensor harmonic analysis for the case of a $D_{3h}$ $O_3$ orbit decorated with valence d-orbitals, Figure 3.2, is displayed in Figure 3.12 using the elliptical projection. The linear

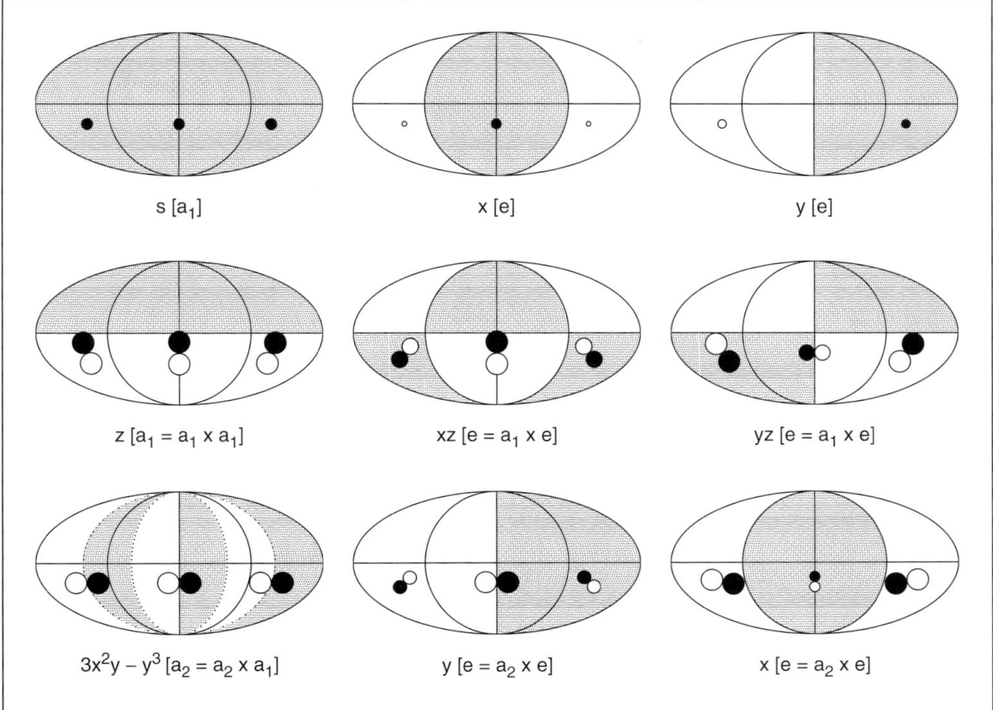


**Figure 3.11** Summary diagrams in the manner of Figure 3.10 identifying the group orbitals on the vertices of the equilateral triangular orbit in a $C_{3v}$ structure such as $NF_3$ with N at the coordinate origin. Note, with reference, to the Character Table for $C_{3v}$ that $\Gamma_z$ is $A_1$ and $\Gamma_\varepsilon$ is $A_2$.

combinations are formed using the tensor harmonics of Table 3.13 and the modulation coefficients of the linear combinations over the d$\delta$-orbital decorations of the orbit vertices are given in their simplest ratios in Table 3.14.

Figure 3.13 summarizes the analysis shown in Figure 3.12 and provides the opportunity to emphasize the generality of the observation made with regard to the similar summary, Figure 3.10, that only the $\sigma$ group orbitals need to be identified formally. We see from the displays in Figure 3.13 that this assertion holds true for the problem of finding the d$\delta$ group orbitals.

The $\delta_{\theta\theta}$ group orbitals can be constructed, simply, by superposition of the $\delta_{\theta\theta}$ local functions for the s local functions in the s linear combinations and then the $\delta_{\theta\phi}$ group orbitals follow on concerted local rotations about the vertex positions.

The difference in the case of the d$\delta$ analysis arises only with regard to the concerted local rotations to turn the linear combinations over the $u^2_{lm}$ resultants in the linear combinations over the $v^2_{lm}$ resultants and so to the identification of the remaining group orbitals in the 3rd row of the figure. These rotations are through $\pi/4$ rather than the $\pi/2$ actions appropriate for the completion of the analysis in Figure 3.10.

If the $O_3$ orbit is displaced from the equatorial great circle the summary results are as shown in Figure 3.14. Again, because of the reduced point symmetry of the orbit in this geometrical arrangement, there is considerable mixing of the various linear combinations,

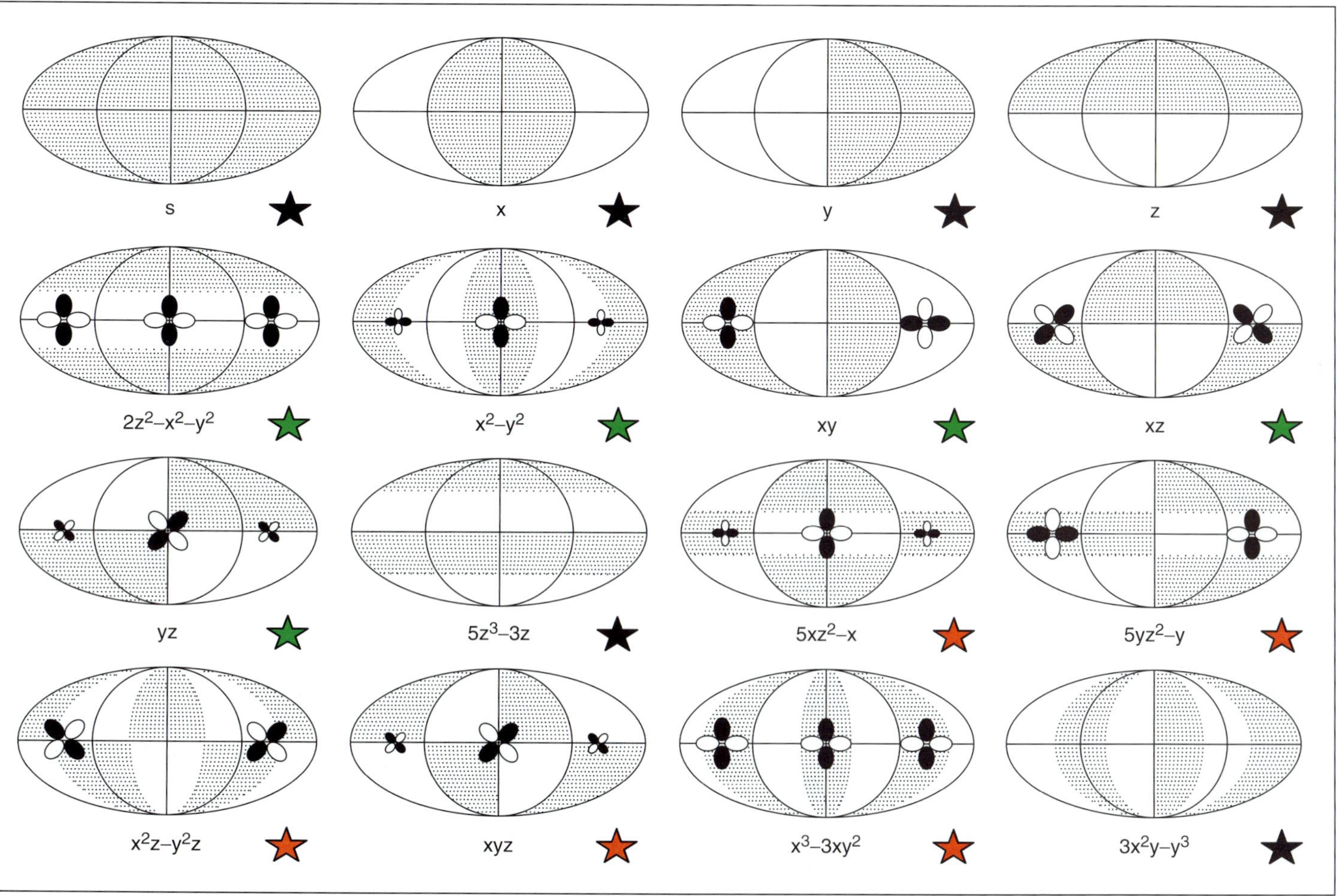


**Figure 3.12** The construction of $\delta$-type group orbitals on the vertices of an $O_3$ structure orbit, Figure 3.2, exhibiting $D_{3h}$ point symmetry, The 'double dumbbells' identify $\delta$-oriented atomic orbital components at the vertices and are sized to reflect equation 3.22 and the values at the orbit vertices of the tensor harmonics of Table 3.13. The motifs ★, ★ and ★ are applied as in Figure 3.8.

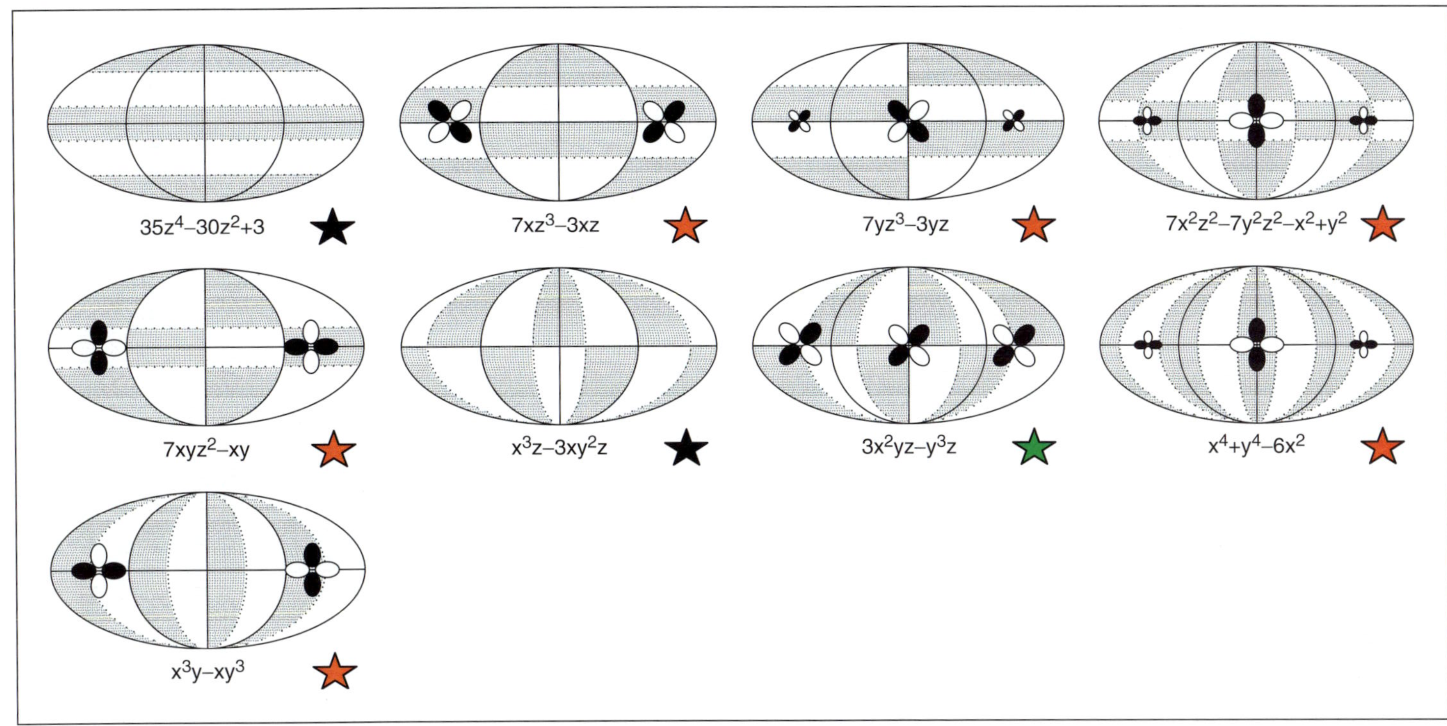


**Figure 3.12** Continued.

**Table 3.14 The modulating coefficients, reduced to their simplest ratios, for the linear combinations over the $u_j^2$ [here $\delta_{\theta\theta}$] and $v_j^2$ [here $\delta_{\theta\phi}$] from equations 3.22 to form the d$\delta$ group orbitals, Figure 3.12, of the $Fe_3$ using the tensor surface harmonics of Table 3.13.**

| $\theta$ | $\pi/2$ | | $\pi/2$ | | $\pi/2$ | |
|---|---|---|---|---|---|---|
| $\phi$ | 0 | | $2\pi/3$ | | $4\pi/3$ | |
| $Y^0_{lm}(\theta,\phi)$ | $U^2_{lm}(\theta,\phi)$ | $V^2_{lm}(\theta,\phi)$ | $U^2_{lm}(\theta,\phi)$ | $V^2_{lm}(\theta,\phi)$ | $U^2_{lm}(\theta,\phi)$ | $V^2_{lm}(\theta,\phi)$ |
| s | 0 | 0 | 0 | 0 | 0 | 0 |
| z | 0 | 0 | 0 | 0 | 0 | 0 |
| y | 0 | 0 | 0 | 0 | 0 | 0 |
| x | 0 | 0 | 0 | 0 | 0 | 0 |
| $z^2$ | 1 | 0 | 1 | 0 | 1 | 0 |
| yz | 0 | 1 | 0 | −½ | 0 | −½ |
| xz | 0 | 0 | 0 | −1 | 0 | 1 |
| xy | 1 | 0 | −½ | 0 | −½ | 0 |
| $x^2 - y^2$ | 0 | 0 | −1 | 0 | 1 | 0 |
| $5z^3 - 3z$ | 0 | 0 | 0 | 0 | 0 | 0 |
| $yz^2$ | 0 | 0 | 1 | 0 | −1 | 0 |
| $xz^2$ | 1 | 0 | −½ | 0 | −½ | 0 |
| $z(x^2 - y^2)$ | 0 | 0 | 0 | 1 | 0 | −1 |
| z(xy) | 0 | 1 | 0 | −½ | 0 | −½ |
| $y(3x^2 - y^2)$ | 0 | 0 | 0 | 0 | 0 | 0 |
| $x(x^2 - 3y^2)$ | 1 | 0 | 1 | 0 | 1 | 0 |

which can be projected onto the unit sphere using the tensor harmonics. Moreover, because the local rotational operations in this case, for $\delta$-oriented interactions, are through only $\pi/4$, the clear distinctions in the group orbitals obtained by the application of the tensor harmonic coefficients in equation 3.22 are less easy to visualize.

Again, the better procedure, in these cases, in which the orbit is of higher symmetry than the surrounding structure defining the overall point group is to construct the different group orbitals for the case of the intrinsic point symmetry of each orbit [e.g. $D_{3h}$ rather than $C_{3v}$ for an $O_3$ orbit].

In summary, therefore, we have seen how to solve the problem to construct group orbitals at different levels of formality in several ways:

1. The mathematics can be applied directly to generate linear combinations over resultant local functions modulated by calculated coefficients in order to construct all the possible group orbitals for an orbit decorated with valence atomic orbitals, or
2. the mathematics can be applied to generate only one half of the number of possible linear combinations over resultant local functions, while the remaining group orbitals are obtained from these by concerted local rotations of the local resultants of the first set, or

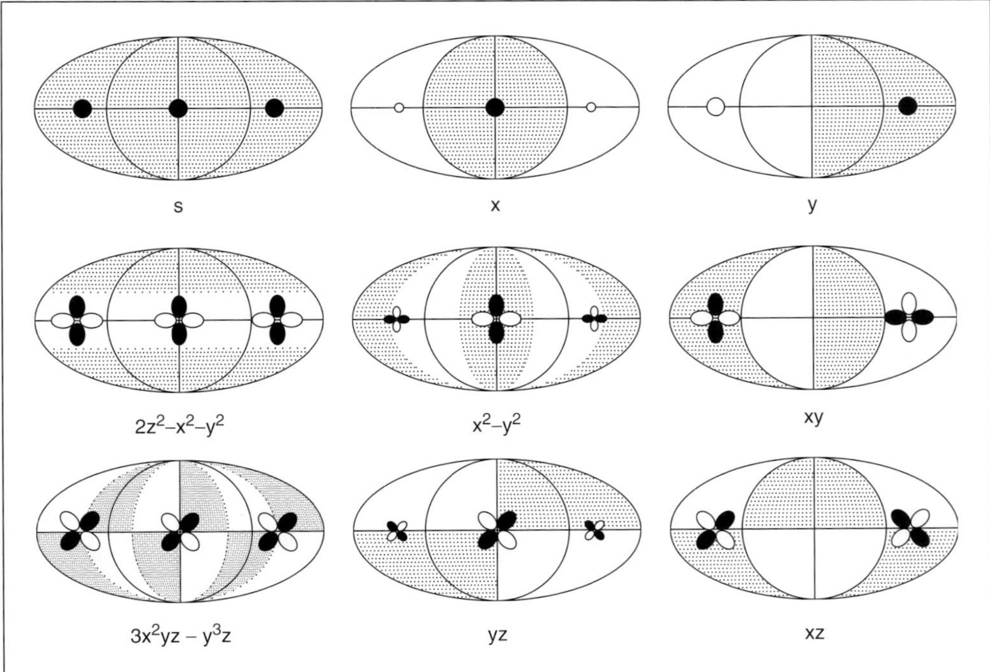


**Figure 3.13** Summary projections of the dδ group orbitals, for example of the $Fe_3$ triangle of Figure 3.2, of s- and p-characters using the superimposition and local rotation procedure. For this occurrence of the $O_3$ orbit there is no complication due to the mixing of z and xz-like components.

3. for cases, in which the local resultants of the valence atomic orbital sets at the vertices can be predicted without calculation, then all the group orbitals can be constructed using only the calculated $\sigma$ group orbitals, or
4. at a pictorial level, it is possible to sketch the group orbitals, for example, for applications in qualitative molecular orbital theory analyses, by choosing resultant local functions to align with the directions on the tangent planes at vertex positions on the unit sphere corresponding to the maxima/minima in the gradients [$\pi$ group orbitals] or the concavities [$\delta$ group orbitals].

### 3.4.2 Group orbitals over the $O_{12}$ orbits of a structure exhibiting either $D_{3h}$ or $D_{6h}$ point symmetry

Inspection of Table 3.2 reveals that an $O_{12}$ structure orbit can be found in objects exhibiting $D_{3h}$ or $D_{6h}$ point symmetries. For the $D_{3h}$ object, the $O_{12}$orbit is the regular orbit of the point group. For an object exhibiting $D_{6h}$ point symmetry, three orbits of 12 vertices, $O_{12v}$, $O_{12d}$ and $O_{12h}$ are possible. From the table we can extract the permutation or $\sigma$-characters as listed in Table 3.15.

The principal symmetry property, which distinguishes the $O_{12}$ orbit in $D_{3h}$ point symmetry, is that the 6-fold rotational symmetry of the three $O_{12}$ orbits is reduced to a 3-fold

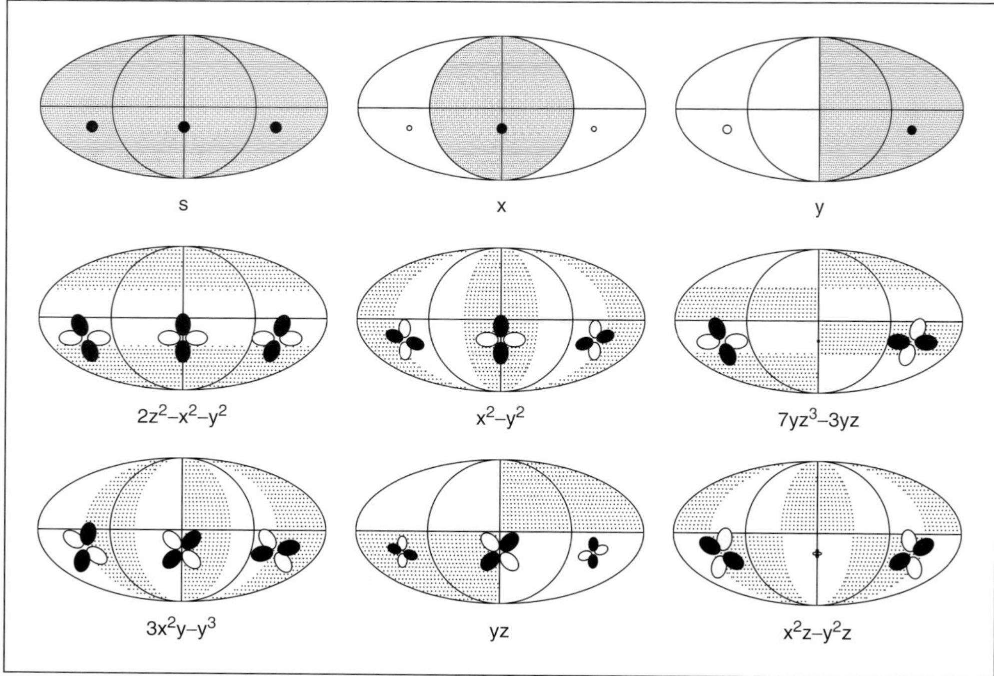


**Figure 3.14** Summary projections to illustrate the superimposition procedure for the construction of the dδ group orbitals on the vertices of the equilateral triangular orbit of $O_3$, in a structure of the lower point symmetry $C_{3v}$. The results are less satisfactory than those displayed in Figure 3.13 for which the full intrinsic $D_{3h}$ of the orbit is used.

symmetry in $D_{3h}$. Thus, as drawn in Figure 2.2, the $O_{12}$ orbit of $D_{3h}$ symmetry is the six-sided prism, in which the hexagon has two distinct bond lengths, and the height of the prism is a third degree of freedom. For the $O_{12}$ orbits, exhibiting $D_{6h}$ point symmetry, the options are to have a twelve-sided polygon with two distinct side lengths, and two orientations of the hexagonal prism, in one orientation with vertices on $\sigma_v$-planes and, in the other, with vertices on the $\sigma_d$-planes.

**Table 3.15 Some examples, taken from Table 3.2, of the occurrence of $O_{12}$ orbits, three in $D_{6h}$ symmetry and one as the regular orbit of a $D_{3h}$ point symmetry object.**

| Group | Orbit | $\Gamma_\sigma$ |
|---|---|---|
| $D_{6h}$ | $O_{12d}$ | $A_{1g} + A_{2u} + B_{1g} + B_{2u} + E_{1g} + E_{1u} + E_{2g} + E_{2u}$ |
| | $O_{12v}$ | $A_{1g} + A_{2u} + B_{1u} + B_{2g} + E_{1g} + E_{1u} + E_{2g} + E_{2u}$ |
| | $O_{12h}$ | $A_{1g} + A_{2g} + B_{1u} + B_{2u} + 2E_{1u} + 2E_{2g}$ |
| $D_{3h}$ | $O_{12}$ | $A_1' + A_1'' + A_2' + A_2'' + 2E' + 2E''$ |

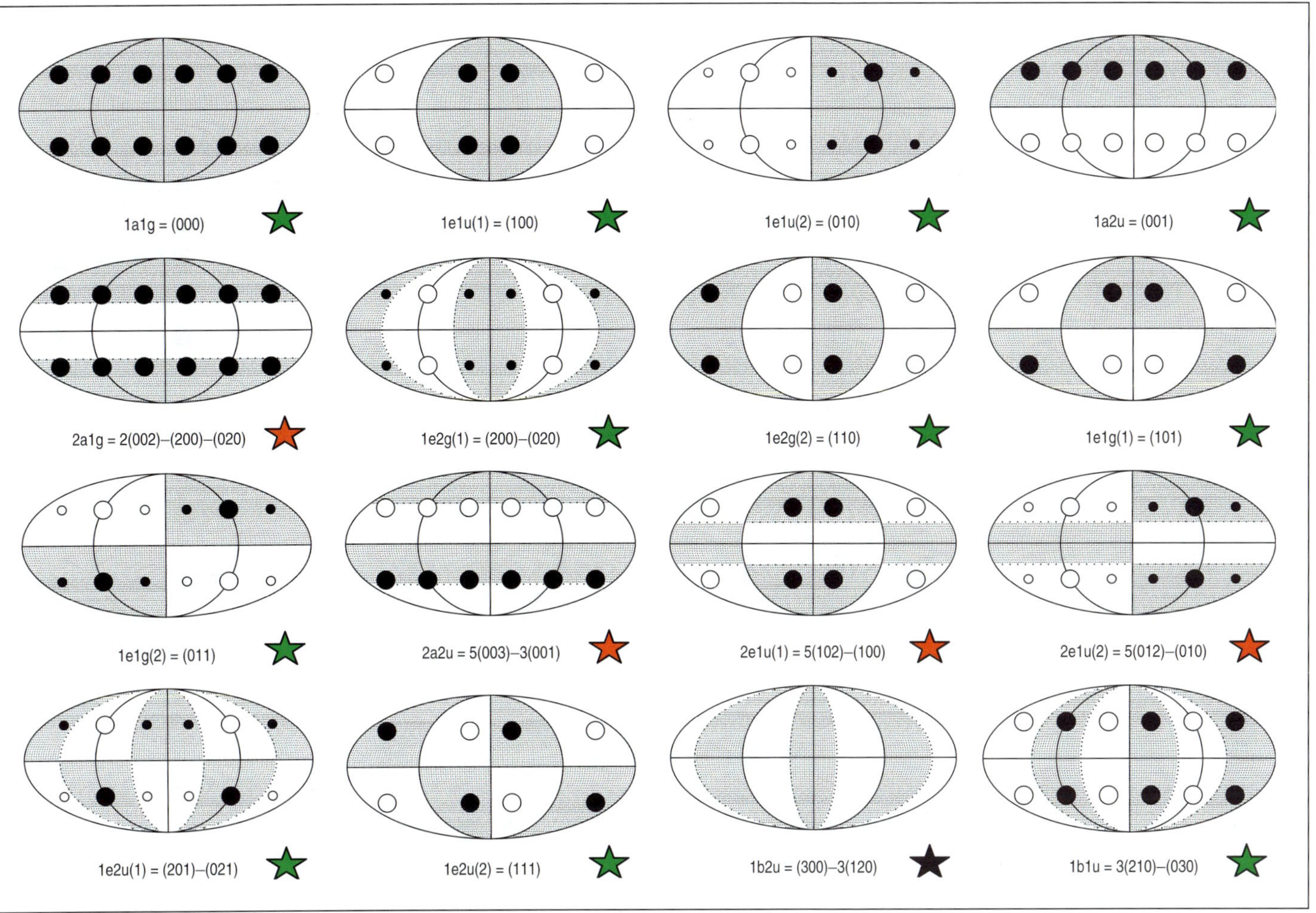


**Figure 3.15** Projections identifying the group orbitals and their irreducible symmetries for local $\sigma$-orbital decoration of the vertices of the $O_{12v}$ orbit, Table 3.15, of a molecular structure with $D_{6h}$ point symmetry. The icons ★, ★ and ★ are applied as in Figure 3.8.

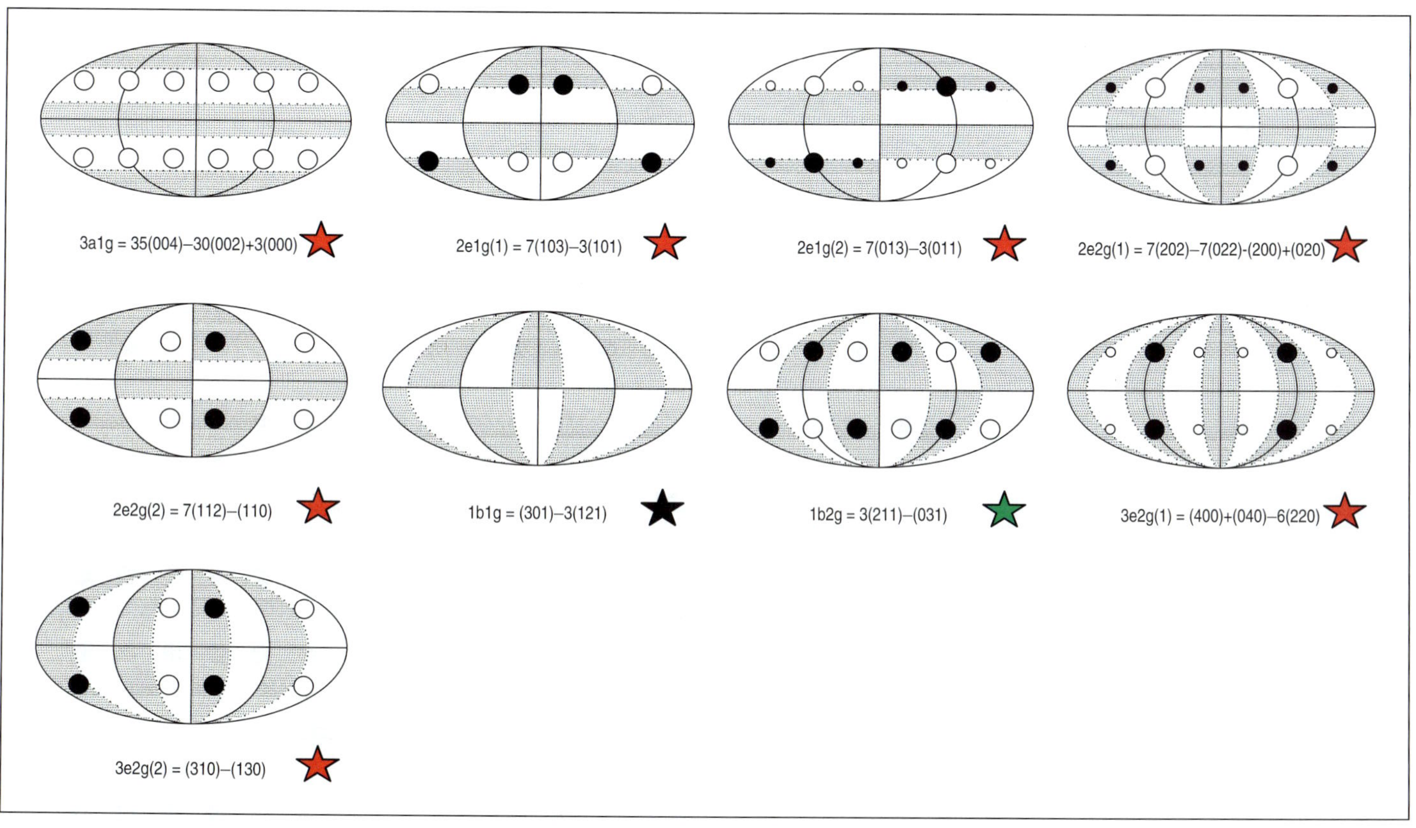


**Figure 3.15** Continued.

The application of the superposition procedure to identify the group orbitals as linear combinations of $\sigma$-type local functions on the vertices of the $O_{12v}$ and $O_{12d}$ structure orbits is set out in Figures 3.15 and 3.16, using the projections of the central general spherical harmonics of Figure 3.7 in the diagrams.

The details in these diagrams highlight the essential features of the method by which one can form the possible group orbitals for a set of $\sigma$-type local functions on the vertices of a structure orbit. In each analysis, we expect to identify 12 unique linear combinations since there are 12 vertices in the structure orbits, $O_{12d}$ and $O_{12v}$, distinguished only in their relative positioning within the Cartesian coordinate system about the central origin. The only difference between the analyses for these cases is that because of the different orientations of the structure orbits, the 1-dimensional components, $A_{2g}$ and $B_{2u}$ of the permutation character of Table 3.15 for the $O_{12d}$ orbit switch labels to $A_{2u}$ and $B_{2g}$ for the $O_{12v}$ orbit. The other effect of the change in orientation is to exclude linear combinations of local functions for the projections of the central functions $1B_{2u}$ and $1B_{1g}$ for the $O_{12v}$ orbit and $1B_{1u}$ with $1B_{2g}$ for the $O_{12d}$ orbit. However, the important point to observe is that there is indeed, in both cases and generally, the hierarchical sequence for $\sigma$-type group orbitals that linear combinations, which are s-, then p-, then d-like and so on about the central origin, until the number of such unique linear combinations equals the number of local functions, which is to say equals the number of vertices in each structure orbit.

The results of the superposition procedure to identify group orbitals using the templates of the projections of the central general spherical harmonics, Figure 3.7, is set out in Figure 3.17 for the case of the $O_{12}$ regular orbit of $D_{3h}$ symmetry. Again 12 linear combinations can be identified and divide as the direct sum components listed for this symmetry in Table 3.15.

However, now the direct sum of the permutation character, Table 3.15, requires the identification of two sets of group orbitals of $e'$ and $e''$ irreducible symmetries and we cannot expect that the simple superposition procedure sampling very few points on the unit sphere amplitude of the central function will return mutually orthogonal linear combinations. These can be obtained as follows.

For linear combinations over the local functions, $\phi$, on the vertices of the orbit,

$$|1\rangle = \sum_{i=1}^{i=12} c_i \phi_i \quad \text{and} \quad |2\rangle = \sum_{j=1}^{j=12} c_j \phi_j \qquad 3.24$$

the overlap integral is the similarity transformation

$$S = \langle 2 | 1 \rangle = \begin{pmatrix} c_1 & \cdots & \cdots & \cdots & c_{12} \end{pmatrix} \begin{pmatrix} \langle \phi_1 | \phi_1 \rangle & \cdots & \cdots & \cdots & \langle \phi_1 | \phi_{12} \rangle \\ \vdots & \ddots & & \iddots & \vdots \\ \vdots & & \langle \phi_6 | \phi_6 \rangle & & \vdots \\ \vdots & \iddots & & \ddots & \vdots \\ \langle \phi_{12} | \phi_1 \rangle & \cdots & \cdots & \cdots & \langle \phi_{12} | \phi_{12} \rangle \end{pmatrix} \begin{pmatrix} c_1 \\ \vdots \\ \vdots \\ \vdots \\ c_{12} \end{pmatrix} \qquad 3.25$$

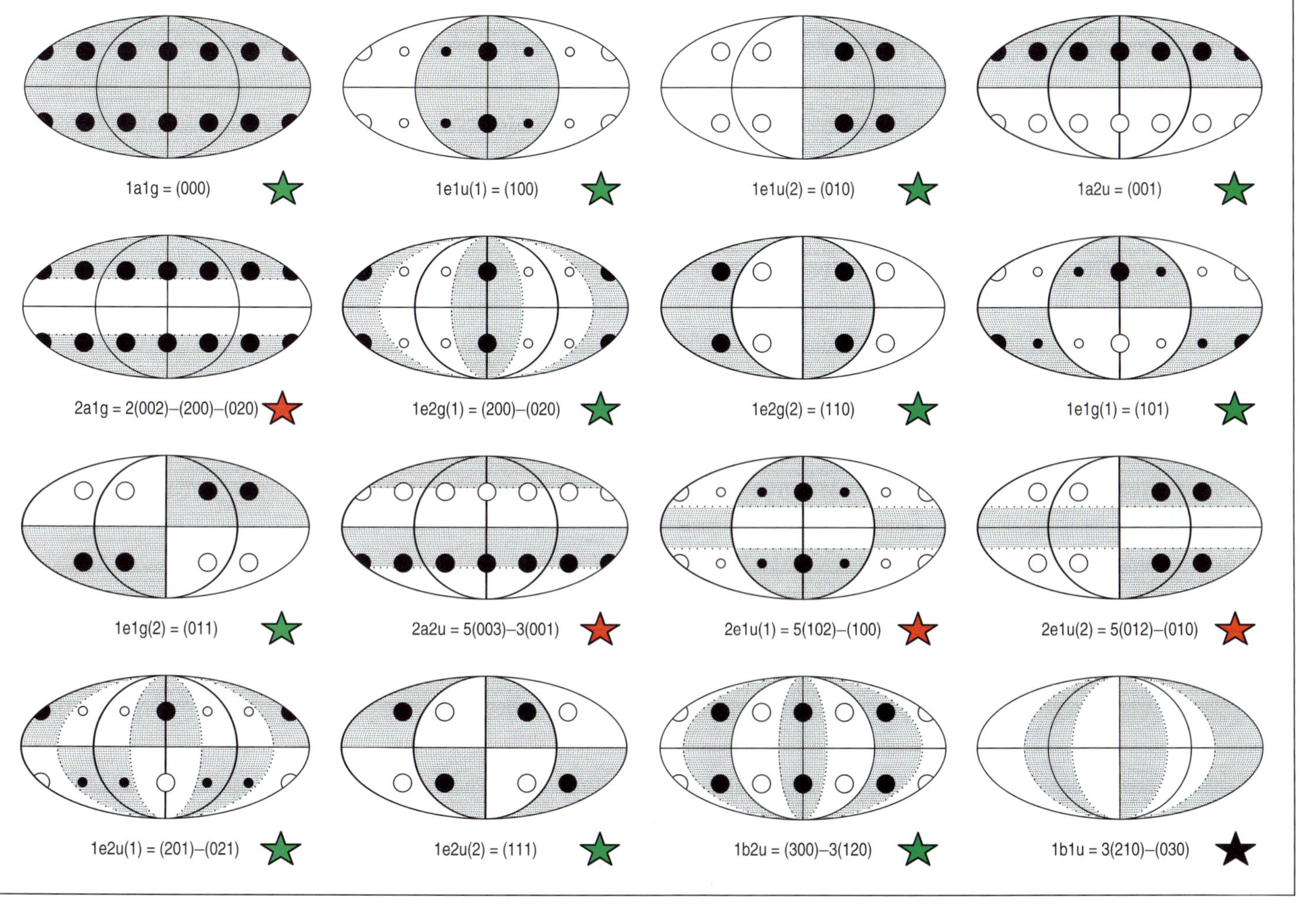


**Figure 3.16** Projections identifying the group orbitals and their irreducible symmetries for 12 local $\sigma$-oriented orbitals on the vertices of the $O_{12d}$ orbit, Table 3.15, of a molecular structure with $D_{6h}$ point symmetry. The icons ★, ★ and ★ are applied as in Figure 3.8.

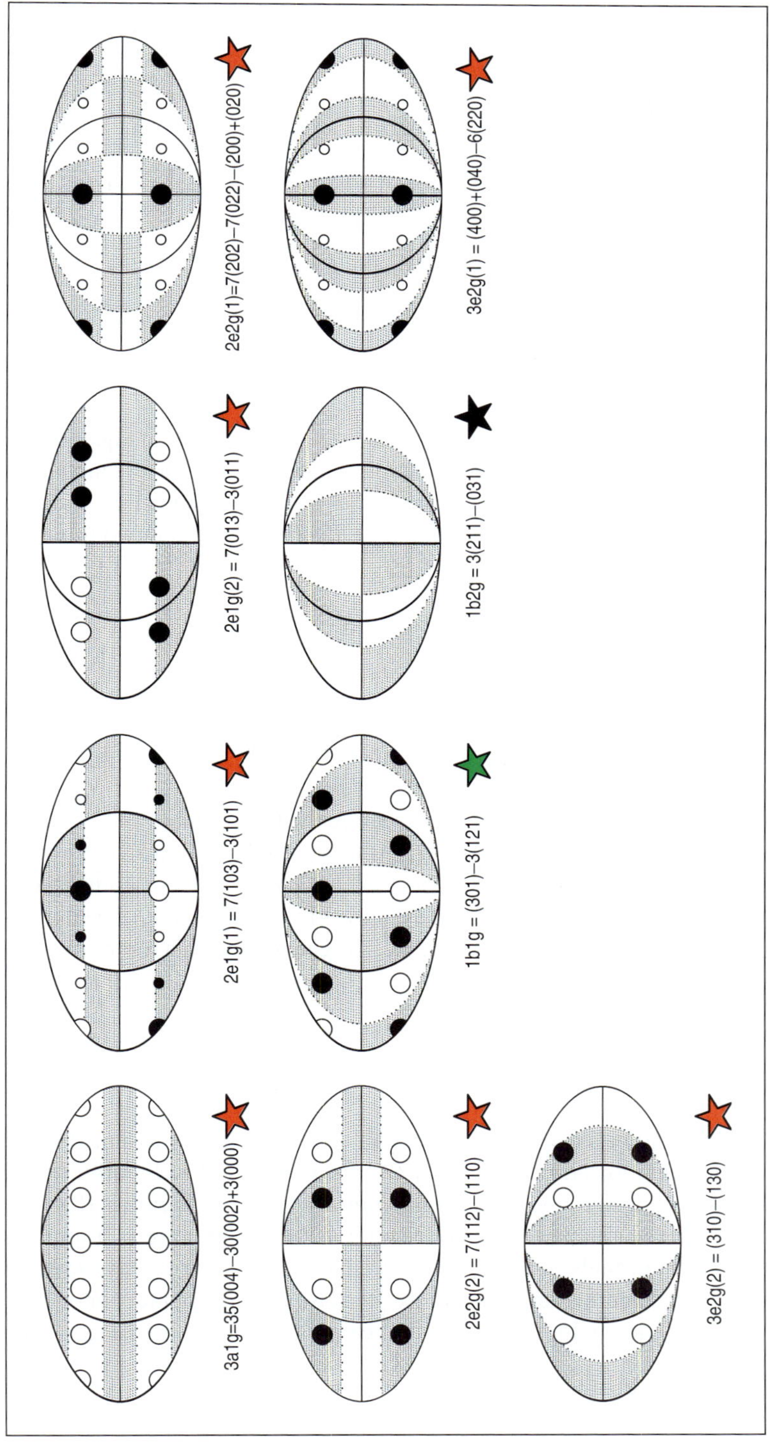


**Figure 3.16** Continued.

**Table 3.16 The coordinates chosen to identify the $O_{12}$ regular orbit of Figure 3.17.**

| # | $\theta$ | $\phi$ | x | y | z | # | $\theta$ | $\phi$ | x | y | z |
|---|---|---|---|---|---|---|---|---|---|---|---|
| 1 | 45 | 15 | 0.683 | 0.183 | 0.707 | 7 | 135 | 15 | 0.683 | 0.183 | −0.707 |
| 2 | 45 | 105 | −0.183 | 0.683 | 0.707 | 8 | 135 | 105 | −0.183 | 0.683 | −0.707 |
| 3 | 45 | 135 | −0.500 | 0.500 | 0.707 | 9 | 135 | 135 | −0.500 | 0.500 | −0.707 |
| 4 | 45 | 225 | −0.500 | −0.500 | 0.707 | 10 | 135 | 225 | −0.500 | −0.500 | −0.707 |
| 5 | 45 | 255 | −0.183 | −0.683 | 0.707 | 11 | 135 | 255 | −0.183 | −0.683 | −0.707 |
| 6 | 45 | 345 | 0.683 | −0.183 | 0.707 | 12 | 135 | 345 | 0.683 | −0.183 | −0.707 |

It is straightforward to determine the overlap integral in the manner of equation 3.25, using EXCEL spreadsheet technology[10], once a suitable model for the local functions has been chosen with which the matrix elements $s_{ij} = \langle\phi_i|\phi_j\rangle$ can be calculated. With this result, Schmidt orthogonalization, to ensure the mutually orthogonality of the two group orbitals of the same symmetry, follows as

$$|2\_orthogonal \geq |2\rangle - [\langle 1|2\rangle/\langle 1|1\rangle]|1\rangle \qquad 3.26$$

since

$$\langle 2\_orthogonal|S|1 \geq 0 \qquad 3.27$$

So, let us assume that the local functions decorating the $O_{12}$ regular orbit , typically, might be Slater 2s orbitals distributed over the vertex positions listed in Table 3.16. Overlap integral formulae for Slater orbitals, based on the Incomplete Gamma Function integrals are to be found in the famous paper by Mulliken and his coworkers[11]. For case of Slater 2s orbitals, with Slater exponent $\zeta$, separated by a distance $R_{ij}$, the general formula for the matrix elements of the overlap integral reduces to

$$\langle\phi_i \mid \phi_j\rangle = e^{-\zeta/R_{ij}}\left[1 + \zeta R_{ij} + \left(\frac{4}{9}\right)(\zeta R_{ij})^2 + \left(\frac{1}{9}\right)(\zeta R_{ij})^3 + \left(\frac{1}{45}\right)(\zeta R_{ij})^4\right] \qquad 3.28$$

Using this approximation for the overlap integral and the geometry assumed in Table 3.16, the mutually orthogonal projections of the central functions onto the $O_{12}$ orbit can be formed using equation 3.26 and lead to the projections shown in row 3 of Figure 3.18.

The listing for the $O_{12h}$ orbit permutation character in Table 3.15 presents a similar problem, since two sets of group orbitals of $E_{1u}$ and $E_{2g}$ irreducible symmetries have to be identified. However, this example identifies an unusual case, because of the large number of orbit vertices in the plane. As can be seen in Figure 3.19, it is necessary to find central functions up to level

[10] Further details of this approach and other applications of EXCEL spreadsheets to a variety of problems in quantum chemistry are to be found in *Computational Quantum Chemistry — an interactive guide to basis set theory*, Charles M. Quinn, Academic Press, New York and London, 2002.

[11] R.S. Mulliken, C.A. Rieke, D. Orloff and H. Orloff, *J. Chem. Physics*, **17** (1949) 1248.

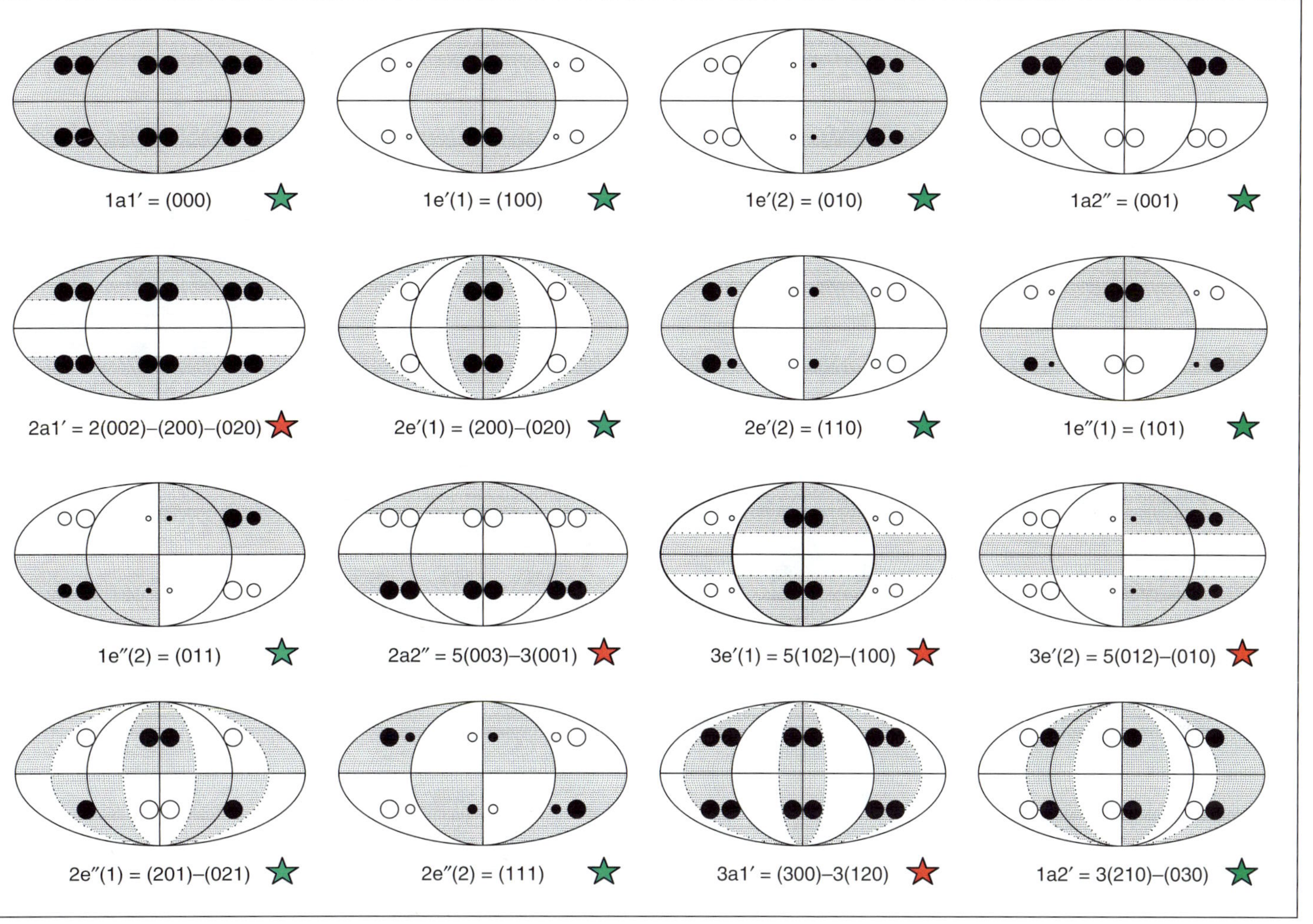


**Figure 3.17** Projections identifying the group orbitals and their irreducible symmetries for 12 local $\sigma$-oriented orbitals on the vertices of the $O_{12}$ regular orbit, Table 3.2, of a molecular structure with $D_{3h}$ point symmetry. The icons, ★, ★ and ★ are applied as in Figure 3.8.

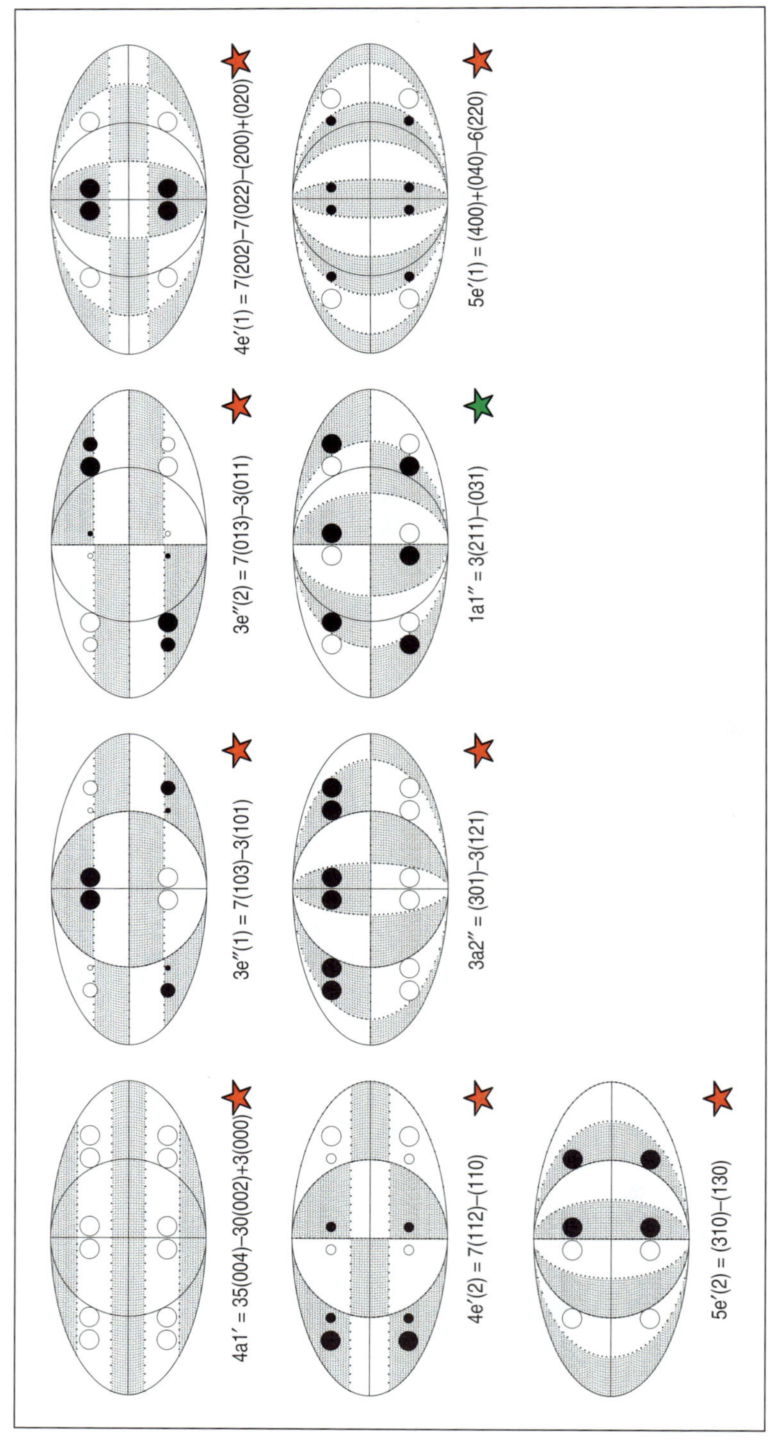


**Figure 3.17** Continued.

$\ell = 5$ [h10 and h11 in the Figure 3.19] to complete the superposition procedure analysis and the standard listings in the tables provide information only to level 4.

Extension of the information in the tables for the dihedral groups is facilitated by application of the information in Table 3.17 in which we identify the spaces spanned by $(x \pm iy)^i$.

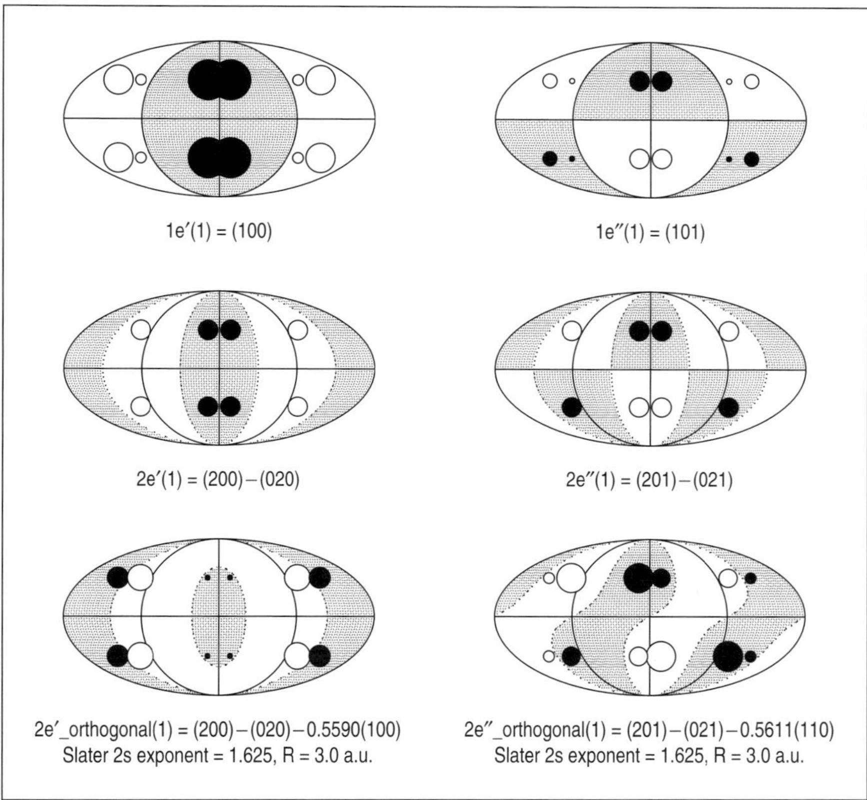


**Figure 3.18** Mutually orthogonal group orbitals of e′ and e″ symmetries [row 3] for the example of carbon 2s orbitals distributed on the vertices of an equilateral $O_{12}$ regular orbit of $D_{3h}$ point symmetry by Schmidt orthogonalization of the functions obtained by simple projection of the central functions $|1e'\rangle$, $|1e''\rangle$ [row 1] and $|2e'\rangle$, $|2e''\rangle$ [row 2]. The 'row 1' and 'row 3' linear combinations are the basis functions for the pairs of irreducible components of these symmetries in the permutation character, Table 3.18, over the vertices of the $O_{12}$ regular orbit of a $D_{3h}$ structure.

For example, consider, first, the application of the table for the case of the group $C_{6v}$. The table columns identify the spaces spanned for different $\ell$ values. The table describes the spaces spanned by $(x \pm iy)^j$ and $z(x \pm iy)^j$. So from the table and breaking each $(x+iy)^j$ into its real and imaginary parts:

$$\Gamma_0 = \left\langle (x \pm iy)^0 \right\rangle \equiv \langle 1 \rangle \leftrightarrow A_1 \qquad 3.29$$

$$\Gamma_0' = \left\langle z\,(x \pm iy)^0 \right\rangle \equiv \langle z \rangle \leftrightarrow A_1 \qquad 3.30$$

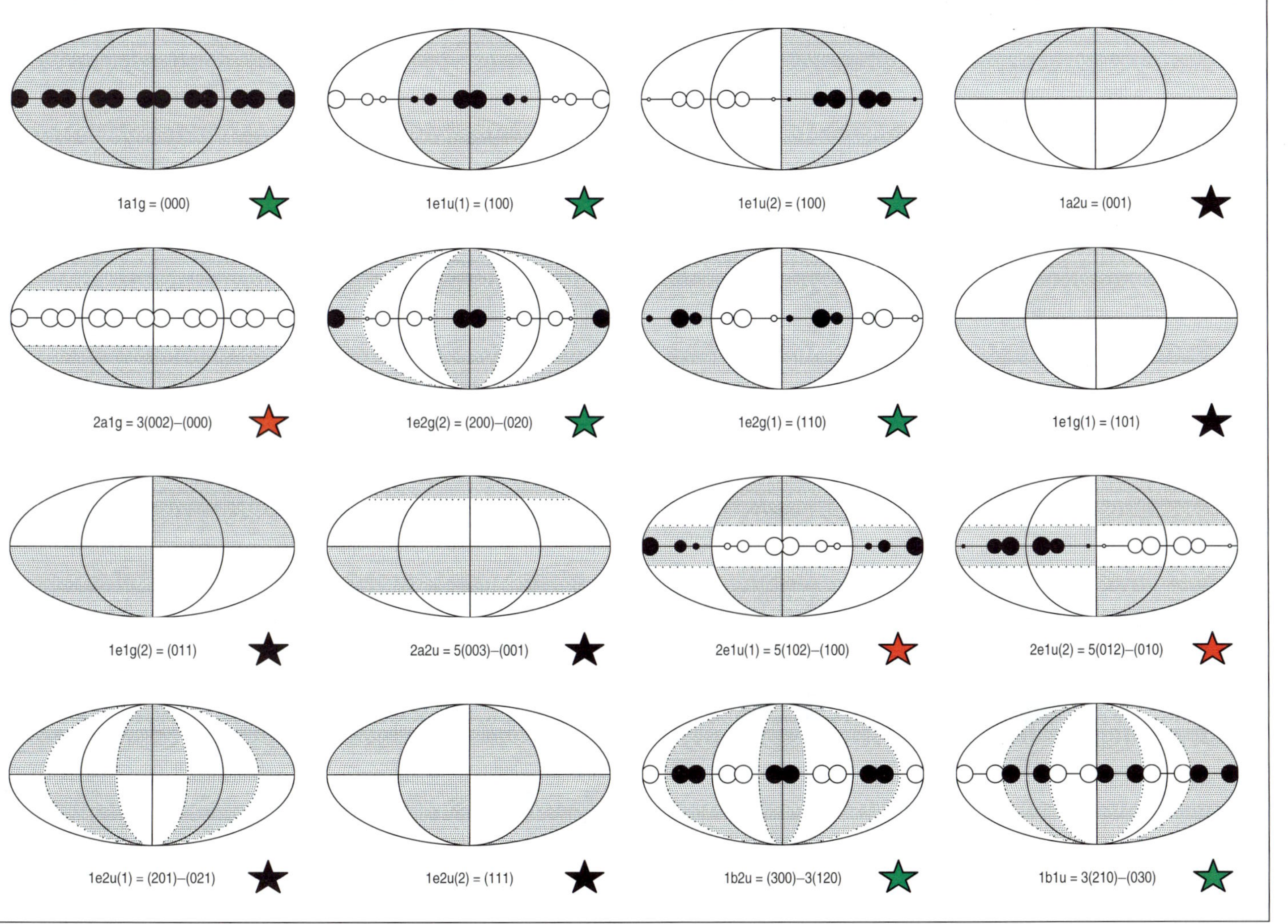


**Figure 3.19** Projections identifying the group orbitals and their irreducible symmetries for the function orbit of 12 local $\sigma$-oriented orbitals on the vertices of the $O_{12h}$ orbit, Table 3.3, of a molecular structure with $D_{6h}$ point symmetry. The icons, ★, ★ and ★ are applied as in Figure 3.8. Note that levels 5 and 6 central functions are required to form the group orbitals of $2_{e1u}$ and $1_{a2g}$ symmetries.

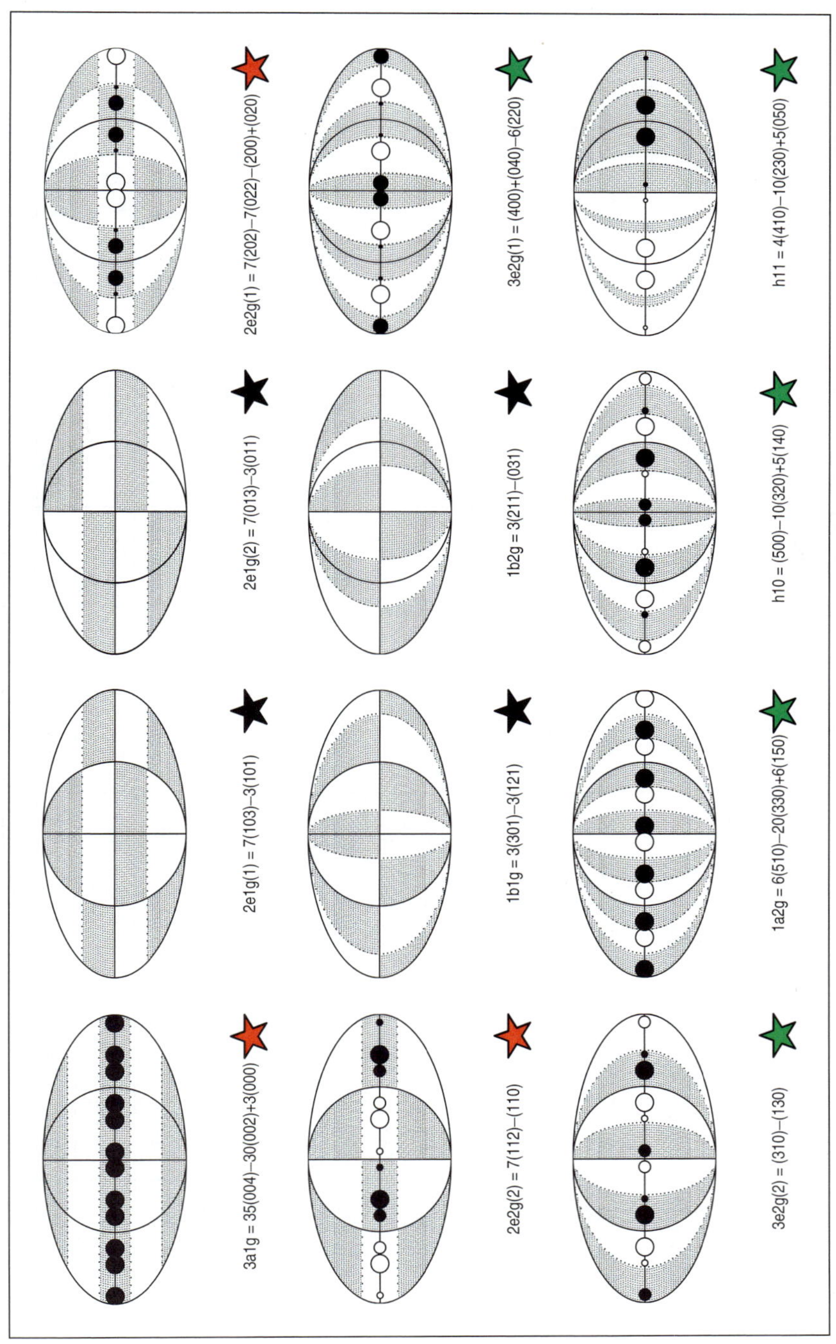


**Figure 3.19** Continued.

**Table 3.17** Mulliken notation for $\Gamma_j$ and $\Gamma_j'$ for the dihedral family of groups. $\Gamma_j$ is the character of the space spanned by $(x \pm iy)^j$ and $\Gamma_j'$ the character spanned by $z(x \pm iy)^j$. Note Re $\Gamma_j$ and Im $\Gamma_j$ mean the spaces spanned by the real and imaginary parts of $(x \pm iy)^j$ and that the column for n/2 is to be included only if n is even. If n is even, it is required that none of the orbit vertices are positioned to lie in the xz plane [this convention is to differentiate between $B_1$ and $B_2$]. In the case that either $\Gamma_j$ or $\Gamma_j'$ is reducible, the first element in the table is the real part and the second is the imaginary part. The Mulliken symbol marked with * identifies the antisymmetric character, $\Gamma_e$ for the group.

| | $\Gamma_0$ | $\Gamma_0'$ | $\Gamma_{n/2}$ | $\Gamma_{n/2}'$ | $\Gamma_n$ | $\Gamma_n'$ | | $\Gamma_i$ | | $\Gamma_i'$ |
|---|---|---|---|---|---|---|---|---|---|---|
| $C_n$ | $A_1^*$ | $A_1$ | $2B_1$ | $2B$ | $2A_1$ | $2A_1$ | $E_i$ | $1 \le i < n/2$ | $E_i$ | $1 \le i \le n/2$ |
| $C_{nh}$ [even] | $A_g$ | $A_u*$ | $2B_g/2B_u$ | $2B_u/2B_g$ | $2A_g$ | $2A_u$ | $E_{ig}/E_{iu}$ | $1 \le i < n/2$ | $E_{ig}/E_{iu}$ | $1 \le i < n/2$ |
| [odd] | $A_1'$ | $A_1''$ | — | — | $2A_1'$ | $2A_1''$ | $E_i'$ | $1 \le i < n/2$ | $E_i'$ | $1 \le i < n/2$ |
| $C_{nv}$ | $A_1$ | $A_1$ | $B_1 + B_2$ | $B_1 + B_2$ | $A_1 + A_2^*$ | $A_1 + A_2$ | $E_i$ | $1 \le i < n/2$ | $E_i$ | $1 \le i < n/2$ |
| $D_2$ | $A^*$ | $B_1$ | $B_3 + B_2$ | $B_2 + B_3$ | $A + B_1$ | $B_1 + A$ | | | | |
| $D_n$ | $A_1^*$ | $A_2$ | $B_1 + B_2$ | $B_2 + B_1$ | $A_1 + A_2$ | $A_2 + A_1$ | $E_i$ | $1 \le i < n/2$ | $E_i$ | $1 \le i < n/2$ |
| $D_{2h}$ | $A_g$ | $B_{1u}$ | $B_{3u} + B_{2u}$ | $B_{2g} + B_{3g}$ | $A_g + B_{1g}$ | $B_{1u} + A_u^*$ | | | | |
| $D_{nh}$ [even] | $A_{1g}$ | $A_{2u}$ | $B_{1g} + B_{2g}/$ $B_{1u} + B_{2u}$ | $B_{2g} + B_{1g}/$ $B_{2u} + B_{1u}$ | $A_{1g} + A_{2g}$ | $A_{2u} + A_{1u}^*$ | $E_{ig}/E_{iu}$ | $1 \le i < n/2$ | $E_{iu}/E_{ig}$ | $1 \le i < n/2$ |
| [odd] | $A_1'$ | $A_2''$ | — | — | $A_1' + A_2'$ | $A_2'' + A_1'$ | $E_i'$ | $1 \le i < n/2$ | $E_i''$ | |
| $S_n$ [n/2 even] | A | $B^*$ | B | 2A | 2A | 2B | $E_i$ | $1 \le i < n/2$ | $E_{n/2-i}$ | |
| [n/2 odd] | $A_g$ | $A_u^*$ | $2A_u$ | $2A_g$ | $2A_{2g}$ | $2A_u$ | $E_{ig}/E_{iu}$ | $1 \le i < n/2$ | $E_{ig}/E_{iu}$ | $1 \le i < n/4$ |
| $D_{n/2d}$ [n/2 even] | $A_1$ | $B_2$ | $B_1^* + B_2$ | $A_2 + A_1$ | $A_1 + A_2$ | $B_2 + B_1$ | $E_i$ | $1 \le i < n/2$ | $E_{n/2-i}$ | |
| [n/2 odd] | $A_{1g}$ | $A_{2u}$ | $A_{1u}^* + A_{2u}$ | $A_{2g} + A_{1g}$ | $A_{1g} + A_{2g}$ | $A_{2u} + A_{1u}$ | $E_{ig}/E_{iu}$ | $1 \le i < n/2$ | $E_{ig}/E_{iu}$ | $1 \le i < n/4$ |

$$\Gamma_1 = \left\langle (x \pm iy)^1 \right\rangle \equiv \langle x, y \rangle \leftrightarrow E_1 \quad 3.31$$

$$\Gamma_2 = \left\langle (x \pm iy)^2 \right\rangle \equiv \left\langle x^2 - y^2, 2xy \right\rangle \leftrightarrow E_2 \quad 3.32$$

$$\Gamma_3 = \left\langle (x \pm iy)^3 \right\rangle \equiv \left\langle x^3 - 3xy^2 \right\rangle \oplus \left\langle 3x^2y - y^3 \right\rangle \leftrightarrow B_2 + B_2 \quad 3.33$$

$$\Gamma_6 = \left\langle (x \pm iy)^6 \right\rangle \equiv \left\langle x^6 - 15x^4y^2 + 15x^2y^4 - y^6 \right\rangle \oplus \left\langle 6x^5y - 20x^3y^3 + 6xy^5 \right\rangle \leftrightarrow B_2 + B_2 \quad 3.34$$

and

$$\Gamma_4 \equiv \Gamma_2, \Gamma_5 \equiv \Gamma_1 \quad 3.35$$

with

$$\Gamma'_j \equiv \Gamma_j \text{ since } \langle z \rangle \leftrightarrow A_1 \quad 3.36$$

For the $O_{12h}$ orbit, the analysis of Figure 3.19 up to level $\ell = 4$ harmonics is not sufficient to identify a second distinct group orbital of $E_{1u}$ symmetry. However, we know that there is a level 4 component of $E_{2g}$ irreducible symmetry and that level 5 components can be generated using the products of the form $z(x \pm iy)^j$. Since 'z' transforms as $A_{2u}$, level 5 harmonics of the required symmetry [$2e_{1u}$] are likely to be found and this is the case for the extended superposition analysis of Figure 3.19.

### 3.4.3 Simplifications in calculations for multi-orbit structures

In multi-orbit molecular structures, the need can arise to identify several LCAO-MOs of the same symmetry, over the structure vertices and this requirement provides a second useful example of the general utility of the orbit by orbit approach to the formation of group orbitals. In this section, the calculation of the electronic structure of the moiety $C_{80}$ as an exercise in Hückel theory, illustrates the benefits, which arise, when the LCAO-MOs are constructed from group orbitals formed on the vertices of the two structure orbits $O_{20}$ and $O_{60}$.

The fullerene, $C_{80}$, exhibits a permutation character over its 80 vertices, which, as set out in Table 3.18, is the sum of the reducible permutation characters [Table 3.4] on the vertices of the $O_{20}$ and $O_{60}$ orbits of the icosahedral point group.

The simplest calculation is to determine the two LCAO-MOs of $a_g$ symmetry and their corresponding Hückel electronic energies. The spherical polar coordinates of the vertices of $C_{80}$ are given in Table 3.19. These coordinates are based on an equilateral $C_{80}$ polyhedron that is then projected onto the sphere and are not the only set that can be generated while maintaining $I_h$ point group symmetry. Another approach would be to insist on equal angular separation for each bond on the sphere. We have chosen to use the coordinates listed in Table 3.19 because the positions are consistent with the neglect of the radial dimension of the polyhedron implied by the Hückel equal–$\beta$ model. Using the coordinates in the table, the trivalent $C_{80}$ cage is drawn in Figure 3.20 as a projection on the unit sphere.

**Table 3.18 The direct sum components of the permutation character over the vertices of the $C_{80}$ fullerene, Table 3.19, formed by summation of the direct sums for the $O_{20}$ and $O_{60}$ orbits of the icosahedral point group $I_h$, Table 3.4.**

| Orbit | $\Gamma_\sigma$ |
|---|---|
| $O_{20}$ | $A_g + T_{1u} + T_{2u} + G_g + G_u + 2H_g + H_u$ |
| $O_{60}$ | $A_g + T_{1g} + 2T_{1u} + T_{2g} + T_{2u} + 2G_g + 2G_u + 3H_g + 3H_u$ |
| $O_{20} + O_{60}$ | $2A_g + T_{1g} + 3T_{1u} + T_{2g} + T_{2u} + 3G_g + 3G_u + 5H_g + 4H_u$ |

The calculation is straightforward, when the 2-orbit approach is adopted. Two linear combinations over the 80-vertex cage can be constructed using only the simple totally symmetric central basis function, $|1a_g\rangle$, of $I_h$. For the first function, we have

$$|1a_g\rangle = \sum_{i=1}^{i=20} \sqrt{\left({}^1\!/_{20}\right)}\sigma_i + \sum_{i=21}^{i=80} 0\sigma_i \qquad 3.37$$

and similarly, for the second function as a group orbital over the $O_{60}$ orbit

$$|1a_g\rangle = \sum_{i=1}^{i=20} 0\sigma_i + \sum_{i=21}^{i=80} \sqrt{{}^1\!/_{60}}\sigma_i \qquad 3.38$$

Solution of the $a_g$ block of the Hückel determinant for $C_{80}$ leads to two linear combinations of $a_g$ irreducible symmetry. The first of these is the linear combination resulting from the projection of the $1a_g$ polynomial of Table 3.20 on the 80 vertices, so that each coefficient is of equal value, while the second linear combination is identical to the one found by projecting the level 6 polynomial of the table onto the 80 vertices and rendering the result orthogonal to the first.

These observations are summarized in Figure 3.21, which displays the projections of linear combinations, which result from the superimposition of the $\sigma$-decorated vertices on the projections of the polynomials of Table 3.20.

Thus, in the first row the totally symmetric Hückel MO is displayed, while in the second row projection on the level 6 function neatly divides the decorations on the vertices into the two orbit sets, with 20 local $\sigma$ orbitals multiplied by a negative coefficient and 60 local $\sigma$ orbitals multiplied by a positive coefficient. The final projection in row 3 of the figure is the result obtained when the linear combination over the level 6 function is rendered orthogonal to that at level 0. In this example, because the overlap is relatively small, little difference is evident on comparison of the row 2 and row 3 results. However, it is clear that without division of the 80-vertexed cage into its $O_{20}$ and $O_{60}$ orbits considerably more algebra is required.

**Table 3.19 The polar coordinates of the vertices of $C_{80}$ on the unit sphere. The first 20 entries identify the vertices of the $O_{20}$ orbit and the remainder, #21–80, the vertices of the $O_{60}$ orbit.**

| # | $\theta$ | $\phi$ | # | $\theta$ | $\phi$ | # | $\theta$ | $\phi$ |
|---|---|---|---|---|---|---|---|---|
| *1* | 37.3774 | 36.0000 | *28* | 50.1740 | 85.0243 | *55* | 99.4400 | 324.0000 |
| *2* | 37.3774 | 108.0000 | *29* | 50.1740 | 130.9757 | *56* | 110.2429 | 18.6336 |
| *3* | 37.3774 | 180.0000 | *30* | 50.1740 | 157.0243 | *57* | 110.2429 | 53.3664 |
| *4* | 37.3774 | 252.0000 | *31* | 50.1740 | 202.9757 | *58* | 110.2429 | 90.6336 |
| *5* | 37.3774 | 324.0000 | *32* | 50.1740 | 229.0243 | *59* | 110.2429 | 125.3664 |
| *6* | 79.1877 | 36.0000 | *33* | 50.1740 | 274.9757 | *60* | 110.2429 | 162.6336 |
| *7* | 79.1877 | 108.0000 | *34* | 50.1740 | 301.0243 | *61* | 110.2429 | 197.3664 |
| *8* | 79.1877 | 180.0000 | *35* | 50.1740 | 346.9757 | *62* | 110.2429 | 234.6336 |
| *9* | 79.1877 | 252.0000 | *36* | 69.7571 | 17.3666 | *63* | 110.2429 | 269.3664 |
| *10* | 79.1877 | 324.0000 | *37* | 69.7571 | 54.6336 | *64* | 110.2429 | 306.6336 |
| *11* | 100.8123 | 0.0000 | *38* | 69.7571 | 89.3664 | *65* | 110.2429 | 341.3664 |
| *12* | 100.8123 | 72.0000 | *39* | 69.7571 | 126.6336 | *66* | 129.8260 | 22.9757 |
| *13* | 100.8123 | 144.0000 | *40* | 69.7571 | 161.3664 | *67* | 129.8260 | 49.0243 |
| *14* | 100.8123 | 216.0000 | *41* | 69.7571 | 198.6336 | *68* | 129.8260 | 94.9757 |
| *15* | 100.8123 | 288.0000 | *42* | 69.7571 | 233.3664 | *69* | 129.8260 | 121.0243 |
| *16* | 142.6226 | 0.0000 | *43* | 69.7571 | 270.6336 | *70* | 129.8260 | 166.9757 |
| *17* | 142.6226 | 72.0000 | *44* | 69.7571 | 302.3664 | *71* | 129.8260 | 193.0243 |
| *18* | 142.6226 | 144.0000 | *45* | 69.7571 | 342.6336 | *72* | 129.8260 | 238.8757 |
| *19* | 142.6226 | 216.0000 | *46* | 80.5600 | 0.0000 | *73* | 129.8260 | 265.0243 |
| *20* | 142.6226 | 288.0000 | *47* | 80.5600 | 72.0000 | *74* | 129.8260 | 310.9757 |
| *21* | 17.1250 | 36.0000 | *48* | 80.5600 | 144.0000 | *75* | 129.8260 | 337.0243 |
| *22* | 17.1250 | 108.0000 | *49* | 80.5600 | 216.0000 | *76* | 162.8750 | 0.0000 |
| *23* | 17.1250 | 180.0000 | *50* | 80.5600 | 288.0000 | *77* | 162.8750 | 72.0000 |
| *24* | 17.1250 | 252.0000 | *51* | 99.4400 | 36.0000 | *78* | 162.8750 | 144.0000 |
| *25* | 17.1250 | 324.0000 | *52* | 99.4400 | 108.0000 | *79* | 162.8750 | 216.0000 |
| *26* | 50.1740 | 13.0243 | *53* | 99.4400 | 180.0000 | *80* | 162.8750 | 288.0000 |
| *27* | 50.1740 | 58.9757 | *54* | 99.4400 | 252.0000 | | | |

### 3.4.4 The application of $\sigma$- and $\pi$-oriented local displacement vectors to analyse the normal mode spectrum of an object with simple $T_d$ symmetry

The GT_calculator CD provides for the identification of the normal mode spectrum of vibrating structures from the reducible character, $\Gamma_{\text{coordinates}}$, formed as the direct sum of the $\ell = 0$

**Table 3.20 Central polynomials of $A_g$ irreducible symmetry for the $I_h$ point group.**

| | |
|---|---|
| $1a_g$ | $x^0y^0z^0$ |
| $2a_g$ | $231z^6 - 315z^4 + 105z^2 - 5 + 42x^5z - 420x^3y^2z + 210xy^4z$ |

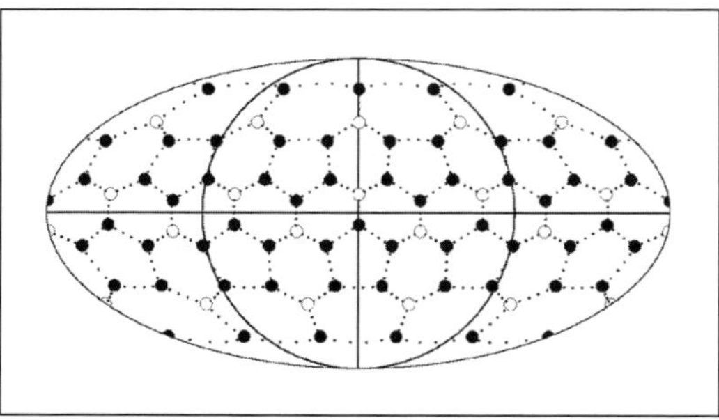

**Figure 3.20** Projection of the $C_{80}$ vertices as listed in Table 3.20, with the $O_{20}$ orbit vertices as open circles and the $O_{60}$ orbit vertices, as the black circles on the unit sphere.

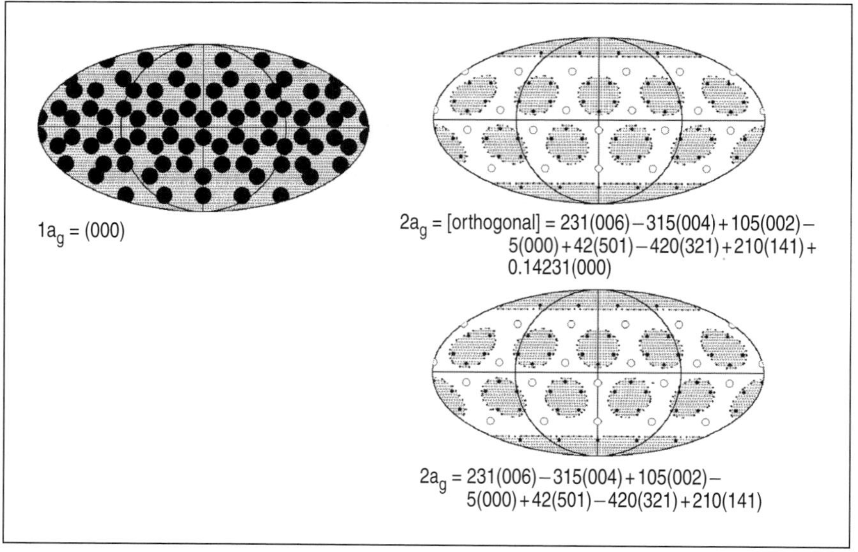


**Figure 3.21** Application of the superposition procedure with orthogonalization to identify the 2 Hückel LCAO-MOs of the $C_{80}$ fullerene, which provide basis functions for the distinct $1a_g$ and $2a_g$ components of the reducible permutation character over the 80 vertices of the molecular structure. Note that the orthogonalization step does not lead to a dramatic change in the $2a_g$ central function, as the integral $\langle 1a_g|2a_g\rangle$ is small.

and $\ell = 1$ characters identified in equation 3.5. The character, $\Gamma_{\text{coordinates}}$,

$$\Gamma_{\text{coordinates}} = \Gamma_{\sigma} + \Gamma_{\pi} = \Gamma_{\text{xyz}} \times \Gamma_{\sigma} \qquad 3.39$$

in turn, can be written in its components representing the distinct motions of translation, rotation and vibration of the structure cage, i.e.

$$\Gamma_{\text{coordinates}} = \Gamma_{\text{xyz}} + \Gamma_{\text{rotation}} + \Gamma_{\text{vibration}} \qquad 3.40$$

With the convenient device of an arrowhead used to reflect infinitesimal motions in the 3D space of the structure, the projections of Figure 3.22 are converted into motion drawings in Figure 3.23. Thus, in diagram 1, of Figure 3.23, the radially disposed $\sigma$-oriented components of Figure 3.22, become the radial concerted radial contractions and extensions of the orbit vertices, which, too, exhibit $A_1$ symmetry.

The remaining diagrams in Figure 3.23 emphasize two features of analyses based on the projection procedure for structures exhibiting cubic and higher symmetries. For dihedral structures the superposition procedure is to replace the local $\sigma$-oriented function with local z and then rotate by $\pi/2$ to identify the complementary set of $\pi_{\phi}$-oriented local functions. For the higher symmetry groups the requirement is to align the local $\pi_{\theta}$-like displacements along the gradient directions in the tangent plane of the appropriate central functions and then

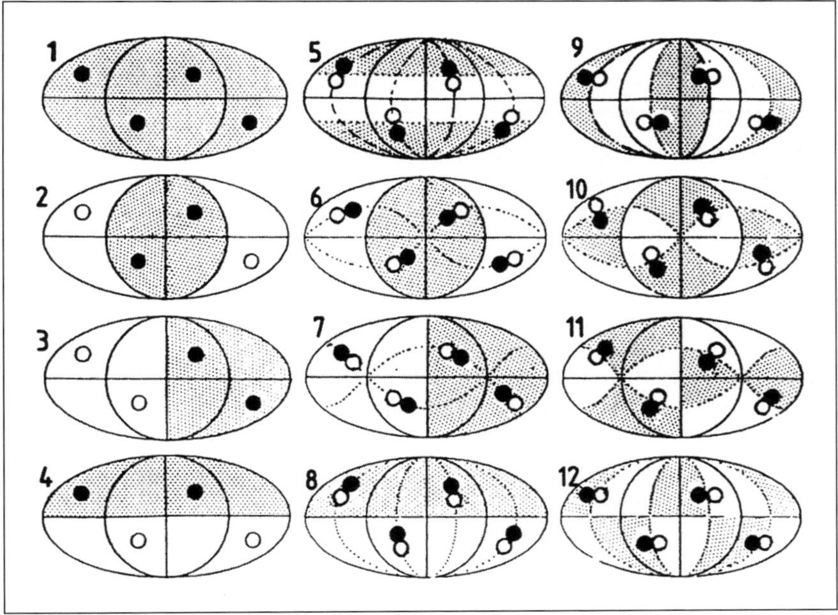


**Figure 3.22** The summary results for the application of the superposition procedure on the tetrahedral, $T_d$, symmetry cage with local $\sigma$- and $\pi$-oriented functions. As explained in the text, these can be applied to investigate the possible normal modes of vibration of the structure.

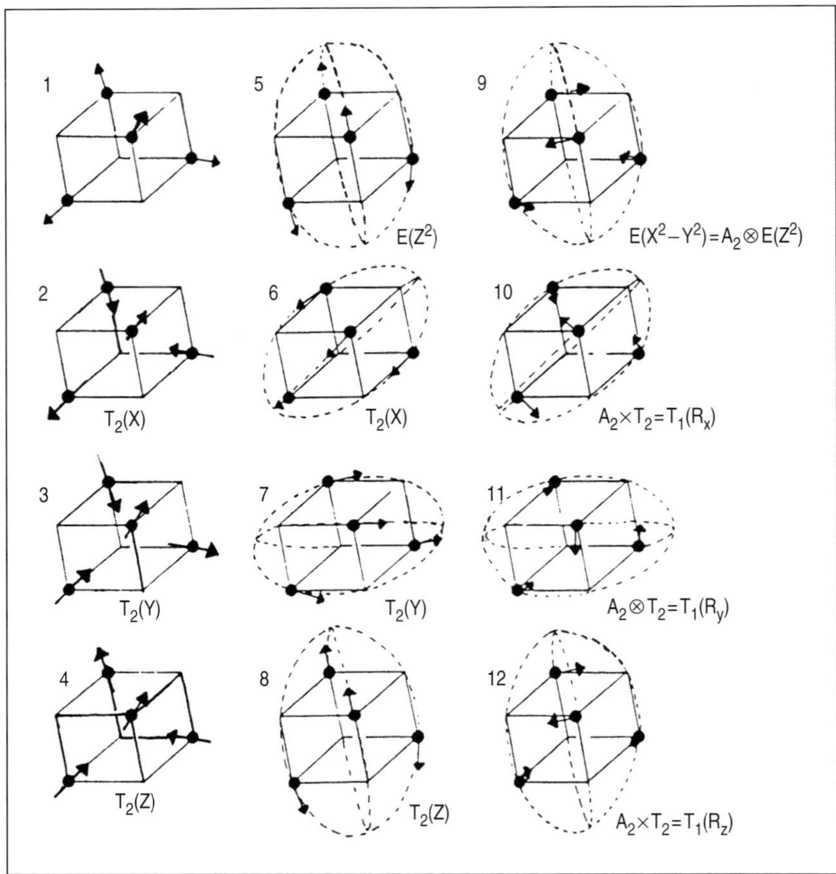


**Figure 3.23** Translation of the summary projection results of Figure 3.33 into motion diagrams, from which in conjunction with the Figure 3.24, all the distinct normal modes of vibration of the $O_4$ orbit cage of a structure exhibiting $T_d$ point symmetry can be identified.

to turn locally, again, by $\pi/2$. As is evident in diagrams 5 to 12 of Figure 3.23, the alternative perspective on these transformations is to recognize that the local functions are rotated to lie as tangents in the planes defined by great circles through the orbit vertices and the turning points of the central functions on the unit sphere, with the complementary set of $\pi_\phi$ functions lying normal to these planes.

Note, the particular result of local rotation on the normal mode components in diagram 5 of Figure 3.23. Diagram 5 identifies one of the pair of the degenerate e-type modes of the $O_4$ orbit, while local rotations on the vectors of diagram 5 lead to diagram 9 and the identification of the second mode of the degenerate pair of this symmetry.

The elementary theory of molecular vibrations identifies the 3n − 5 and 3n − 6 rules for vibrating objects defined by sets of vertices forming linear and polyhedral shaped structures.

For the $O_4$ orbit of a $T_d$ symmetry structure, there will be 3 translations, 3 rotations and 6 [3*4 − 6] distinct vibrations. Diagrams 2, 3 and 4 and 6, 7 and 8 of Figure 3.23 correspond to motions of the $O_4$ structure, which are of $T_2$ symmetry, but which are neither pure translation nor pure vibration, whereas it is straightforward to assign the natures of the motions represented in the other diagrams in the figure.

The deficiency is rectified in Figure 3.24 by taking suitable sums and differences of the motions of Figure 3.23. For molecular motions these transformations correspond to the orthogonalization of the like symmetry linear combinations, which occur as repetitions. Note that this particular example is an especially favourable one, as it is a single-orbit problem and is also one where the vibrational character contains not more than one copy of any irreducible symmetry, so the forms of all the vibrational modes are entirely determined by symmetry. In a more general case, symmetry considerations provide a basis for the normal modes rather than the modes themselves.

The final result, then is that there are 3 rotational motions of $T_1$ symmetry and that the obviously occurring 3 translations of $T_2$ symmetry can be rendered distinct from the same symmetry vibrations, in the direct sum, for the normal mode character.

$$\Gamma_{\text{vibrations}} = A_1 + E + T_2 \qquad 3.41$$

Analysis of motions of a structure using $\sigma$ and $\pi$ representations has many applications beyond chemistry. For example, some interesting results come from applications of these ideas to macroscopic, engineering-scale structures. In structural mechanics, Maxwell's rule gives a necessary though not sufficient condition for rigidity of a 3D bar-and-joint assembly. It is

$$m - s = 3j - b - 6$$

where m is the number of mechanisms, s is the number of states of self-stress, j is the number of joints and b the number of bars. A symmetry-extended version[12] uses characters for sets of points at the positions of the joints and the centres of the bars and is

$$\Gamma_{(m)} - \Gamma_{(s)} = \Gamma_{\sigma(j)} \times \Gamma_{xyz} - \Gamma_{\sigma(b)} - \Gamma_{xyz} \times (\Gamma_0 - \Gamma_\varepsilon) \qquad 3.42$$

As this rule is obeyed separately under every symmetry operation, rather than just under the identity as in the pure counting version, it can reveal extra information about the mechanisms and states of self stress, giving a count m – s for every irreducible representation of the group. Results such as the fact that the vibrations of a fully triangulated polyhedron are all edge stretches (showing the generic rigidity of deltahedra)[13], and that every fully

[12] P.W. Fowler and S.D. Guest, *Int. J. Solids and Structures*, **37** (2000) 1793–1804 A symmetry extension of Maxwell's rule for rigidity of frames.

[13] A. Ceulemans and P.W. Fowler, *Nature*, **353** (1991) 52–54 Extension of Euler's theorem to the symmetry properties of polyhedra.

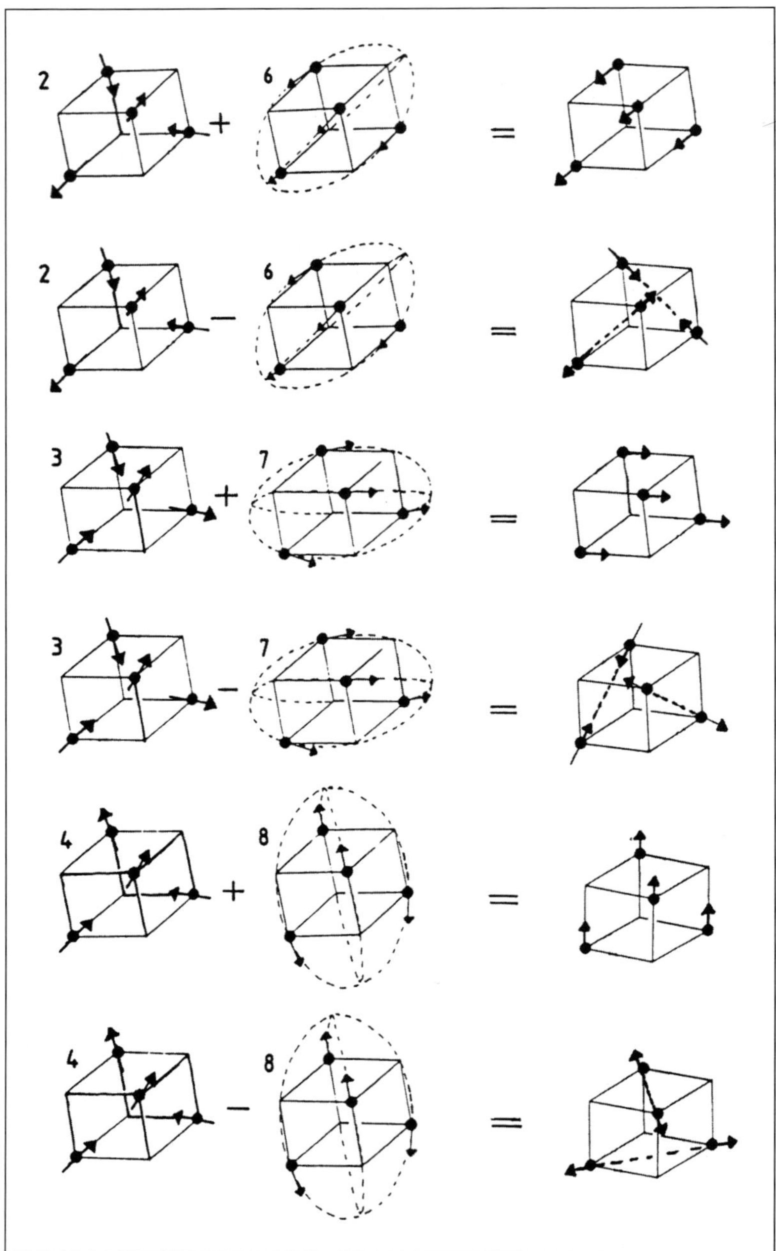


**Figure 3.24** Deconvolution of the same symmetry components of the motion analysis displayed in Figure 3.34 into orthogonal combinations of the local vectors which distinguish pure vibrations from pure translations.

triangulated torus has six states of self stress of particular symmetries[14], can be derived from the symmetry-extended Maxwell Rule. Other engineering rules, such as the 'mobility criterion' which gives the count m–s for a set of n bodies linked by g joints with residual freedoms at each joint, can be given a symmetry-extended form[15], with similar improvement in the detailed understanding of structures and mechanisms.

[14] P.W. Fowler and S.D. Guest, *Int. J. Solids and Structures*, **39** (2002) 4385–4393 Symmetry and states of self stress in triangulated toroidal frames.

[15] S.D. Guest and P.W. Fowler, *Mechanism and Machine Theory*, **40** (2005) 1002–1014. A symmetry-extended mobility rule.

# 4

# Symmetrized powers and their applications

This chapter deals with a number of applications of group theory to molecular properties and structure, all connected by the idea of symmetrized powers of representations. The GT_calculator has a general facility for calculation of these powers, and specialised routines for their application to angular momentum, molecular electric properties and isomer counting.

In this chapter you will learn:

1. how to calculate the first six fully symmetric and fully antisymmetric powers of reducible and irreducible representations;
2. how to count independent components of electric multipole moments and polarisabilities and molecular force fields; and
3. how to count chiral and achiral derivatives of symmetrical molecules.

Symmetrized powers are useful in various areas of spectroscopy and quantum mechanics, and arise from some basic considerations about sets and products. Consider a set of three variables {x, y, z}, denoting the coordinates of a point in 3D space with respect to three orthogonal axes. Any *linear* function of position can be expressed by taking combinations $ax+by+cz$. Suppose now we wish to represent a function that depends on the second powers of coordinates. From the three quantities x, y and z we can form nine terms of second degree:

$$\{xx, xy, xz, yx, yy, yz, zx, zy, zz\} \quad 4.1$$

Of course, the fact that x, y and z are numbers implies that $xy = yx$, $xz = zx$, $yz = zy$ and so the nine can be reduced to six:

$$\{x^2, xy, xz, y^2, yz, z^2\} \quad 4.2$$

Suppose now that we have two different vectors $\{x_1, y_1, z_1\}$ and $\{x_2, y_2, z_2\}$, perhaps representing coordinates of two points or two sets of quantum mechanical operators that have the symmetries of coordinates. We can rewrite the nine products

$$\{x_1x_2, x_1y_2, x_1z_2, y_1x_2, y_1y_2, y_1z_2, z_1x_2, z_1y_2, z_1z_2\} \quad 4.3$$

in a way that shows their symmetry under a permutation $P_{12}$ that exchanges subscripts 1 and 2.

The nine products comprise a *permutationally symmetric* set of six combinations $x_1x_2$, ½ $\{x_1y_2 + y_1x_2\}$, ½ $\{x_1z_2 + z_1x_2\}$, $y_1y_2$, ½ $\{y_1z_2 + z_1y_2\}$, $z_1z_2$, which are each unchanged under the operation of $P_{12}$, e.g.

$$P_{12}\left[\frac{1}{2}\{x_1y_2 + y_1x_2\}\right] = \frac{1}{2}\{x_2y_1 + y_2x_1\} = \frac{1}{2}\{x_1y_2 + y_1x_2\} \qquad 4.4$$

and a *permutationally antisymmetric* set of three combinations

$$\frac{1}{2}\{x_1y_2 - y_1x_2\},\ \frac{1}{2}\{x_1z_2 - z_1x_2\},\ \frac{1}{2}\{y_1z_2 - z_1y_2\} \qquad 4.5$$

which each change sign under the operation of $P_{12}$, e.g.

$$P_{12}\left[\frac{1}{2}\{x_1y_2 - y_1x_2\}\right] = \frac{1}{2}\{x_2y_1 - y_2x_1\} = -\frac{1}{2}\{x_1y_2 - y_1x_2\} \qquad 4.6$$

When subscripts 1 and 2 are suppressed, we have the simple case of products of {x, y, z}, and the six permutationally symmetric products reduce to (4.2), with the three permutationally antisymmetric products vanishing identically when x, y, z are numbers but not necessarily when they stand for (non-commuting) operators.

The process of taking products can be extended to three, four, ... sets $\{x_i, y_i, z_i\}$, corresponding to higher powers, and the size of the starting set from which the products are generated can also be generalized. A starting set might be only two objects (e.g. a pair of orthogonal molecular orbitals spanning a doubly degenerate representation of a group such as $C_{3v}$), or arbitrarily large (e.g. the 3N-6 vibrational normal modes of a non-linear polyatomic molecule). The possibilities for permutational symmetry become more varied as the power increases. A given combination of product terms may be symmetric under some permutations, antisymmetric under others, or may be mixed with other products by the action of the permutation. However, there will always be one set of *permutationally totally symmetric* combinations, defined by their preservation under *every* pairwise permutation of subscripts. If $d$ is the dimension of the set and and $p$ is the power in the product, a simple counting argument gives the number of totally symmetric products as

$$n(\text{symm}) = (p + d - 1)!/p!(d - 1)! \qquad 4.7$$

In the limit of $d = 1$, there is just one symmetric combination ($x^p$) at all $p$.

When the power is small enough, there is also a set of fully antisymmetric combinations (*determinantally* antisymmetric), defined by the property that they are reversed in sign by every pairwise permutation of indices. For a set of given dimension $d$, combinations of this type exist for all those powers that lie between 2 and $d$, i.e. $(d + 1) > p > 1$. The number of totally antisymmetric combinations is

$$n(\text{anti}) = d!/p!(d - p)! \qquad 4.8$$

Note that, for every set of order $> 1$, there is always at least one power $p$ for which there is an antisymmetric combination.

How does this concept of powers of a set connect with point-group theory? If the starting sets $\{x_i, y_i, z_i\}$ each span some character [or representation] $\Gamma$, then the full sets of products of 2, 3, 4, ... sets span $\Gamma \times \Gamma = \Gamma^2, \Gamma \times \Gamma \times \Gamma = \Gamma^3, \Gamma \times \Gamma \times \Gamma \times \Gamma = \Gamma^4, \ldots$ Thus, for example, in the group $I_h$, the three translations span $\Gamma_{xyz} = T_{1u}$ and the full set of 9 second degree products $\{x_1x_2, x_1y_2, x_1z_2, y_1x_2, y_1y_2, y_1z_2, z_1x_2, z_1y_2, z_1z_2\}$ spans $T_{1u} \times T_{1u} = A_g + T_{1g} + H_g$. We can define the character of symmetric and antisymmetric sets of the *pth* power products as parts of the full power $\Gamma^p$. The character formula for each of these can be worked out in terms of the theory of permutation groups[1] and becomes increasingly cumbersome as $p$ increases. For the first few cases, the symmetric power results are

$$\chi_{[2]}(R) = \frac{1}{2}\left\{\chi^2(R) + \chi(R^2)\right\} \tag{4.9}$$

$$\chi_{[3]}(R) = \frac{1}{6}\left\{\chi^3(R) + 3\chi(R)\chi(R^2) + 2\chi(R^3)\right\} \tag{4.10}$$

$$\chi_{[4]}(R) = \frac{1}{24}\left\{\chi^4(R) + 8\chi(R)\chi(R^3) + 6\chi^2(R)\chi(R^2) + 3\chi^2(R^2) + 6\chi(R^4)\right\} \tag{4.11}$$

$$\chi_{[5]}(R) = \frac{1}{120}\left\{\begin{array}{l}\chi^5(R) + 30\chi(R)\chi(R^4) + 15\chi(R)\chi^2(R^2) + 20\chi(R^2)\chi(R^3)\\ +20\chi^2(R)\chi(R^3) + 10\chi^3(R)\chi(R^2) + 24\chi(R^5)\end{array}\right\} \tag{4.12}$$

and the antisymmetric power results are:

$$\chi_{\{2\}}(R) = \frac{1}{2}\left\{\chi^2(R) - \chi(R^2)\right\} \tag{4.13}$$

$$\chi_{\{3\}}(R) = \frac{1}{6}\left\{\chi^3(R) - 3\chi(R)\chi(R^2) + 2\chi(R^3)\right\} \tag{4.14}$$

$$\chi_{\{4\}}(R) = \frac{1}{24}\left\{\chi^4(R) + 8\chi(R)\chi(R^3) - 6\chi^2(R)\chi(R^2) + 3\chi^2(R^2) - 6\chi(R^4)\right\} \tag{4.15}$$

$$\chi_{\{5\}}(R) = \frac{1}{120}\left\{\begin{array}{l}\chi^5(R) - 30\chi(R)\chi(R^4) + 15\chi(R)\chi^2(R^2) - 20\chi(R^2)\chi(R^3)\\ +20\chi^2(R)\chi(R^3) - 10\chi^3(R)\chi(R^2) + 24\chi(R^5)\end{array}\right\} \tag{4.16}$$

where $\chi(R)$, $\chi_{[p]}(R)$, $\chi_{\{p\}}(R)$ are the characters of the starting set and of its *symmetrized* and *antisymmetrized pth* powers. By convention, we define $\chi_{[1]}(R) = \chi_{\{1\}}(R) = \chi(R)$. Notice that addition of $\chi_{[2]}(R)$ and $\chi_{\{2\}}(R)$ recovers the character of the full square, but for higher powers $\chi_{[p]}(R) + \chi_{\{p\}}(R) \neq \chi^p(R)$, as $p > 2$ there are other types of permutational behaviour beyond the fully symmetric and antisymmetric.

The formulae 4.9 to 4.16 apply to both irreducible and reducible characters; if the starting set spans $\Gamma = \sum_a n_a \Gamma_a$ and all the expressions apply directly with this sum. Similarly, expressions can be derived for the various powers in terms of the component characters and their powers.

[1]For example, L.L. Boyle, *Int. J. Quant. Chem.*, **6** (1972) 725.

If $\Gamma$ is a sum, then (with indices a, b, c, d ... all distinct)

$$[\Gamma^2] = \sum_{\mathrm{a}} [\Gamma_{\mathrm{a}}^2] + \sum_{\mathrm{a<b}} \Gamma_{\mathrm{a}} \times \Gamma_{\mathrm{b}} \tag{4.17}$$

$$[\Gamma^3] = \sum_{\mathrm{a}} [\Gamma_{\mathrm{a}}^2] + \sum_{\mathrm{a,b}} [\Gamma_{\mathrm{b}}^2] + \Gamma_{\mathrm{a}} + \sum_{\mathrm{a<b<c}} \Gamma_{\mathrm{a}} + \Gamma_{\mathrm{b}} + \Gamma_{\mathrm{c}} \tag{4.18}$$

$$\begin{aligned}[\Gamma^4] = &\sum_{\mathrm{a}} [\Gamma_{\mathrm{a}}^4] + \sum_{\mathrm{a,b}} \Gamma_{\mathrm{a}} + \Gamma[_{\mathrm{b}}^3] + \sum_{\mathrm{a<b}} [\Gamma_{\mathrm{a}}^2] + [\Gamma_{\mathrm{b}}^2] \\ &+ \sum_{\mathrm{a<b,c}} \Gamma_{\mathrm{a}} + \Gamma_{\mathrm{b}} + \Gamma_{\mathrm{c}} + \sum_{\mathrm{a<b<c<d}} \Gamma_{\mathrm{a}} + \Gamma_{\mathrm{b}} + \Gamma_{\mathrm{c}} + \Gamma_{\mathrm{d}}\end{aligned} \tag{4.19}$$

$$\begin{aligned}[\Gamma^5] = &\sum_{\mathrm{a}} [\Gamma_{\mathrm{a}}^5] + \sum_{\mathrm{a,b}} [\Gamma_{\mathrm{a}}^3] \times [\Gamma_{\mathrm{b}}^2] + \sum_{\mathrm{a,b}} [\Gamma_{\mathrm{a}}^4] \times \Gamma_{\mathrm{b}} \\ &+ \sum_{\mathrm{a<b,c}} [\Gamma_{\mathrm{a}}^2] \times [\Gamma_{\mathrm{b}}^2] \times \Gamma_{\mathrm{c}} + \sum_{\mathrm{b<c}} [\Gamma_{\mathrm{a}}^3] \times \Gamma_{\mathrm{b}} \times \Gamma_{\mathrm{c}} \\ &+ \sum_{\mathrm{b<c<d}} [\Gamma_{\mathrm{a}}^2] \times \Gamma_{\mathrm{b}} \times \Gamma_{\mathrm{c}} \times \Gamma_{\mathrm{d}} + \sum_{\mathrm{a<b<c<d<e}} \Gamma_{\mathrm{a}} \times \Gamma_{\mathrm{b}} \times \Gamma_{\mathrm{c}} \times \Gamma_{\mathrm{d}} \times \Gamma_{\mathrm{e}}\end{aligned} \tag{4.20}$$

where $[\Gamma^p]$ is the symmetrized *pth* power of $\Gamma$. If every term in square brackets is replaced by its antisymmetric counterpart, $\{\Gamma^p\}$, the same equations apply to the *antisymmetrized* powers of $\Gamma$. One potentially confusing feature of these formulae occurs when a coefficient $n_a$ is negative, as it can be in formal expressions for reducible characters. For example, for vibrations $\Gamma_{vib} = \Gamma_{3N} - \Gamma_T - \Gamma_R$ results by subtraction of the rigid-body motions from the degrees of freedom of a molecule and the roles of symmetric and antisymmetric powers reverse, since $[(-\Gamma_{\mathrm{a}})]^{\mathrm{b}} = (-1)^p\{\Gamma_{\mathrm{a}}^{\mathrm{b}}\}$ and $\{(-\Gamma_{\mathrm{a}})^p\} = (-1)^p[\Gamma_{\mathrm{a}}^p]$. If there is doubt, it is always correct to use the formulae 4.9 to 4.16 directly. In any event the GT_calculator returns the correct result irrespective of the signs of the coefficients $n_{\mathrm{a}}$.

While cubes, fourth and fifth powers of degenerate irreducible characters of the common point groups are available in tables[1], these results are extended here for both symmetrized and antisymmetrized powers of reducible and irreducible characters up to $p = 6$. For one dimensional $\Gamma_{\mathrm{a}}$, the powers are trivially already permutationally symmetric, with $[\Gamma_{\mathrm{a}}^p] = \Gamma_0$ for even $p$ and $[\Gamma_{\mathrm{a}}^p] = \Gamma_{\mathrm{a}}$ for odd $p$.

The remainder of this chapter is devoted to discussion of some applications of symmetrized powers and their calculation with the GT_calculator.

## 4.1 Symmetrized Squares, Electronic States and the Jahn-Teller Effect

The symmetric and antisymmetric squares have special prominence in molecular spectroscopy as they give information about some of the simplest open-shell electronic states. A closed-shell configuration has a totally symmetric space function, arising from multiplication of all occupied orbital symmetries, one per electron. The required antisymmetry of the space/spin wavefunction as a whole is satisfied by the exchange-antisymmetric spin function, which returns ${}^1\Gamma_0$ as the term symbol. In open-shell molecules belonging to a group without degeneracies, the ground state is not in general spatially totally symmetric, but would have term symbol ${}^2\Gamma_a$, where $\Gamma_a$ is the symmetry of the singly occupied orbital.

If an electronic state corresponds to occupation of a doubly degenerate orbital, say of symmetry $\Gamma(e)$, the full square $\Gamma(e) \times \Gamma(e)$ gives the full set of possible symmetries of the space parts of the electronic wavefunction. The square divides into symmetric and antisymmetric components: the symmetric component $[\Gamma(e)^2]$ is associated with singlet states (permutationally symmetric space part, permutationally antisymmetric spin part), and the antisymmetric component $\{\Gamma(e)^2\}$ with triplet states (antisymmetric space part, symmetric spin part). When the two electrons are in orbitals belonging to different degenerate $e$ pairs, the full set of symmetries in $\Gamma(e) \times \Gamma(e)$ is accessible as both singlets and triplets.

The symmetric square also has a role in the theory of the Jahn-Teller effect. The Jahn-Teller theorem asserts the existence of a spontaneous symmetry breaking in degenerate electronic states of non-linear molecules. Jahn-Teller distortion is governed by the *epikernel* principle, which states that symmetry breaking proceeds down any particular subgroup chain only as far as is needed to lift the degeneracy causing the distortion[2]. According to the rule for Jahn-Teller instability, the distortion coordinates for a degenerate electronic state that corresponds to an irreducible character $\Gamma$ must span the space of the character $\Lambda$ (both in the original higher group) belonging to the non-totally symmetric part of the symmetric square of $\Gamma$, i.e. $\Lambda = [\Gamma^2] - \Gamma_0$. If $\Gamma$ has dimension $d$, $\Lambda$ has dimension $\mathrm{d(d+1)/2-1}$. $\Lambda$ gives information on the possible Jahn-Teller activity of vibrations, the active vibrations being those in the overlap of $\Lambda$ and $\Gamma_{vib}$, the reducible character of the total set of vibrations.

## 4.2 Electric and Magnetic Properties of Molecules

A standard question in spectroscopy is: does a given molecule have a permanent dipole moment? This can usually be decided immediately by inspection, but the formal group-theoretical answer is that a molecule can exhibit a component of permanent dipole for each occurrence of the totally symmetric $\Gamma_0$ in the dipole character $\Gamma(\mu) \equiv \Gamma_{xyz} \equiv \Gamma_T$. Similar reasoning can be extended to other multipole moments, polarisabilities and corresponding magnetic properties.

The energy of a molecule in its non-degenerate ground state immersed in a non-uniform electric field $F$ can be developed as a Taylor series in terms of the components of field $F_\alpha$ and its gradients $F_{\alpha\beta} = (\partial F_\alpha/\partial\beta)$, $F_{\alpha\beta\gamma} = (\partial^2 F_\alpha/\partial\beta\partial\gamma)$, ... evaluated at some fixed origin, and sets of molecular properties: the dipole, quadrupole, octopole, ... moments $\mu_\alpha$, $\Theta_{\alpha\beta}$, $\Omega_{\alpha\beta\gamma}$, ... , polarisabilities $\alpha_{\alpha\beta}$, $\mathrm{C}_{\alpha\beta,\gamma\delta}$, ... and various cross- and hyper-polarisabilities. In the convention where Greek subscripts $\alpha$, $\beta$, ... each take all values x, y, z, and summation is applied for any repeated subscript on the RHS of an equation, the energy change is:

$$\begin{aligned} W = &-\mu_\alpha F_\alpha - \frac{1}{3}\Theta_{\alpha\beta}F_{\alpha\beta} - \frac{1}{15}\Omega_{\alpha\beta\gamma}F_{\alpha\beta\gamma} - \cdots \\ &- \frac{1}{2}\alpha_{\alpha\beta}F_\alpha F_\beta - \frac{1}{6}\beta_{\alpha\beta\gamma}F_\alpha F_\beta F_\gamma - \frac{1}{24}\gamma_{\alpha\beta\gamma}F_\alpha F_\beta F_\gamma F_\delta - \cdots \\ &- \frac{1}{6}\mathrm{C}_{\alpha\beta}F_{\alpha\beta}F_{\gamma\delta} - \frac{1}{3}\mathrm{A}_{\alpha,\beta\gamma}F_\alpha F_{\beta\gamma} - \frac{1}{6}\mathrm{B}_{\alpha\beta,\gamma\delta}F_\alpha F_\beta F_{\gamma\delta} - \cdots \end{aligned} \tag{4.21}$$

[2] A. Ceulemans and L.G. Vanquickenborne, The epikernel principle, *Structure and Bonding*, **71** (1989) 125–159.

Each molecular property is defined by an energy derivative taken at the zero-field limit e.g. dipole moment:

$$\mu_\alpha = -(\partial W/\partial F_\alpha)_0 \qquad 4.22$$

quadrupole moment:

$$\Theta_{\alpha\beta} = -(\partial W/\partial F_{\alpha\beta})_0 \qquad 4.23$$

dipole polarisability:

$$\alpha_{\alpha\beta} = -(\partial^2 W/\partial F_\alpha \partial F_\beta)_0 \qquad 4.24$$

first dipole hyperpolarisability:

$$\beta_{\alpha\beta\gamma} = -(\partial^3 W/\partial F_\alpha \partial F_\beta \partial F_\gamma)_0 \qquad 4.25$$

Moments and polarisabilities have permutational symmetry arising both from the exchange of like field derivatives (e.g. $F_\alpha F_\beta = F_\beta F_\alpha$) and their invariance to the order of differentiation (e.g. $F_{\alpha\beta\gamma} = F_{\alpha\gamma\beta} = F_{\beta\alpha\gamma} = F_{\beta\gamma\alpha} = F_{\gamma\alpha\beta} = F_{\gamma\beta\alpha}$). They also have a property of being traceless: any moment of polarisability vanishes when contracted (summed) over two indices of a field gradient. This follows from the Laplace condition on an electrostatic potential arising from distant charges, $\nabla^2\phi = 0$. Thus,

$$\Theta_{\alpha\alpha} = \sum_{\beta=\alpha} \Theta_{\alpha\beta} = \Theta_{xx} + \Theta_{yy} + \Theta_{zz} = 0 \qquad 4.26$$

$$\Omega_{\alpha\beta\beta} = \sum_{\beta=\gamma} \Omega_{\alpha\beta\gamma} = \Omega_{\alpha xx} + \Omega_{\alpha yy} + \Omega_{\alpha zz} = 0 \qquad 4.27$$

$$A_{\alpha,\beta\beta} = \sum_{\beta=\gamma} A_{\alpha,\beta\gamma} = A_{\alpha,xx} + A_{\alpha,yy} + A_{\alpha,zz} = 0 \qquad 4.28$$

and so on.

Each field index corresponds to a set of cartesian directions, and so each derivative has the symmetry of a power of the set {x, y, z}, symmetrized and with the trace(s) subtracted. The symmetries spanned by the moments and polarisabilities can therefore be deduced from their indices as, for the moments

$$\begin{aligned} \Gamma(\mu) &= \Gamma_{xyz} \\ \Gamma(\Theta) &= [\Gamma^2_{xyz}] - \Gamma_0 \\ \Gamma(\Omega) &= [\Gamma^3_{xyz}] - \Gamma_{xyz} \end{aligned} \qquad 4.29$$

with general term for the *pth* multipole moment

$$\Gamma(\xi^p) = [\Gamma^p_{xyz}] - [\Gamma^{p-2}_{xyz}] \qquad 4.30$$

and

$$\Gamma(\alpha) = [\Gamma^2_{xyz}], \Gamma(\beta) = [\Gamma^3_{xyz}], \ \Gamma(\gamma) = [\Gamma^4_{xyz}], \ldots \qquad 4.31$$

$$\Gamma(A) = \Gamma(\mu) \times \Gamma(\Theta) = \Gamma_{xyz} \times [\Gamma^2_{xyz}] - \Gamma_{xyz}, \ \Gamma(C) = [\Gamma(\Theta)^2], \ldots \qquad 4.32$$

For each property $P$, the number of independent components needed to specify the whole property tensor is given by the number of copies of $\Gamma_0$ in $\Gamma(P)$. This information is available from the GT_calculator. Example: use the calculator to show that $A_{\alpha,\beta\gamma}$ has 1 independent component for a molecule of $T_d$ symmetry but no non-vanishing components for a molecule of $O_h$ symmetry. [This illustrates the general rule that electric properties with odd numbers of subscripts vanish identically for any centrosymmetric system].

## 4.3 Counting Molecular Force Constants

Within the Born-Oppenheimer approximation, the electronic and nuclear motions of molecules can be decoupled, and vibrational motion of a molecule takes place on a potential surface, $V(s_i)$, representing the variation of the sum of electronic energy and nuclear repulsion energy with the internal coordinates $\{s_i\}$. In this simple, but remarkably useful picture, the electrons are assumed to keep up perfectly with the motion of the more ponderous nuclei. The internal coordinates would typically include bond lengths, bond angles, dihedral and torsional angles, referred to their values at the equilibrium geometry about which the vibrations are take place. The symmetry spanned by any non-redundant set of coordinates capable of describing the vibrations is $\Gamma_{vib}$, the vibrational character

$$\Gamma_{vib} = \Gamma_{3N} - \Gamma_T - \Gamma_R = \Gamma_\sigma \times \Gamma_{xyz} - \Gamma_{xyz} - \Gamma_{xyz} \times \Gamma_\varepsilon \qquad 4.33$$

Vibrations typically correspond to small excursions from the equilibrium geometry, and the energy scale of the potential surface is defined with respect to a zero at that geometry, where by definition all first derivatives $(\partial V/\partial s_i)_e$ vanish. The higher derivatives $(\partial^2 V/\partial s_i \partial s_j)_e$, $(\partial^3 V/\partial s_i \partial s_j \partial s_k)_e, \ldots$ are the quadratic, cubic, ... force constants. As coefficients in the Taylor expansion

$$V = \sum_{i,j} \frac{1}{2}(\partial^2 V/\partial s_i \partial s_j) s_i s_j + \sum_{i,j,k} \frac{1}{6}(\partial^3 V/\partial s_i \partial s_j \partial s_k) s_i s_j s_k$$

$$\sum_{i,j,k,l} \frac{1}{24}(\partial^4 V/\partial s_i \partial s_j \partial s_k \partial s_l) s_i s_j s_k s_l + \cdots \qquad 4.34$$

the force constants have permutational symmetry and span symmetrized powers of the symmetry of the set of internal coordinates, i.e. $\Gamma_{vib}$. Thus the sets of quadratic, cubic, quartic ... force constants span $[\Gamma^2_{vib}]$, $[\Gamma^3_{vib}]$, $[\Gamma^4_{vib}]$ ...and the numbers of independent force constants at each level are found by counting the copies of $\Gamma_0$.

The observations provide an opportunity to demonstrate one aspect of the utility of the GT_calculator files on the CDROM. It is straightforward to calculate that while the $C_{2v}$ point symmetry $H_2O$ molecule exhibits 4 independent quadratic, 6 independent cubic and

(a)

| README | $O_1$ | $O_{2xz}$ | $O_{2yz}$ | $O_4$ |
|---|---|---|---|---|
| # | 1 | 0 | 1 | 0 |
| $C_{2v}$ | E | $C_2$ | $\sigma_v(yz)$ | $\sigma_v(xz)$ |
| $\Gamma$ | 3 | 1 | 1 | 3 |
| $C_{2v}$ | $A_1$ | $A_2$ | $B_1$ | $B_2$ |
| # | 2 | 0 | 0 | 1 |
| $\Gamma_{nm}$ | 2 | 0 | 0 | 1 |
| $\Gamma^2$ | 4 | 0 | 0 | 2 |
| $\Gamma^3$ | 6 | 0 | 0 | 4 |
| $\Gamma^4$ | 9 | 0 | 0 | 6 |

(b)

| README | $O_1$ | $O_{12}$ | $O_{20}$ | $O_{30}$ | $O_{60}$ | $O_{120}$ | | | | |
|---|---|---|---|---|---|---|---|---|---|---|
| # | 0 | 0 | 0 | 0 | 1 | 0 | | | | |
| $I_h$ | E | $12C_5$ | $12C_5^2$ | $20C_3$ | $15C_2$ | i | $12S_{10}$ | $12S_{10}^3$ | $20S_6$ | $15\sigma$ |
| $\Gamma_\sigma$ | 60 | 0 | 0 | 0 | 0 | 0 | 0 | 0 | 0 | 4 |
| $I_h$ | $A_g$ | $T_{1g}$ | $T_{2g}$ | $G_g$ | $H_g$ | $A_u$ | $T_{1u}$ | $T_{2u}$ | $G_u$ | $H_u$ |
| # | 1 | 1 | 1 | 2 | 3 | 0 | 2 | 2 | 2 | 2 |
| $\Gamma^2$ | 24 | 38 | 38 | 62 | 86 | 14 | 46 | 46 | 60 | 74 |
| $\Gamma^3$ | 335 | 929 | 929 | 1264 | 1589 | 302 | 962 | 962 | 1264 | 1556 |
| $\Gamma^4$ | 5116 | 14755 | 14755 | 19871 | 24987 | 4928 | 14912 | 14912 | 19840 | 24768 |

**Figure 4.1** Composite of the GT_Calculator displays for the calculations of the quadratic, cubic and quartic force constants for the examples of $H_2O$ and $C_{60}$ discussed in the text.

9 independent quartic force constants, the $I_h$ point symmetry $C_{60}$ molecule exhibits 24, 335 and 5116 force constants at these orders. From the permutation characters, Figure 4.1a, the number of vibrational modes, follow from the option button, as in Figure 4.1b. Then, these results for $H_2O$ and $C_{60}$ are used as input data for the symmetric powers analysis in the remainder of Figure 4.1.

The same ideas extend straightforwardly to deal with property surfaces describing the dependence of a molecular property on geometry, for example, the dipole moment and polarisability derivatives that control the activity of a vibrational mode in IR and Raman spectroscopy. Extension to the case of redundant internal coordinates, the typical situation for polyatomic molecules, is also straightforward.

## 4.4 Symmetries of Central Functions with Arbitrarily High Angular Momentum

Spherical harmonic functions are important in many problems in Chemistry and Physics. Spherical harmonic functions are central in discussions of rotation, motion in a central potential, multipole expansion, cluster bonding, spherical wave expansions and many more topics. The calculation of symmetrized powers of representations give a way of obtaining the symmetry $\Gamma(J)$ spanned in any particular point group by the set of spherical harmonic functions corresponding to a given total angular momentum $J$. The required relation is identical, apart from notation, with that used already for multipole moments. In fact, $\Gamma(\xi^J)$ and $\Gamma(J)$ are identical

$$\Gamma(J) = [\Gamma(1)^J] - [\Gamma(1)^{J-2}] \qquad 4.35$$

where $\Gamma(1) \equiv \Gamma(J = 1) \equiv \Gamma(\mu)$ and $\Gamma(0)$ is the trivial character $\Gamma_0$. In the implementation of the GT_Calculator, symmetrized powers are programmed up to $p = 6$, placing a corresponding limit on the values of $J$ accessible by this route.

In view of the importance of spherical harmonics, an alternative calculator facility is provided, allowing explicit calculation up to $\Gamma(60)$ and, indirectly, calculation to any value of $J$ for all point groups of practical interest. The alternative method is based on the multiplication property

$$\Gamma(J) \times \Gamma(1) = \Gamma(J+1) + \Gamma(J) \times \Gamma_\varepsilon + \Gamma(J-1) \qquad 4.36$$

which follows from the rules of addition of angular momentum, but can easily be proved directly from the characters, $\Gamma(J)$, with

$$\begin{aligned} \chi_J(E) &= 2J+1, \\ \chi_J(C_\phi) &= \sin(J+\frac{1}{2})\phi / \sin\frac{1}{2}\phi, \\ \chi_J(i) &= (-1)^J \chi_J(E), \\ \chi_J(S_\phi) &= (-1)^J \chi_J(C_\phi) \end{aligned} \qquad 4.37$$

Given $\Gamma(J)$ and $\Gamma(1)$, we can use the multiplication in the form

$$\Gamma(J) = \Gamma(J-1) \times \Gamma(1) - \Gamma(J-1) \times \Gamma_\varepsilon - \Gamma(J-2) \qquad 4.38$$

to bootstrap to high $J$. The GT_calculator is set up to deliver $\Gamma(J)$ up to $\Gamma(60)$.

Extension to arbitrary $J$ follows from the fact that, for any finite point group, there is an integer $K$ such that $\Gamma(J+K) - \Gamma(J)$ is a constant (reducible) character depending (at most) on the value of $J$ mod $K$. The constant term is either the regular character of the group, or, for centrosymmetric groups, is alternately the g and u 'halves' of the regular character.

For example, in $C_{3v}$, $\Gamma(J)$ obeys a pattern that repeats every $K = 3$ steps, i.e.

$$\begin{aligned} \Gamma(3n) &= A_1 + n(A_1 + A_2 + 2E) \\ \Gamma(3n+1) &= A_1 + E + n(A_1 + A_2 + 2E) \\ \Gamma(3n+1) &= A_1 + 2E + n(A_1 + A_2 + 2E) \end{aligned} \qquad 4.39$$

| $I_h$ | $A_g$ | $T_{1g}$ | $T_{2g}$ | $G_g$ | $H_g$ | $A_u$ | $T_{1u}$ | $T_{2u}$ | $G_u$ | $H_u$ |
|---|---|---|---|---|---|---|---|---|---|---|
| 0 | 1 | | | | | | | | | |
| 1 | 0 | 0 | 0 | 0 | 0 | 0 | 1 | 0 | 0 | 0 |
| 2 | 0 | 0 | 0 | 0 | 1 | 0 | 0 | 0 | 0 | 0 |
| 3 | 0 | 0 | 0 | 0 | 0 | 0 | 0 | 1 | 1 | 0 |
| 4 | 0 | 0 | 0 | 1 | 1 | 0 | 0 | 0 | 0 | 0 |
| 5 | 0 | 0 | 0 | 0 | 0 | 0 | 1 | 1 | 0 | 1 |
| 6 | 1 | 1 | 0 | 1 | 1 | 0 | 0 | 0 | 0 | 0 |
| 7 | 0 | 0 | 0 | 0 | 0 | 0 | 1 | 1 | 1 | 1 |
| 8 | 0 | 0 | 1 | 1 | 2 | 0 | 0 | 0 | 0 | 0 |
| 9 | 0 | 0 | 0 | 0 | 0 | 0 | 1 | 1 | 2 | 1 |
| 10 | 1 | 1 | 1 | 1 | 2 | 0 | 0 | 0 | 0 | 0 |
| 11 | 0 | 0 | 0 | 0 | 0 | 0 | 2 | 1 | 1 | 2 |
| 12 | 1 | 1 | 1 | 2 | 2 | 0 | 0 | 0 | 0 | 0 |
| 13 | 0 | 0 | 0 | 0 | 0 | 0 | 1 | 2 | 2 | 2 |
| 14 | 0 | 1 | 1 | 2 | 3 | 0 | 0 | 0 | 0 | 0 |
| 15 | 0 | 0 | 0 | 0 | 0 | 1 | 2 | 2 | 2 | 2 |
| 16 | 1 | 2 | 1 | 2 | 3 | 0 | 0 | 0 | 0 | 0 |
| 17 | 0 | 0 | 0 | 0 | 0 | 0 | 2 | 2 | 2 | 3 |
| 18 | 1 | 1 | 2 | 3 | 3 | 0 | 0 | 0 | 0 | 0 |
| 19 | 0 | 0 | 0 | 0 | 0 | 0 | 2 | 2 | 3 | 3 |
| 20 | 1 | 2 | 2 | 2 | 4 | 0 | 0 | 0 | 0 | 0 |
| 21 | 0 | 0 | 0 | 0 | 0 | 1 | 3 | 2 | 3 | 3 |
| 22 | 1 | 2 | 2 | 3 | 4 | 0 | 0 | 0 | 0 | 0 |
| 23 | 0 | 0 | 0 | 0 | 0 | 0 | 2 | 3 | 3 | 4 |
| 24 | 1 | 2 | 2 | 4 | 4 | 0 | 0 | 0 | 0 | 0 |
| 25 | 0 | 0 | 0 | 0 | 0 | 1 | 3 | 3 | 3 | 4 |
| 26 | 1 | 3 | 2 | 3 | 5 | 0 | 0 | 0 | 0 | 0 |
| 27 | 0 | 0 | 0 | 0 | 0 | 1 | 3 | 3 | 4 | 4 |
| 28 | 1 | 2 | 3 | 4 | 5 | 0 | 0 | 0 | 0 | 0 |
| 29 | 0 | 0 | 0 | 0 | 0 | 0 | 3 | 3 | 4 | 5 |
| 30 | 2 | 3 | 3 | 4 | 5 | 0 | 0 | 0 | 0 | 0 |
| 31 | 0 | 0 | 0 | 0 | 0 | 1 | 4 | 3 | 4 | 5 |

**Figure 4.2** Composite of the GT_Calculator display for the direct sum components in $I_h$ point symmetry of centrally placed spherical harmonics up to level 31.

whereas in $D_{3d}$ the repeat length is $K = 6$, with $\Gamma(J+2) - \Gamma(J) = A_{1g} + A_{2g} + 2E_g$ for even $J$ and $\Gamma(J+2) - \Gamma(J) = A_{1u} + A_{2u} + 2E_u$ for odd $J$. In every case a run of $K + 2$ values establishes all $\Gamma(J)$. The integers $K$ are m for $C_m$, $C_{mv}$, $C_{mh}$ and $S_m$; 2m for $D_{2m}$ and $D_{2mh}$; 4m for $D_{2md}$; 4m + 2 for $D_{(2m+1)}$, $D_{(2m+1)h}$ and $D_{(2m+1)d}$; 6 for $T$, $T_h$; 12 for $T_d$, $O$, $O_h$; 15 for I and 30 for $I_h$.

The GT_Calculator displays for the direct sum components of the spherical harmonics about a central origin in a structure of $I_h$ point symmetry are shown in Figure 4.2. These results demonstrate the validity of equations 4.36 to 4.40 for this example.

The data in the figure illustrate that for the $I_h$ direct sum the orders of the first three non-vanishing multipole moments of the $C_{60}$ molecule by finding the first three $\Gamma(J)$ to have a component transforming as the totally symmetric representation that are found at levels 1, 6 and 10.

## 4.5 Isomer Counting using Point Group Symmetry

Counting the isomers arising by addition to, or substitution in, a basic framework is a mathematical problem with many practical applications in chemistry. In classical organic chemistry, for example, the number of derivatives of a compound was often cited as proof or disproof of structure. Point group theory that uses concepts familiar to most chemists and is easy to apply when the number of addends/substituents is small provides a unified method for deciding, for example, the number of dihydrides $C_{70}H_2$ of fullerene $C_{70}$, or the number of trihalo-derivatives $C_{20}H_{17}FClBr$ of dodecahedrane. All that is needed to determine such matters is the availability of the permutation character, $\Gamma_\sigma$, of the atoms in the parent molecule.

$\Gamma_\sigma$ is, in general, reducible and exhibits traces, $\chi(R)$, equal to the number of members of the set that are left unshifted by the operations R. $\Gamma_\sigma$ has been tabulated for all point-set orbits of the common groups and is available from the GT_calculator orbit-by-orbit or from inspection of the tables in Chapter 3. In what follows, the direct sum irreducible is written

$$\Gamma_\sigma = \sum_i a_i \Gamma_i \tag{4.40}$$

wherein each $\Gamma_i$ is an irreducible representation of G with degeneracy $g_i$.

For isomer counting, the problem to be solved is to determine the number of distinct isomers corresponding to decoration of atoms in a parent molecule. Substitution of an atom by another isotope of the same element or by a different chemical species, addition of a structureless ligand or functional group oriented so as to preserve the local site symmetry, all are to be treated as aspects of the same decoration process.

It is possible too, to choose whether or not to distinguish enantiomers. The basic equations all count each enantiomeric pair as a single structural isomer. An option for counting the number of such pairs, hence distinguishing chiral and achiral isomers, is also available in the GT_Calculator. The main problem is solved from the perspective that decoration of *n* vertices of a graph is analogous to selecting terms from the *n*-fold direct product of the vertex set with itself. Therefore, symmetry consequences can be established and isomers can be counted by taking appropriate functions of the powers of the permutation representation, $\Gamma_\sigma$.

(i) *Single replacements*: This case is trivial. Replacement of any one member of an orbit gives the same result as replacement of any other and so the set of mono-substituted isomers spans the same permutation representation as the vertices:

$$\Gamma(X) = \Gamma_\sigma \tag{4.41}$$

and, *n*(X), the number of distinct single-replacement isomers is just $a_0$, i.e. the number of copies of $\Gamma_0$ in $\Gamma_\sigma$.

(ii) *Double replacements*: All possible replacements of two atoms by identical substituents generate a set of isomers spanning the permutation representation $\Gamma(X_2)$. The isomers are generated by taking all distinct pairs of vertices, excluding pairing of a vertex to itself, and hence span

$$\Gamma(X_2) = [\Gamma_\sigma^2] - \Gamma_\sigma \tag{4.42}$$

The number of distinct isomers, $n(X_2)$, is the number of copies of $\Gamma_0$ in $\Gamma(X_2)$ and it follows from the definition of $\Gamma_\sigma$ in reducible form that this is

$$n(X_2) = \sum_i \frac{a_i(a_i+1)}{2} - a_0 = \frac{a_0(a_0-1)}{2} + \sum_{i\neq 0} \frac{a_i(a_i+1)}{2} \qquad 4.43$$

When the two substituents are distinguishable, the full square replaces its symmetric part in the permutation representation:

$$\Gamma(XY) = \Gamma_\sigma^2 - \Gamma_\sigma \qquad 4.44$$

and

$$n(XY) = a_0(a_0-1) + \sum_{i\neq 0} a_i^2 \qquad 4.45$$

The difference between $\Gamma(XY)$ and $\Gamma(X_2)$ is just the antisymmetric part of the square of $\Gamma_\sigma$, with dimension $\Sigma_i a_i(a_i - 1)/2$. In both cases, the number of isomers is immediately available once the reduction of the permutation representation of the vertices $\Gamma_\sigma$ is known.

(iii) *Triple replacements*: Three types of triple replacement can be envisaged, giving rise to isomer permutation representations $\Gamma(X_3)$, $\Gamma(X_2Y)$ and $\Gamma(XYZ)$. For three identical substituents, the isomers can be imagined as constructed by taking all distinct triples of vertices and excluding all those with a repeated vertex. The full set of triplets spans the symmetric cube $[\Gamma_\sigma^3]$ and the excluded cases span $\Gamma(XY)$ (pairs consisting of one unique and one doubled vertex) plus $\Gamma(X)$ (one tripled vertex), and so

$$\Gamma(X_3) = [\Gamma_\sigma^3] - \Gamma_\sigma^2 \qquad 4.46$$

The dimension of $\Gamma(X_3)$ is $N(N+1)(N+2)/6 - N^2 = N(N-1)(N-2)/6$. The number of distinct isomers is $n(X_3)$, which is the number of copies of $\Gamma_0$ in $\Gamma(X_3)$.
To count the isomers with 2 substituents of one kind and 1 of another, consider first all possible placings of $X_2$ and then add Y anywhere to each of them, excluding the case where Y falls on an X site. The permutation representation $\Gamma(X_2Y)$ follows as

$$\Gamma(X_2Y) = \Gamma(X) \times \Gamma(X_2) - \Gamma(XY) = \Gamma_\sigma \times [\Gamma_\sigma^2] - 2\Gamma_\sigma^2 + \Gamma_\sigma \qquad 4.47$$

For three distinct substituents X, Y and Z, consider all possible placings of XY and add Z anywhere except on an X site (where it would produce a heteropair XY) or a Y site (where it would produce a different heteropair XZ). The permutation representation $\Gamma(XYZ)$ is

$$\Gamma(XYZ) = \Gamma(X) \times \Gamma(XY) - 2\Gamma(XY) = \Gamma_\sigma^3 - 3\Gamma_\sigma^2 + 2\Gamma_\sigma \qquad 4.48$$

(iv) *Quadruple replacements:* Allowing a total of four substituents leads to decoration patterns $X_4$, $X_3Y$, $X_2Y_2$, $X_2YZ$ and WXYZ. Construction of the relevant fourth-order products and exclusion of degenerate special cases leads to the following expressions for the various isomer permutation representations

$$\begin{aligned}\Gamma[X_4] &= [\Gamma_\sigma^4] - \Gamma(X_2Y) - \Gamma(XY) - \Gamma(X_2) - \Gamma(X)\\ &= [\Gamma_\sigma^4] - \Gamma_\sigma \times \left[\Gamma_\sigma^2\right] + \Gamma_\sigma^2 - [\Gamma_\sigma^2]\end{aligned} \quad 4.49$$

$$\begin{aligned}\Gamma(X_3Y) &= \Gamma(X) \times \Gamma(X_3) - \Gamma(X_2Y)\\ &= \Gamma_\sigma \times [\Gamma_\sigma^3] - \Gamma_\sigma^3 - \Gamma_\sigma x[\Gamma_\sigma^2] + 2\Gamma_\sigma^2 - \Gamma_\sigma\end{aligned} \quad 4.50$$

$$\begin{aligned}\Gamma(X_2Y_2) &= \Gamma(X_2) \times \Gamma(X_2) - \Gamma(XYZ) - \Gamma(X_2)\\ &= [\Gamma_\sigma^2] \times [\Gamma_\sigma^2] - 2\Gamma_\sigma \times [\Gamma_\sigma^2] - \Gamma_\sigma^3 + 4\Gamma_\sigma^2 - [\Gamma_\sigma^2] - \Gamma_\sigma\end{aligned} \quad 4.51$$

$$\begin{aligned}\Gamma(X_2YZ) &= \Gamma(X) \times \Gamma(X_2Y) - \Gamma(X_2Y) - \Gamma(XYZ)\\ &= \Gamma(XY) \times \Gamma(X_2) - 2\Gamma(XYZ) - \Gamma(XY)\\ &= [\Gamma_\sigma^2] \times [\Gamma_\sigma^2] - 3\Gamma_\sigma^3 - \Gamma_\sigma \times [\Gamma_\sigma^2] + 6\Gamma_\sigma^2 - 3\Gamma_\sigma\end{aligned} \quad 4.52$$

$$\begin{aligned}\Gamma(WXYZ) &= \Gamma(X) \times \Gamma(XYZ) - 3\Gamma(XYZ)\\ &= \Gamma_\sigma^4 - 6\Gamma_\sigma^3 + 11\Gamma_\sigma^2 - 6\Gamma_\sigma\end{aligned} \quad 4.53$$

with the expected dimensions $\alpha/24$, $\alpha/6$, $\alpha/4$, $\alpha/2$ and $\alpha$, respectively, where $\alpha = N(N-1)(N-2)(N-3) = N^4 - 6N^3 + 11N^2 - 6N$. The isomers for each type of substitution follow from the number of copies of $\Gamma_0$ in each.

(v) *Quintuple and higher replacements:* Notice that when all $n$ substituents are different, i.e. in $\Gamma(XY)$, $\Gamma(XYZ)$, $\Gamma(WXYZ)$ …, the permutation representation can be obtained by formally replacing N in the polynomial expansion of $N!/(N-n)!$ by the vertex representation $\Gamma_\sigma$. Five and six distinct substituents give

$$\Gamma(VWXYZ) = \Gamma_\sigma^5 - 10\Gamma_\sigma^4 + 35\Gamma_\sigma^3 - 50\Gamma_\sigma^2 + 24\Gamma_\sigma \quad 4.54$$

$$\Gamma(UVWXYZ) = \Gamma_\sigma^6 - 15\Gamma_\sigma^5 + 85\Gamma_\sigma^4 - 225\Gamma_\sigma^3 + 274\Gamma_\sigma - 120\Gamma_\sigma \quad 4.55$$

Some recursion formulae for mixed substitutions/additions are

$$\Gamma(X_nY) = \Gamma(X) \times \Gamma(X_n) - \Gamma(X_{n-1}Y) \quad 4.56$$

$$\Gamma(X_nY_2) = \Gamma(X_2) \times \Gamma(X_n) - \Gamma(X_{n-1}YZ) - \Gamma(X_{n-2}Y_2) \quad 4.57$$

$$\Gamma(X_nYZ) = \Gamma(XY) \times \Gamma(X_n) - 2\Gamma(X_{n-1}YZ) - \Gamma(X_{n-2}YZ) \quad 4.58$$

The bootstrap process can be continued indefinitely, but clearly becomes increasingly cumbersome. The cases for five and six identical substituents are

$$\begin{aligned}\Gamma(X_5) &= [\Gamma_\sigma^5] - \Gamma(X_3Y) - 2\Gamma(X_2Y) - 2\Gamma(XY) - \Gamma(X) \\ &= [\Gamma_\sigma^5] - \Gamma_\sigma x[\Gamma_\sigma^3]x\Gamma_\sigma^3 - \Gamma_\sigma x[\Gamma_\sigma^2]\end{aligned} \tag{4.59}$$

$$\begin{aligned}\Gamma(X_6) &= [\Gamma_\sigma^6] - \Gamma(X_4Y) - \Gamma(X_3Y) - \Gamma(X_2Y) \\ &\quad - \Gamma(XYZ) - \Gamma(X_3) - 2\Gamma(XY) - \Gamma(X_2) - \Gamma(X) \\ &= [\Gamma_\sigma^6] - \Gamma_\sigma x[\Gamma_\sigma^4] + \Gamma_\sigma^2 x[\Gamma_\sigma^2] \\ &\quad - [\Gamma_\sigma^2]x[\Gamma_\sigma^2] = 2\Gamma_\sigma x[\Gamma_\sigma^2] - [\Gamma_\sigma^3] - [\Gamma_\sigma^3]\end{aligned} \tag{4.60}$$

In the special case of a molecule without symmetry all atoms are inequivalent and the isomer counts are purely combinatorial. When the molecular point group is the trivial group $C_1$ the permutation representation is $\Gamma_\sigma = N\Gamma_0$, where N is the number of atoms. The numbers $n(X)$, $n(X_2)$, $n(XY) \ldots n(X_xY_y \ldots Z_z) \ldots$ are then equal to the dimensions of the respective permutation representations $\Gamma(X)$, $\Gamma(X_2)$, $\Gamma(XY) \ldots \Gamma(X_xY_y \ldots Z_z) \ldots$ i.e. N, N(N−1)/2, N(N−1), ... $N!/(N-n)!x!y! \ldots z!$ where $x + y + \cdots + z = n$.

**Chiral isomers:** Although assignment of isomers to individual point groups can be laborious, division into chiral and achiral classes is straightforward. Decoration of an achiral framework *can* lead to chiral isomers, but, decoration of a chiral framework *will* lead to chiral isomers. In the first case, in the achiral group G, the left- and right-handed forms of such isomers will be exchanged by improper operations, so that the $n_\varepsilon$ left and $n_\varepsilon$ right enantiomers form a single orbit of dimension $2n_\varepsilon$. If improper formations were removed, by descent to the pure rotational half of G, the left and right enantiomers would split into two separate orbits of dimension $n_\varepsilon$ and each copy of $\Gamma_e$, the antisymmetric representation in G, would then become totally symmetric in the smaller group. The number of pairs of enantiomers is therefore found without leaving G by counting the copies of $\Gamma_e$ in $\Gamma(X_xY_y \ldots)$ or equivalently the copies of $\Gamma_0$ in $\Gamma_\varepsilon$ times $\Gamma(X_xY_y \ldots)$. The number of achiral isomers follows by subtraction. If G, itself, is a pure rotational group, the original framework is chiral and all derivatives are chiral, $\Gamma_\varepsilon = \Gamma_0$, and the two enantiomers of any particular derivative arise from distinct enantiomers of the parent.

As an example of chirality assignment, consider vertex decoration of the icosahedral [60]-fullerene. Here the vertices span

$$\Gamma_\sigma = A_g + T_{1g} + T_{2g} + 2G_g + 3H_g + 2T_{1u} + 2T_{2u} + 2G_u + 2H_u \tag{4.61}$$

wherein $\Gamma_0 = A_g$ and $\Gamma_\varepsilon = A_u$. The $X_2$ and XY isomers span

$$\begin{aligned}\Gamma(X_2) &= 23A_g + 37T_{1g} + 37T_{2g} + 60G_g + 83H_g \\ &\quad + 14A_u + 44T_{1u} + 44T_{2u} + 58G_u + 72H_u\end{aligned} \tag{4.62}$$

$$\begin{aligned}\Gamma(XY) &= 31A_g + 87T_{1g} + 87T_{2g} + 118G_g + 149H_g \\ &\quad + 28A_u + 90T_{1u} + 90T_{2u} + 118G_u + 146H_u\end{aligned} \tag{4.63}$$

These symmetries imply that the 23 distinct positional isomers of $C_{60}X_2$ comprise 9 achiral structures and 14 enantiomeric pairs whilst the 31 distinct positional isomers of $C_{60}XY$ split into 3 achiral structures and 28 enantiomeric pairs.

As an example of the operation of the GT_Calculator, in this application, consider the composite displays for isomers calculations of decorations of the four of a regular tetrahedron. All vertices are equivalent, belonging to a single orbit of $C_{3v}$ site symmetry. Figures 4.3 to 4.7 displays the GT_calculator results confirming that

$$\Gamma(X) = \Gamma(X_3) = \Gamma(X_3Y) = A_1 + T_2 \tag{4.64}$$

$$\Gamma(X_2) = \Gamma(X_2Y_2) = A_1 + E + T_2 \tag{4.65}$$

$$\Gamma(XY) = \Gamma(X_2Y) = \Gamma(X_2YZ) = A_1 + E + T_1 + 2T_2 \tag{4.66}$$

$$\Gamma(XYZ) = \Gamma(WXYZ) = A_1 + A_2 + 2E + 3T_1 + 3T_2 \tag{4.67}$$

$$\Gamma(X_4) = A_1 \tag{4.68}$$

These results are consistent with the familiar facts that substitution of one vertex of a tetrahedron gives a single isomer ($C_{3v}$), substitution of two gives unique $C_{2v}(X_2)$ and $C_s(XY)$ isomers, and substitution with three distinct atoms gives one positional isomer of $C_1$ symmetry that can exist in enantiomeric forms.

| README | $O_1$ | $O_4$ | $O_6$ | $O_{12}$ | $O_{24}$ |
|---|---|---|---|---|---|
| # | 0 | 1 | 0 | 0 | 0 |
| $T_d$ | E | $8C_3$ | $3C_2$ | $6C_4$ | $6\sigma_d$ |
| $\Gamma_\sigma$ | 4 | 1 | 0 | 0 | 2 |
| $T_d$ | $A_1$ | $A_2$ | E | $T_1$ | $T_2$ |
| # | 1 | 0 | 0 | 0 | 1 |
| Decoration | # | | | | |
| x | 1 | | | | |
| $T_d$ | E | $8C_3$ | $3C_2$ | $6C_4$ | $6\sigma_d$ |
| $\Gamma_\sigma$ | 4 | 1 | 0 | 0 | 2 |
| $T_d$ | $A_1$ | $A_2$ | E | $T_1$ | $T_2$ |
| # | 1 | 0 | 0 | 0 | 1 |
| xxx | 1 | | | | |
| $T_d$ | $A_1$ | $A_2$ | E | $T_1$ | $T_2$ |
| # | 1 | 0 | 0 | 0 | 1 |
| xxxy | 1 | | | | |
| $T_d$ | $A_1$ | $A_2$ | E | $T_1$ | $T_2$ |
| # | 1 | 0 | 0 | 0 | 1 |

**Figure 4.3** Composite of the GT_Calculator displays illustrating equation 4.64.

| README | $O_1$ | $O_4$ | $O_6$ | $O_{12}$ | $O_{24}$ |
|---|---|---|---|---|---|
| # | 0 | 1 | 0 | 0 | 0 |
| $T_d$ | E | $8C_3$ | $3C_2$ | $6C_4$ | $6\sigma_d$ |
| $\Gamma_\sigma$ | 4 | 1 | 0 | 0 | 2 |
| $T_d$ | $A_1$ | $A_2$ | E | $T_1$ | $T_2$ |
| # | 1 | 0 | 0 | 0 | 1 |
| Decoration | # | | | | |
| xx | 1 | | | | |
| $T_d$ | E | $8C_3$ | $3C_2$ | $6C_4$ | $6\sigma_d$ |
| $\Gamma_\sigma$ | 6 | 0 | 2 | 0 | 2 |
| $T_d$ | $A_1$ | $A_2$ | E | $T_1$ | $T_2$ |
| # | 1 | 0 | 1 | 0 | 1 |
| xxyy | 1 | | | | |
| $T_d$ | $A_1$ | $A_2$ | E | $T_1$ | $T_2$ |
| # | 1 | 0 | 1 | 0 | 1 |

**Figure 4.4** Composite of the GT_Calculator displays illustrating equation 4.65.

| README | $O_1$ | $O_4$ | $O_6$ | $O_{12}$ | $O_{24}$ |
|---|---|---|---|---|---|
| # | 0 | 1 | 0 | 0 | 0 |
| $T_d$ | E | $8C_3$ | $3C_2$ | $6C_4$ | $6\sigma_d$ |
| $\Gamma_\sigma$ | 4 | 1 | 0 | 0 | 2 |
| $T_d$ | $A_1$ | $A_2$ | E | $T_1$ | $T_2$ |
| # | 1 | 0 | 0 | 0 | 1 |
| Decoration | # | | | | |
| xy | 1 | | | | |
| $T_d$ | E | $8C_3$ | $3C_2$ | $6C_4$ | $6\sigma_d$ |
| $\Gamma_\sigma$ | 12 | 0 | 0 | 0 | 2 |
| $T_d$ | $A_1$ | $A_2$ | E | $T_1$ | $T_2$ |
| # | 1 | 0 | 1 | 1 | 2 |
| xxyz | 1 | | | | |
| $T_d$ | $A_1$ | $A_2$ | E | $T_1$ | $T_2$ |
| # | 1 | 0 | 1 | 1 | 2 |
| xxy | 1 | | | | |
| $T_d$ | $A_1$ | $A_2$ | E | $T_1$ | $T_2$ |
| # | 1 | 0 | 1 | 1 | 2 |

**Figure 4.5** Composite of the GT_Calculator displays illustrating equation 4.66.

| README | $o_1$ | $o_4$ | $o_6$ | $o_{12}$ | $o_{24}$ |
|---|---|---|---|---|---|
| # | 0 | 1 | 0 | 0 | 0 |
| $T_d$ | E | $8C_3$ | $3C_2$ | $6C_4$ | $6\sigma_d$ |
| $\Gamma_\sigma$ | 4 | 1 | 0 | 0 | 2 |
| $T_d$ | $A_1$ | $A_2$ | E | $T_1$ | $T_2$ |
| # | 1 | 0 | 0 | 0 | 1 |
| Decoration | # | | | | |
| xyz | 1 | | | | |
| $T_d$ | E | $8C_3$ | $3C_2$ | $6C_4$ | $6\sigma_d$ |
| $\Gamma_\sigma$ | 24 | 0 | 0 | 0 | 0 |
| $T_d$ | $A_1$ | $A_2$ | E | $T_1$ | $T_2$ |
| # | 1 | 1 | 2 | 3 | 3 |
| XYZW | 1 | | | | |
| $T_d$ | $A_1$ | $A_2$ | E | $T_1$ | $T_2$ |
| # | 1 | 1 | 2 | 3 | 3 |

**Figure 4.6** Composite of the GT_Calculator displays illustrating equation 4.67.

| README | $o_1$ | $o_4$ | $o_6$ | $o_{12}$ | $o_{24}$ |
|---|---|---|---|---|---|
| # | 0 | 1 | 0 | 0 | 0 |
| $T_d$ | E | $8C_3$ | $3C_2$ | $6C_4$ | $6\sigma_d$ |
| $\Gamma_\sigma$ | 4 | 1 | 0 | 0 | 2 |
| $T_d$ | $A_1$ | $A_2$ | E | $T_1$ | $T_2$ |
| # | 1 | 0 | 0 | 0 | 1 |
| Decoration | # | | | | |
| xxxx | 1 | | | | |
| $T_d$ | E | $8C_3$ | $3C_2$ | $6C_4$ | $6\sigma_d$ |
| $\Gamma_\sigma$ | 1 | 1 | 1 | 1 | 1 |
| $T_d$ | $A_1$ | $A_2$ | E | $T_1$ | $T_2$ |
| # | 1 | 0 | 0 | 0 | 0 |

**Figure 4.7** Composite of the GT_Calculator displays illustrating equation 4.68.

# Appendix 1

## The icosahedral harmonics

While the appropriate linear combinations of the spherical harmonics known as the *kubic* harmonics, which transform in cubic symmetry, have been known for many years from the seminal works of Bethe and van Vleck, we believe that the icosahedral harmonics available in the basis functions lists in the files Ih.xls and I.xls on the CDROM have not been identified, in this form, in the literature. Thus, for the record, this appendix contains the complete list of such harmonic functions, classified as basis functions for the irreducible representations of the groups I and $I_h$.

These functions have the appropriate transformation properties for applications in calculation, as illustrated on the CDROM in the 'BonusPack', where sets of Hückel theory calculations can be found for the energy spectra of regular orbits of I and $I_h$ point symmetry decorated with $H_{1s}$ atomic orbitals. It is to be noted that, while the continuous functions are mutually orthonormal on the unit sphere, this property is not maintained in their discrete samplings on the 60 and 120-vertex orbit cages, and so a further orthogonalization transformation is required to restore orthogonality.

As explained in Chapter 1, page 24, the polynomials are listed using Elert's compact notation.

**Table A1.1 The icosahedral harmonics, fashioned to be basis functions for all the irreducible representations of the regular orbit cage of $I_h$ point symmetry. The polynomial functions are written in Elert's notation, as described in Chapter 1 and their irreducible properties under $I_h$ and I are identified, in columns 1 and 2 of the table, using Mulliken symbols.**

| $I_h$ | I | Basis Functions |
|---|---|---|
| $1A_g$ | 1A | (000) |
| $1T_{1g[a]}$ | $3T_{1[a]}$ | 5(411) − 10(231) + (051) |
| $1T_{1g[b]}$ | $3T_{1[b]}$ | 66(105) − 60(103) + 10(101) − 11(402) + 66(222) − 11(042) + (400) − 6(220) + (040) + (600) − 15(420) + 15(240) − (060) |
| $1T_{1g[c]}$ | $3T_{1[c]}$ | 33(015) − 30(013) + 5(011) + 22(312) − 22(132) − 2(310) + 2(130) + 3(510) − 10(330) + 3(150) |
| $2T_{1g[a]}$ | – | 4845(415) − 9690(235) + 969(055) − 2550(413) + 5100(233) − 510(053) + 225(411) − 450(231) + 45(051) − 85(910) + 1020(730) − 2142(550) + 1020(370) − 85(190) |

| | | |
|---|---|---|
| $2T_{1g[b]}$ | – | 14535(424) + 612(721) + 54(220) – 2142(541) + 1428(361) – 17(901) – 9(600) + 9(060) – 17442(226) + 2907(406) – 2295(404) + 2907(046) – 306(062) + 969(064) + 306(602) + 4590(242) – 4590(422) – 14535(244) – 2430(222) + 13770(224) + 405(042) – 969(604) – 2295(044) + 405(402) + 63(101) + 4199(109) – 7956(107) + 4914(105) – 1092(103) – 153(181) + 135(420) – 135(240) – 9(400) – 9(040) |
| $2T_{1g[c]}$ | – | 1620(132) + 2142(451) – 612(271) – 1620(312) + 17(091) – 5814(514) + 19380(334) – 5814(154) + 1836(512) – 6120(332) + 1836(152) + 4199(019) – 7956(017) + 4914(015) – 1092(013) + 63(011) + 9180(314) + 11628(136) – 11628(316) + 36(310) – 9180(134) – 36(130) + 153(811) – 1428(631) – 54(510) + 180(330) – 54(150) |
| $3T_{1g[a]}$ | – | 28980(415) – 57960(235) – 4200(413) + 5796(055) – 840(053) – 15(910) + 180(730) – 300(231) + 150(411) + 30(051) + 10350(059) – 13800(057) – 15(190) + 8400(233) – 378(550) + 180(370) + 51750(419) – 103500(239) – 69000(417) + 138000(237) – 3375(914) + 40500(734) – 85050(554) + 40500(374) – 3375(194) + 750(912) – 9000(732) + 18900(552) – 9000(372) + 750(192) |
| $3T_{1g[b]}$ | – | –2(040) – 3003(A40) – 1(060) + 1(600) – 3024(541) – 24(901) + 84(062) – 84(602) – 966(064) + 864(721) + 966(604) + 3220(066) – 3220(606) – 3105(068) + 3105(608) – 33600(363) + 12236(046) + 12236(406) + 15960(224) – 73416(226) – 1260(242) + 1260(422) + 14490(244) – 2660(404) – 14490(424) – 48300(246) + 48300(426) + 46575(248)400(903) – 46575(428) – 1080(905) + 190(042) + 131100(228) – 78660(22A) – 44(1A1) + 190(402) – 2660(044) + 660(381) – 21850(048) – 1848(561) + 1320(741) – 220(921) – 21850(408) + 13110(04A) + 13110(40A) + 396(1A3) – 5940(383) + 16632(563) + 4(B01) – 11880(743) + 3600(183) + 1980(923) – 216(181) + 2016(361) – 36(B03) – 1140(222) + 50400(543) – 14400(723) – 9720(185) + 90720(365) – 136080(545) + 38880(725) + 12(220) + 15(240) – 15(420) – 273(2C0) – 9009(680) + 3003(4A0) + 273(C20) + 9009(860) – 3(E00) + 3(0E0) – 2(400) |
| $3T_{1g[c]}$ | – | 87400(318) – 87400(138) + 52440(13A) – 52440(31A) – 4(0E1) – 16632(653) – 8(130) + 8(310) + 18630(518) – 760(312) + 760(132) + 10640(314) – 10640(134) – 660(831) + 220(291) + 5940(833) – 1980(293) + 36(0E3) + 11880(473) + –396(A13) + 3024(451) – 864(271) + 24(091) + 216(811) + 14400(273) – 2016(631) – 3600(813) – 50400(453) + 33600(633) + 136080(455) – 38880(275) + 9720(815) – 90720(635) + 1080(095) – 504(512) – 504(152) + 1680(332) – 19320(334) + 5796(154) + 64400(336) – 19320(156) + 5796(514) – 62100(338) + 18630(158) – 19320(516) – 48944(316) + 48944(136) + 1848(651) – 1320(471) + 44(A11) + 6(150) – 20(330) + 6(510) + 6006(590) – 1092(3E0) + 42(D10) – 10296(770) – 1092(B30) + 42(1D0) + 6006(950) – 400(093) |
| $1T_{2g[a]}$ | – | 25(413) – 50(233) + 5(053) – 5(411) + 10(231) – (051) |

| | | |
|---|---|---|
| $1T_{2g[b]}$ | – | 143(206) − 143(026) − 143(204) + 143(024) + 33(202) − 33(022) − (200) + (020) − 39(305) + 117(125) + 26(303) − 78(123) − 3(301) + 9(121) − 3(701) + 63(521) − 105(341) + 21(161) − (800) + 28(620) − 70(440) + 28(260) − (080) |
| $1T_{2g[c]}$ | – | 286(116) − 286(114) + 66(112) − 2(110) + 117(215) − 39(035) − 78(213) + 26(033) + 9(211) − 3(031) − 21(611) + 105(431) − 63(251) + 3(071) + 8(710) − 56(530) + 56(350) − 8(170) |
| $2T_{2g[a]}$ | – | 6460(415) − 12920(235) + 1292(055) − 3400(413) + 6800(233) − 680(053) + 300(411) − 600(231) + 60(051) + 95(910) − 1140(730) + 2394(550) − 1140(370) + 95(190) |
| $2T_{2g[b]}$ | – | 336(121) − 112(301) + 3724(262) − 15504(127) + 10374(523) − 494(703)+7(800)+7(080)+3458(163)−17290(343)+78(701)− 5712(305) − 546(161) + 2730(341) − 9310(442) − 196(620) + 490(440) − 196(260) + 17136(125) + 3724(622) − 133(802) + 7(200) − 7(020) − 5040(123) + 1680(303) + 4199(208) − 133(082) + 6188(026) + 2730(204) − 2730(024) − 364(202) + 364(022) − 4199(028) − 6188(206) − 1638(521) + 5168(307) |
| $2T_{2g[c]}$ | – | 17136(215) − 5040(213) − 5712(035) − 12376(116) + 336(211) + 5460(114) + 8398(118) − 112(031) + 1638(251) + 7448(352) + 14(110) − 56(710) + 392(530) − 392(350) + 56(170) − 78(071) + 494(073) − 15504(217) + 5168(037) − 3458(613) + 17290(433) − 10374(253) + 546(611) + 1680(033) − 7448(532) − 1064(172) − 728(112) + 1064(712) − 2730(431) |
| $3T_{2g[a]}$ | – | 6555(417) − 13110(237) + 1311(057) − 5985(415) + 11970(235) − 1197(055) + 1425(413) − 2850(233) + 285(053) − 75(411) + 150(231) − 15(051) − 115(912) + 1380(732) − 2898(552) + 1380(372) − 115(192) + 5(910) − 60(730) + 5(910) − 60(730) + 126(550) − 60(370) + 5(190) |
| $T_{2g[b]}$ | – | 3(020) − 3(200) − 238(620) + 595(440) − 238(260) + 17/2(800) − 23/2(0C0) + 759(A20) − 11385/2(840) + 10626(660) − 11385/2(480) + 759(2A0) − 23/2(C00) + 95795(444) + 2040(303)+7429(20A)−69768(127)−357(802)+2737/2(804)+ 9690(206) − 6120(123) − 7429(02A) + 225(202) − 14535(208) − 11628(305) − 9690(026) − 24990(442) − 14858(309) + 14535(028) + 23256(307) + 9996(262) + 44574(129) − 3920(163) + −3920(163) + 34884(125) − 11760(523) − 40(701) + 19600(343) + 9016(165) + 280(161) + 560(703) − 2550(204) + 840(521) + 9996(262) + 270(121) + 2550(024) + 27048(525) − 1400(341) − 90(301) − 38318(264) − 38318(624) − 1288(705) − 45080(345) + 2737/2(084) − 357(082) − 225(022) + 17/2(080) |
| $3T_{2g[c]}$ | – | 2856(712) − 14858(039) − 6120(213) − 476(350) + 68(170) + 476(530) + 23256(037) − 68(710) + 44574(219) + 9108(570) − 9108(750) + 2530(930) − 138(B10) + 270(211) + 138(1B0) − 2530(390) + 76636(534) − 840(251) − 76636(354) + 10948(174) + 40(071) + 11760(253) − 280(611) + 76636(534) − 840(251) − 76636(354) + 10948(174) + 40(071) + 11760(253) − 280(611) + 1400(431) − 560(073) + 1288(075) − 27048(255) + 45080(435) + 2040(033) − 90(031) − 9016(615) − 11628(035) + 34884(215) + 450(112) + 14858(11A) − 2856(172) − 29070(118) − 19992(532) − 10948(714) + 3920(613)− 19600(433) + 19380(116) − 6(110) − 5100(114) − 69768(217) |

| | | |
|---|---|---|
| $1G_{g[a]}$ | $2G_{[a]}$ | (400) − 6(220) + (040) + 7(103) − 3(101) |
| $1G_{g[b]}$ | $2G_{[b]}$ | 4(310) − 4(130) − 7(013) + 3(011) |
| $1G_{g[c]}$ | $2G_{[c]}$ | (301) − 3(121) − 7(202) + 7(022) + (200) − (020) |
| $1G_{g[d]}$ | $2G_{[d]}$ | 3(211) − (031) + 14(112) − 2(110) |
| $2G_{g[a]}$ | $3G_{[a]}$ | 264(105) − 240(103) + 40(101) + 66(402) − 396(222) + 66(042) − 6(400) + 36(220) − 6(040) − 11(600) + 165(420) − 165(240) + 11(060) |
| $2G_{g[b]}$ | $3G_{[b]}$ | 132(015) − 120(013) + 20(011) − 132(312) + 132(132) + 12(310) − 12(130) − 33(510) + 110(330) − 33(150) |
| $2G_{g[c]}$ | $3G_{[c]}$ | 33(204) − 33(024) − 18(202) + 18(022) + (200) − (020) − 44(303) + 132(123) + 12(301) − 36(121) |
| $2G_{g[d]}$ | $3G_{[d]}$ | 33(114) − 18(112) + (110) + 66(213) − 22(033) − 18(211) + 6(031) |
| $3G_{g[a]}$ | – | 1430(107) − 2002(105) + 770(103) − 70(101) − 455(404) + 2730(224) − 455(044) + 182(402) − 1092(222) + 182(042) − 7(400) + 42(220) − 7(040) − 195(602) + 2925(422) − 2925(242) + 195(062) + 13(600) − 195(420) + 195(240) − 13(060) |
| $3G_{g[b]}$ | – | 1430(017) − 2002(015) + 770(013) − 70(011) + 1820(314) − 1820(134) − 728(312) + 728(132) + 28(310) − 28(130) − 1170(512) + 3900(332) − 1170(152) + 78(510) − 260(330) + 78(150) |
| $3G_{g[c]}$ | – | 1001(206) − 1001(026) − 1001(204) + 1001(024) + 231(202) − 231(022) − 7(200) + 7(020) + 1092(305) − 3276(125) − 728(303) + 2184(123) + 84(301) − 252(121) − 26(701) + 546(521) − 910(341) + 182(161) + 13(800) − 364(620) + 910(440) − 364(260) + 13(080) |
| $3G_{g[d]}$ | – | 2002(116) − 2002(114) + 462(112) − 14(110) − 3276(215) + 1092(035) + 2184(213) − 728(033) − 252(211) + 84(031) − 182(611) + 910(431) − 546(251) + 26(071) − 104(710) + 728(530) − 728(350) + 104(170) |
| $4G_{g[a]}$ | – | 48620(600) + 5408312(903) + 663390(220) − 110565(040) + 729300(240) − 729300(420) + 428549940(224) + 290990700(244) − 290990700(424) − 446185740(246) + 446185740(426) − 1199939832(226) + 985664862(228) − 45110520(222) − 6348888(181) + 59256288(361) − 88884432(541) + 25395552(721) + 48674808(183) − 454298208(363) + 681447312(543) − 194699232(723) − 12252240(105) + 7518420(042) + 7518420(402) + 1801800(103) + 29745716(066) − 29745716(606) − 72072(101) + 199989972(046) + 199989972(406) − 164277477(048) − 164277477(408) − 705432(901) − 19399380(064) + 19399380(604) − 38798760(109) + 16224936(10B) − 48620(060) + 2771340(062) − 41570100(242) + 41570100(422) − 2771340(602) − 71424990(404) − 71424990(044) + 33256080(107) − 110565(400) − 71424990(044) + 33256080(107) − 110565(400) |

| | | |
|---|---|---|
| $4G_{g[b]}$ | – | +157080(015) − 23100(013) + 497420(019) − 426360(017) − 1515516(138) + 1515516(318) − 1020(130) + 1020(310) + 720(150) − 2400(330) + 720(510) − 208012(01B) + 924(011) − 294360(471) + 412104(651) + 9812(A11) − 892(0B1) + 49060(291) − 147180(831) + 96768(451) − 27648(271) + 768(091) + 6912(811) − 5888(093) + 211968(273) − 64512(631) − 52992(813) + 494592(633) − 741888(453) + 136800(332) − 41040(152) − 957600(334) + 287280(154) − 41040(512) − 440496(156) + 1468320(336) + 287280(514) + 69360(132) − 440496(516) − 658920(134) − 69360(312) − 1844976(316) + 658920(314) + 1844976(136) |
| $4G_{g[c]}$ | – | 226(800) − 326876(20A) + 326876(02A) + 639540(208) − 639540(028) − 426360(206) + 426360(026) + 112200(204) − 112200(024) − 9900(202) + 9900(022) + 267444(129) − 418608(127) + 209304(125) − 36720(123) + 1620(121) − 89148(309) + 139536(307) − 69768(305) + 12240(303) − 540(301) − 6328(620) − 6328(260) + 15820(440) + 40572(705) − 17640(703)+1260(701)+36386(804)+36386(084)−9492(802)− 9492(082)+132(200)−132(020)−852012(525)+1420020(345)− 284004(165) + 370440(523) − 617400(343) + 123480(163) − 26460(521) + 44100(341) − 8820(161) − 1018808(624) + 2547020(444) − 1018808(264) + 265776(622) − 664440(442) + 265776(262) + 281(0C0) + 81(C00) + 226(080) − 259644(660) − 18546(2A0) + 139095(480) − 18546(A20) + 139095(840) |
| $4G_{g[d]}$ | – | 617400(433) − 123480(613) + 852012(255) − 1420020(435) − 370440(253) + 26460(251) − 44100(431) + 8820(611) + 2037616(534) − 291088(714) − 75936(172) + 531552(352) − 531552(532) + 75936(712) + 291088(174) − 2037616(354) + 17640(073) − 1260(071) + 284004(615) − 40572(075) + 1808(170) − 12656(350) + 12656(530) − 1808(710) + 264(110) − 222552(570) − 3372(1B0) + 61820(390) + 222552(750) + 3372(B10) − 61820(930) + 267444(219) − 418608(217) + 209304(215) − 36720(213) + 1620(211) − 89148(039) + 139536(037) − 69768(035) − 540(031) − 653752(11A) + 1279080(118) − 852720(116) + 224400(114) − 19800(112) + 12240(033) |
| $1H_{g[a]}$ | $1H_{[a]}$ | 2(002) − (200) − (020) |
| $1H_{g[b]}$ | $1H_{[b]}$ | (101) |
| $1H_{g[c]}$ | $1H_{[c]}$ | (011) |
| $1H_{g[d]}$ | $1H_{[d]}$ | (200) − (020) |
| $1H_{g[e]}$ | $1H_{[e]}$ | (110) |
| $2H_{g[a]}$ | $2H_{[a]}$ | 35(004) − 30(002) + 3(000) |
| $2H_{g[b]}$ | $2H_{[b]}$ | 7(400) − 42(220) + 7(040) − 56(103) + 24(101) |
| $2H_{g[c]}$ | $2H_{[c]}$ | 7(310) − 7(130) + 14(013) − 6(011) |
| $2H_{g[d]}$ | $2H_{[d]}$ | 14(301) − 42(121) + 7(202) − 7(022) − (200) + (020) |
| $2H_{g[e]}$ | $2H_{[e]}$ | 21(211) − 7(031) − 7(112) + (110) |

| | | |
|---|---|---|
| $3H_{g[a]}$ | $4H_{[a]}$ | 231(006) − 315(004) + 105(002) − 5(000) − 33(501) + 330(321) − 165(141) |
| $3H_{g[b]}$ | $4H_{[b]}$ | 66(105) − 60(103) + 10(101) + 99(402) − 594(222) + 99(042) − 9(400) + 54(220) − 9(040) + 11(600) − 165(420) + 165(240) − 11(060) |
| $3H_{g[c]}$ | $4H_{[c]}$ | 33(015) − 30(013) + 5(011) − 198(312) + 198(132) + 18(310) − 18(130) + 33(510) − 110(330) + 33(150) |
| $3H_{g[d]}$ | $4H_{[d]}$ | 33(204) − 33(024) − 18(202) + 18(022) + (200) − (020) + 11(303) − 33(123) − 3(301) + 9(121) |
| $3H_{g[e]}$ | $4H_{[e]}$ | 66(114) − 36(112) + 2(110) − 33(213) + 11(033) + 9(211) − 3(031) |
| $4H_{g[a]}$ | – | 6435(008) − 12012(006) + 6930(004) − 1260(002) + 35(000) − 1080(503) + 10800(323) − 5400(143) + 216(501) − 2160(321) + 1080(141) |
| $4H_{g[b]}$ | – | 195(404) − 1170(224) + 195(044) − 78(402) + 468(222) − 78(042) + 3(400) − 18(220) + 3(040) − 30(602) + 450(422) − 450(242) + 30(062) + 2(600) − 30(420) + 30(240) − 2(060) |
| $4H_{g[c]}$ | – | 780(314) − 780(134) − 312(312) + 312(132) + 12(310) − 12(130) + 180(512) − 600(332) + 180(152) − 12(510) + 40(330) − 12(150) |
| $4H_{g[d]}$ | – | 1716(206) − 1716(026) − 1716(204) + 1716(024) + 396(202) − 396(022) − 12(200) + 12(020) − 2028(305) + 6084(125) + 1352(303) − 4056(123) − 156(301) + 468(121) + 84(701) − 1764(521) + 2940(341) − 588(161) + 33(800) − 924(620) + 2310(440) − 924(260) + 33(080) |
| $4H_{g[e]}$ | – | 3432(116) − 3432(114) + 792(112) − 24(110) + 6084(215) − 2028(035) − 4056(213) + 1352(033) + 468(211) − 156(031) + 588(611) − 2940(431) + 1764(251) − 84(071) − 264(710) + 1848(530) − 1848(350) + 264(170) |
| $5H_{g[a]}$ | – | 92378(00A) − 218790(008) + 180180(006) − 60060(004) + 6930(002) − 126(000) − 247(A00) + 11115(820) − 51870(640) + 51870(460) − 11115(280) + 247(0A0) |
| $5H_{g[b]}$ | – | −516(220) − 1368(721) + 23220(222) − 131580(224) + 166668(226) + 86(400) + 86(040) + 111435(424) − 111435(244) − 35190(422) + 4788(541) − 3192(361) + 2346(602) + 21930(404) − 27778(406) − 3870(042) − 27778(046) + 342(181) + 35190(242) − 2346(062) + 38(901) − 7429(604) + 7429(064) + 21930(044) − 3870(402) + 29484(105) − 6552(103) + 378(101) + 25194(109) − 47736(107) − 69(600) + 69(060) + 1035(420) − 1035(240) |
| $5H_{g[c]}$ | – | 344(130) − 344(310) − 414(150) + 1380(330) − 414(510) + 111112(316) − 4788(451) + 1368(271) − 38(091) + 3192(631) + 148580(334) − 46920(332) + 14076(152) − 44574(154) + 14076(512) + 15480(312) − 15480(132) − 44574(514) − 111112(136) + 87720(134) − 87720(314) − 342(811) − 6552(013) + 29484(015) − 47736(017) + 25194(019) + 378(011) |
| $5H_{g[d]}$ | – | −67(800) − 3234(161) + 16170(341) − 9702(521) + 20482(163) − 102410(343) + 61446(523) − 35644(262) + 89110(442) − 35644(622) − 2926(703) + 462(701) + 24570(204) + 55692(026) − 55692(206) − 3276(202) + 3276(022) − 37791(028) + 37791(208) − 7980(303) − 24548(307)+ |

| | | |
|---|---|---|
| | | 27132(305) + 532(301) − 24570(024) − 4690(440) + 1876(260) + 1876(620) − 81396(125) − 1596(121) + 23940(123) + 73644(127) + 63(200) − 63(020) + 1273(082) − 67(080) + 1273(802) |
| $H_{g[e]}$ | – | 126(110) + 73644(217) − 81396(215) + 23940(213) − 1596(211) − 24548(037) + 27132(035) − 7980(033) + 532(031) + 536(710) − 3752(530) + 3752(350) − 536(170) + 75582(118) − 111384(116) + 49140(114) − 6552(112) + 2926(073) − 20482(613) + 102410(433) − 61446(253) + 3234(611) − 16170(431) + 9702(251) − 10184(712) + 71288(532) − 71288(352) + 10184(172) − 462(071) |
| $1A_u$ | – | 187920(375) + 150075(41A) − 300150(23A) − 232875(418) + 465750(238) + 120750(416) + 3(0F0) − 15660(195) − 3(050) − 7560(551) + 3600(371) − 64800(733) + 5400(913) + 30(230) + 136080(553) − 300(191) + 30015(05A) − 46575(058) + 24150(056) − 4830(054) + 315(052) + 4095(4B0) + 1365(C30) + 19305(870) − 9009(A50) − 45(E10) − 24150(414) + 315(052) + 4095(4B0) + 1365(C30) + 19305(870) − 9009(A50) − 45(E10) − 24150(414) − 315(2D0) − 15015(690) − 15(410) − 394632(555) + 187920(735) + 3600(731) |
| $1T_{1u[a]}$ | $1T_{1[a]}$ | (001) |
| $1T_{1u[b]}$ | $1T_{1[b]}$ | (100) |
| $1T_{1u[c]}$ | $1T_{1[c]}$ | (010) |
| $2T_{1u[a]}$ | $2T_{1[a]}$ | 126(005) − 140(003) + 30(001) + 7(500) − 70(320) + 35(140) |
| $2T_{1u[b]}$ | $2T_{1[b]}$ | 21(104) − 14(102) + (100) − 7(401) + 42(221) − 7(041) |
| $2T_{1u[c]}$ | $2T_{1[c]}$ | 21(014) − 14(012) + (010) + 28(311) − 28(131) |
| $3T_{1u[a]}$ | – | 12155(009) − 25740(007) + 18018(005) − 4620(003) + 315(001) − 8415(504) + 84150(324) − 315(2D0) − 15015(690) − 15(410) − 394632(555) + 187920(735) + 3600(731) |
| $3T_{1u[b]}$ | – | 83160(223) + 396(601) + 2244(063) − 2244(603) + 1386(041) + 1386(401) − 13860(043) − 141372(225) + 23562(405) + 5940(241) − 5940(421) − 33660(243) + 33660(423) − 8316(221) − 13860(403) + 23562(045) − 396(061) + 63(100) − 1683(180) + 15708(360) − 23562(540) + 6732(720) + 18018(104) − 36036(106) + 21879(108) − 2772(102) − 187(900) |
| $3T_{1u[c]}$ | – | +21879(018) − 36036(016) + 18018(014) − 2772(012) + 63(010) − 94248(315) + 94248(135) + 55440(313) − 55440(133) − 5544(311) + 5544(131) − 13464(513) + 44880(333) − 13464(153) + 2376(511) − 7920(331) + 2376(151) + 1683(810) − 15708(630) + 23562(450) − 6732(270) + 187(090) |
| $1T_{2u[a]}$ | $1T_{2[a]}$ | 5(003) − 3(001) |
| $1T_{2u[b]}$ | $1T_{2[b]}$ | (300) − 3(120) + 3(201) − 3(021) |
| $1T_{2u[c]}$ | $1T_{2[c]}$ | 3(210) − (030) − 6(111) |
| $2T_{2u[a]}$ | $2T_{2[a]}$ | 63(005) − 70(003) + 15(001) − 9(500) + 90(320) − 45(140) |

| | | |
|---|---|---|
| $2T_{2u[b]}$ | $2T_{2[b]}$ | 18(203) − 18(023) − 6(201) + 6(021) − 9(302) + 27(122) + (300) − 3(120) |
| $2T_{2u[c]}$ | $2T_{2[c]}$ | 36(113) − 12(111) + 27(212) − 9(032) − 3(210) + (030) |
| $3T_{2u[a]}$ | – | 3432(007) − 5544(005) + 2520(003) − 280(001) − 1001(502) + 10010(322) − 5005(142) + 77(500) − 770(320) + 385(140) |
| $3T_{2u[b]}$ | – | 2002(205) − 2002(025) − 1540(203) + 1540(023) + 210(201) − 210(021) + 7007(304) − 21021(124) − 3234(302) + 9702(122) + 147(300) − 441(120) − 143(700) + 3003(520) − 5005(340) + 1001(160) |
| $3T_{2u[c]}$ | – | 4004(115) − 3080(113) + 420(111) − 21021(214) + 7007(034) + 9702(212) − 3234(032) − 441(210) + 147(030) − 1001(610) + 5005(430) − 3003(250) + 143(070) |
| $1G_{u[a]}$ | $1G_{[a]}$ | 5(102) − (100) |
| $1G_{u[b]}$ | $1G_{[b]}$ | 5(012) − (010) |
| $1G_{u[c]}$ | $1G_{[c]}$ | (300) − 3(120) − 2(201) + 2(021) |
| $1G_{u[d]}$ | $1G_{[d]}$ | 3(210) − (030) + 4(111) |
| $2G_{u[a]}$ | $2G_{[a]}$ | 429(106) − 495(104) + 135(102) − 5(100) + 286(403) − 1716(223) + 286(043) − 66(401) + 396(221) − 66(041) − 11(601) + 165(421) − 165(241) + 11(061) |
| $2G_{u[b]}$ | $2G_{[b]}$ | 429(016) − 495(014) + 135(012) − 5(010) − 1144(313) + 1144(133) + 264(311) − 264(131) − 66(511) + 220(331) − 66(151) |
| $2G_{u[c]}$ | $2G_{[c]}$ | 429(205) − 429(025) − 330(203) + 330(023) + 45(201) − 45(021) − 286(304) + 858(124) + 132(302) − 396(122) − 6(300) + 18(120) − 11(700) + 231(520) − 385(340) + 77(160) |
| $2G_{u[d]}$ | $2G_{[d]}$ | 858(115) − 660(113) + 90(111) + 858(214) − 286(034) − 396(212) + 132(032) + 18(210) − 6(030) − 77(610) + 385(430) − 231(250) + 11(070) |
| $3G_{u[a]}$ | – | 102(405) − 612(225) + 102(045) − 60(403) + 360(223) − 60(043) + 6(401) − 36(221) + 6(041) + 34(603) − 510(423) + 510(243) − 34(063) − 6(601) + 90(421) − 90(241) + 6(061) + (900) − 36(720) + 126(540) − 84(360) + 9(180) |
| $3G_{u[b]}$ | – | 408(315) − 408(135) − 240(313) + 240(133) + 24(311) − 24(131) − 204(513) + 680(333) − 204(153) + 36(511) − 120(331) + 36(151) + 9(810) − 84(630) + 126(450) − 36(270) + (090) |
| $3G_{u[c]}$ | – | 6(120) + 3(700) − 1326(126) + 1170(124) − 234(122) + 1071(522) − 1785(342) + 357(162) − 42(201) + 42(021) + 442(306) − 390(304) + 78(302) − 51(702) + 1638(025) + 546(203) − 546(023) + 1326(207) − 1326(027) − 1638(205) − 21(160) − 2(300) + 105(340) − 63(520) − 12(081) + 336(621) − 840(441) − 12(801) + 336(261) |
| $3G_{u[d]}$ | – | 2652(117) − 3276(115) + 1092(113) − 84(111) − 1326(216) + 442(036) + 1170(214) − 390(034) − 234(212) + 78(032) + 6(210) − 2(030) − 357(612) + 1785(432) − 1071(252) + 51(072) + 21(610) − 105(430) + 63(250) − 3(070) + 96(711) − 672(531) + 672(351) − 96(171) |

| | | |
|---|---|---|
| $4G_{u[a]}$ | – | 741150(421) – 307440(221) + 5226480(223) + 51240(041) + 19860624(227)3(070) + 96(711) – 672(531) + 672(351) – 96(171) + 6133428(542) – 1752408(722) – 741150(241) – 871080(403) – 871080(043) + 51240(401) – 49410(601) + 49410(061) – 1281(100) – 20862(180) + 194712(360) – 292068(540) + 83448(720) – 625860(063) – 1338645(380) – 8113(B00) + 89243(1A0) + 3748206(560) – 2677290(740) + 446215(920) – 2318(900) – 9387900(423) + 9387900(243) + 625860(603) – 19860624(225) + 3310104(045) + 3310104(405) + 83265(102) + 1792973(10A) – 3310104(407) – 3842085(108) – 3310104(047) + 48678(902) + 1314306(065) + 2831010(106) – 832650(104) – 1314306(605) |
| $4G_{u[b]}$ | – | –160260459(090) + 84296919(010) + 13461878556(630) – 20192817834(450) + 5769376524(270) – 1442344131(810) – 93033084375(830) + 260492636250(650) – 186066168750(470) + 31011028125(290) + 6202205625(A10) – 117987587627(01A) + 252830544915(018) – 186296190990(016) + 54792997350(014) – 5479299735(012) + 3365469639(092) – 282699449676(632) + 545214008052(155) – 259625718120(513) + 865419060400(333) – 259625718120(153) + 20496767220(511) – 68322557400(331) + 20496767220(151) + 30289226751(812) + 424049174514(452) – 121156907004(272) + 920044890016(137) + 920044890016(315) – 920044890016(135) – 242117076320(313) + 242117076320(133) + 14242180960(311) – 14242180960(131) + 545214008052(515) – 1817380026840(335) – 920044890016(317) |
| $4G_{u[c]}$ | – | 315(340) – 63(160) – 189(520) – 42(201) – 840(023) + 840(203) + 4284(025) – 4284(205) – 7752(027) + 7752(207) + 4522(029) + 42(021) + 63(120) + 32130(124) – 81396(126) + 61047(128) + 1260(302) – 10710(304) + 27132(306) – 20349(308) – 3780(122) – 4522(209) – 21(300) – 342(702) + 1064(083) + 1064(803) – 152(801) – 152(081) + 1197(704) – 11970(342) + 41895(344) + 2394(162) + 4256(621) – 10640(441) – 25137(524) – 8379(164) + 7182(522) + 4256(261) – 29792(623) + 74480(443) – 29792(263) + 9(700) |
| $4G_{u[d]}$ | – | –9(070) – 10710(034) + 1260(032) + 61047(218) – 81396(216) + 32130(214) – 3780(212) – 20349(038) + 27132(036) + 63(610) – 315(430) + 189(250) – 21(030) – 8568(115) + 1680(113) + 15504(117) – 84(111) – 9044(119) + 342(072) – 7182(252) – 8512(713) + 59584(533) – 59584(353) + 8512(173) + 1216(711) – 8512(531) + 8512(351) – 1216(171) – 1197(074) + 8379(614) + 1216(711) – 8512(531) + 8512(351) – 1216(171) – 1197(074) + 8379(614) |
| $1H_{u[a]}$ | $3H_{[a]}$ | 5(410) – 10(230) + (050) |
| $1H_{u[b]}$ | $3H_{[b]}$ | 21(104) – 14(102) + (100) + 3(401) – 18(221) + 3(041) |
| $1H_{u[c]}$ | $3H_{[c]}$ | 21(014) – 14(012) + (010) – 12(311) + 12(131) |

| | | |
|---|---|---|
| $1H_{u[d]}$ | $3H_{[d]}$ | 12(203) − 12(023) − 4(201) + 4(021) + 9(302) − 27(122) − (300) + 3(120) |
| $1H_{u[e]}$ | $3H_{[e]}$ | 24(113) − 8(111) − 27(212) + 9(032) + 3(210) − (030) |
| $2H_{u[a]}$ | $5H_{[a]}$ | 65(412) − 130(232) + 13(052) − 5(410) + 10(230) − (050) |
| $2H_{u[b]}$ | $5H_{[b]}$ | 429(106) − 495(104) + 135(102) − 5(100) − 104(403) + 624(223) − 104(043) + 24(401) + 1216(711) − 8512(531) + 8512(351) − 1216(171) − 1197(074) + 8379(614) |
| $2H_{u[c]}$ | $5H_{[c]}$ | 429(016) − 495(014) + 135(012) − 5(010) + 416(313) − 416(133) − 96(311) + 96(131) − 156(511) + 520(331) − 156(151) |
| $2H_{u[d]}$ | $5H_{[d]}$ | 858(205) − 858(025) − 660(203) + 660(023) + 90(201) − 90(021) + 143(304) − 429(124) − 66(302) + 198(122) + 3(300) − 9(120) + 13(700) − 273(520) + 455(340) − 91(160) |
| $2H_{u[e]}$ | $5H_{[e]}$ | 1716(115) − 1320(113) + 180(111) − 429(214) + 143(034) + 198(212) − 66(032) − 9(210) + 3(030) + 91(610) − 455(430) + 273(250) − 13(070) |
| $3H_{u[a]}$ | – | 425(414) − 850(234) + 85(054) − 150(412) + 300(232) − 30(052) + 5(410) − 10(230) + (050) |
| $3H_{u[b]}$ | – | 1260(241) − 1260(421) + −4004(106) + 2431(108) − 308(102) + 2002(104) + 7(100) + 238(405) + 238(045) − 140(403) − −140(043)+14(401)+14(041)−476(603)+476(063)+84(601)− 1428(225) + 840(223) − 84(221) + 7140(423) − 7140(243) − 612(720) − 1428(360) + 153(180) + 2142(540) + 17(900) − 84(061) |
| $3H_{u[c]}$ | – | 2431(018) − 4004(016) + 2002(014) − 308(012) + 7(010) − 952(315) + 952(135) + 560(313) − 560(133) − 56(311) + 56(131) − 2856(513) + 9520(333) − 2856(153) + 504(511) − 1680(331) + 504(151) − 153(810) + 1428(630) − 2142(450) + 612(270) − 17(090) |
| $3H_{u[d]}$ | – | 2(700)−21(120)+4641(126)−4095(124)+819(122)+714(522)− 1190(342) + 238(162) − 28(201) + 28(021) − 1547(306) + 1365(304) − 273(302) − 34(702) + 1092(025) + 364(203) − 364(023)+884(207)−884(027)−1092(205)−14(160)+7(300)+ 70(340) − 42(520) + 17(081) + 1190(441) + 17(081) − 476(621) |
| $3H_{u[e]}$ | – | 1768(117) − 2184(115) + 728(113) − 56(111) + 4641(216) − 1547(036) − 4095(214) + 1365(034) + 819(212) − 273(032) − 21(210) + 7(030) − 238(612) + 1190(432) − 714(252) + 34(072) + 14(610) − 70(430) + 42(250) − 2(070) − 136(711) + 952(531) − 952(351) + 136(171) |
| $4H_{u[a]}$ | – | 11305(416) − 22610(236) + 2261(056) − 8075(414) + 16150(234) − 1615(054) + 1275(412) − 2550(232) + 255(052) − 25(410) + 50(230) − 5(050) |
| $4H_{u[b]}$ | – | 675(421) − 45(601) − 570(063) + 45(061) + 17955(425) − 2268(182) + 11628(407) + 1197(065) + 9072(722) + 8550(243) − 180(041) − 8550(423) + 570(603) − 675(241) + 21168(362) − 17955(245) − 18360(223) − 180(401) − 11628(405) + 3060(043) + 11628(047) − 11628(045) − 31752(542) − 252(902) + 1080(221) + 69768(225) + 3060(403) − 1197(605) − 69768(227) − 432(720)+ |

| | | |
|---|---|---|
| | | 1512(540) – 1008(360) + 108(180) + 12474(560) – 4455(380) + 297(1A0) – 27(B00) + 1485(920) – 8910(740) + 12(900) |
| $4H_{u[c]}$ | – | 270(150) – 27(0B0) + 7182(515) – 900(331) + 270(511) – 3420(153) + 11400(333) – 3420(513) + 9072(272) – 31752(452) + 21168(632) – 2268(812) + 12(090) – 432(270) + 1512(450) – 1008(630) + 108(810) – 12240(133) – 720(311) + 720(131) – 23940(335) + 7182(155) – 252(092) – 4455(830) + 12474(650) + 46512(317) – 46512(137) – 46512(315) + 46512(135) + 12240(313) + 8910(470) + 1485(290) + 297(A10) |
| $4H_{u[d]}$ | – | 1530(124) – (300) + 7980(342) – 1596(162) – 1890(441) – 180(122) – 27930(344) – 4788(522) – 5292(623) – 3876(126) + 756(261) + 2907(128) + 13230(443) + 16758(524) + 5586(164) + 756(621) – 5292(263) + 11628(027) – 6(700) – 27(801) – 11628(207) + 60(302) + 1292(306) – 969(308) – 63(021) + 1260(023) + 63(201) – 27(081) + 6783(209) + 6426(205) – 6783(029) – 6426(025) + 189(803) – 798(704) – 1260(203) + 189(083) + 228(702) – 510(304) + 3(120) + 42(160) + 126(520) – 210(340) |
| $4H_{u[e]}$ | – | 1292(036) – 969(038) – (030) + 3(210) – 216(171) – 2520(113) – 1512(531) + 1512(351) + 4788(252) + 1512(173) + 216(711) + 798(074) – 228(072) + 1596(612) – 16758(254) – 7980(432) + 27930(434) – 5586(614) – 510(034) + 60(032) + 1530(214) – 180(212) – 3876(216) + 2907(218) – 23256(117) + 126(111) + 12852(115) + 13566(119) + 10584(533) + 6(070) + 210(430) – 42(610) – 126(250) – 10584(353) – 1512(713) |
| $5H_{u[a]}$ | – | 54625(418) – 109250(238) + 10925(058) – 61180(416) + 122360(236) – 12236(056) + 19950(414) – 39900(234) + 3990(054) – 1900(412) + 3800(232) – 380(052) + 25(410) – 50(230) + 5(050) |
| $5H_{u[b]}$ | – | 330(740) – 462(560) – 756(540) + 504(360) – 54(180) – 40(601) + 4600(607) + 40(061) + 840(603) + 276(902) – 840(063) – 4600(067) – 3864(605) + 3864(065) – 1150(904) – 25(B02) + 21850(049) + 21850(409) – 10350(184) – 8250(742) – 57960(245) – 12600(423) + 12600(243) + 1375(922) + 69000(247) – 144900(544) + 34776(542) + 57960(425) + 41400(724) – 600(241) + 600(421) – 9936(722) + 2484(182) + 96600(364) – 69000(427) – 23184(362) + 275(1A2) + 11550(562) – 4125(382) – 131100(229) + 165(380) – 11(1A0) – 55(920) + 216(720) + (B00) – 6(900) – 86184(225) + 13680(223) – 540(221) + 188784(227) – 2280(403) + 14364(045) – 2280(043) + 90(041) + 90(401) – 31464(047) – 31464(407) + 14364(405) |
| $5H_{u[c]}$ | – | 6(090) – (0B0) – 41400(274) + 25(0B2) – 276(092) – 11550(652) + 4125(832) – 34776(452) + 9936(272) – 1375(292) – 2484(812) + 144900(454) + 1150(094) – 92000(337) + 27600(157) – 23184(515) + 77280(335) – 23184(155) + 5040(513) – 16800(333) + 5040(153) – 240(511) + 800(331) – 240(151) + 10350(814) + 27600(517) – 87400(319) + 87400(139) + 125856(317) – 125856(137) – 57456(315) + 57456(135) + 9120(313) – 9120(133) – 360(311) + |

| | | |
|---|---|---|
| | | 360(131) + 756(450) − 216(270) + 54(810) − 504(630) + 11(A10) − 165(830) + 462(650) + 55(290) − 330(470) + 8250(472) − 96600(634) + 23184(632) − 275(A12) |
| $5H_{u[d]}$ | – | 169050(526) − 33810(164) + 169050(344) − 101430(524) + 56350(166) − 281750(346) − 1771000(445) + 708400(625) + 4410(162) − 22050(342) + 13230(522) − 660(A21) + 18480(261) − 46200(441) + 18480(621) − 283360(263) + 708400(443) − 283360(623) + 708400(265) − 660(2A1) + 4950(481) − 9240(66) + 4950(841) − 8050(706) − 630(702) + 4830(704) + 10120(083) + 10120(803) − 25300(085) − 660(081) − 660(801) + 10(0C1) + 10(C01) − 1365(1C0) + 30030(3A0) − 135135(580) + 180180(760) − 105(D00) + 8190(B20) − 75075(940) − 334305(308) + 203490(306) − 48450(304) + 3825(302) + 185725(30A) + 145350(124) − 11475(122) + 1002915(128) − 610470(126) − 557175(12A) − 45(300) − 210(520) + 350(340) − 70(160) + 10(700) + 135(120) − 25300(805) |
| $5H_{u[e]}$ | – | −10(070) − 4410(612) − 56350(616) + 33810(614) − 557175(21A) + 1002915(218) − 610470(216) + 145350(214) − 11475(212) + 70(610) − 350(430) + 210(250) − 30030(A30) − 180180(670) + 105(0D0) + 1365(C10) − 8190(2B0) − 169050(434) + 22050(432) − 1416800(535) + 1416800(355) − 45(030) − 5280(171) + 120(B11) − 120(1B1) − 2200(931) + 2200(391) − 7920(571) + 7920(751) + 80960(173) + 5280(711) − 36960(531) + 36960(351) − 202400(175) − 80960(713) + 566720(533) − 566720(353) + 202400(715) − 8050(076) − 4830(074) + 630(072) + 281750(436) + 135(210) + 75075(490) + 135135(850) − 169050(256) − 13230(252) + 101430(254) − 334305(038) + 185725(03A) + 3825(032) − 48450(034) + 203490(036) |

# Appendix 2

## Quantum Chemistry on an EXCEL® spreadsheet

In this Appendix, Hückel theory calculations are used to demonstrate that the polynomials of Appendix 1 can be applied as basis functions for the irreducible subspaces of a Hamiltonian invariant under icosahedral point symmetry, while extended Hückel theory calculations on 'cubium' cages of cubic point symmetry are used to demonstrate the same result for the *kubic* harmonics, since single bond-length regular orbits are not possible in all cases.

For an approximate molecular orbital, $\varphi(\mathbf{r})$, formed by linear combination of appropriately weighted atomic orbitals, $\phi(\mathbf{r})$,

$$\varphi(\mathbf{r}) = \sum_{m} c_i \phi_i (\mathbf{r} - \mathbf{R}_i) \qquad \text{A2.1}$$

the expectation value of the 1-electron Hamiltonian for the molecular structure defined by the m atomic positions $\mathbf{R}_i$ is

$$\varepsilon = \frac{\langle \varphi | \mathbf{H} | \varphi \rangle}{\langle \varphi \mid \varphi \rangle} = \frac{\left\langle \sum_m c_i \phi_i (\mathbf{r} - \mathbf{R}_i) \middle| \mathbf{H} \middle| \sum_m c_i \phi_i (\mathbf{r} - \mathbf{R}_i) \right\rangle}{\left\langle \sum_m c_i \phi_i (\mathbf{r} - \mathbf{R}_i) \middle| \sum_m c_i \phi_i (\mathbf{r} - \mathbf{R}_i) \right\rangle} \qquad \text{A2.2}$$

For our present purpose[1] it is convenient to rewrite equation A2.2 as the similarity transforms

$$(c_1 \quad \cdots \quad c_m) \begin{pmatrix} H_{11} & \cdots & H_{1m} \\ \vdots & \ddots & \vdots \\ H_{m1} & \cdots & H_{mm} \end{pmatrix} \begin{pmatrix} c_1 \\ \vdots \\ c_6 \end{pmatrix} = \varepsilon \times (c_1 \quad \cdots \quad c_m) \begin{pmatrix} S_{11} & \cdots & S_{1m} \\ \vdots & \ddots & \vdots \\ S_{m1} & \cdots & S_{mm} \end{pmatrix}$$

$$= \varepsilon \begin{pmatrix} c_1 \\ \vdots \\ c_m \end{pmatrix} N^2 \qquad \text{A2.3}$$

[1] *See also* Charles M. Quinn, *The electronic structure of the water molecule* in 'The spreadsheet at 25 — the invention that changed the world', 25 amazing EXCEL® examples, Bill Jelen [Holy Macro Books, Ohio, 2005] and Charles M. Quinn, Quantum Chemistry on an Excel® spreadsheet, 2nd Int. Conference on Chemistry and its Applications, University of Qatar, December 2003.

thereby identifying the familiar Coulomb, Resonance and Overlap integrals of LCAO-MO theory

$$\begin{aligned}
H_{ii} &= \langle \phi_i (\mathbf{r} - \mathbf{R}_i)|\, H \,|\phi_i (\mathbf{r} - \mathbf{R}_i)\rangle \\
H_{ij} &= \langle \phi_i (\mathbf{r} - \mathbf{R}_i)|\, H \,|\phi_j (\mathbf{r} - \mathbf{R}_j)\rangle = \langle \phi_j (\mathbf{r} - \mathbf{R}_j)|\, H \,|\phi_i (\mathbf{r} - \mathbf{R}_i)\rangle \\
S_{ij} &= \langle \phi_i (\mathbf{r} - \mathbf{R}_i) \mid \phi_j (\mathbf{r} - \mathbf{R}_j)\rangle = \langle \phi_j (\mathbf{r} - \mathbf{R}_j) \mid \phi_i (\mathbf{r} - \mathbf{R}_i)\rangle
\end{aligned} \qquad \text{A2.4}$$

and the normalization constant, N, for the linear combination of equation A2.1.

It is straightforward to programme equation A2.3 onto an EXCEL® spreadsheet and to employ the SOLVER®[2] macro to fulfil the Variation Principle optimization of the eigenvalue calculation to identify the first energy value of the eigenvalue spectrum. Then, the other energies and corresponding molecular orbitals follow in subsequent applications of SOLVER® with the additional constraints that each of the remaining molecular orbitals be orthogonal to all of those previously calculated.

The orthogonality tests in the BonusPack files on the CDROM are examples of this approach to Quantum Chemistry on a spreadsheet. Other examples of SOLVER's utility in Quantum Chemistry are to be found elsewhere[3].

In Hückel theory, the integrals of equation A2.3 are simplified by the assumptions

$$\begin{aligned}
H_{ii} &= \langle \pi_i | H | \pi_i \rangle = \alpha \\
H_{ij} &= H_{ji} = \langle \pi_i | H | \pi_j \rangle = \langle \pi_j | H | \pi_i \rangle = \beta \\
s_{ii} &= \langle \pi_i | \pi_i \rangle = 1,\ s_{ij} = \langle \pi_i | \pi_i \rangle = \langle \pi_j | \pi_i \rangle = 0,\ \text{i.e. } s_{ij} = \delta_{ij}
\end{aligned} \qquad \text{A2.5}$$

and the similarity transform in the coefficient matrix and its transpose, for example, for the Hückel theory of the $\pi$-electronic structure of benzene, can be written

$$\begin{pmatrix} c_1 & c_2 & c_3 & c_4 & c_5 & c_6 \end{pmatrix}
\begin{pmatrix}
\frac{\alpha-\varepsilon}{\beta} & 1 & 0 & 0 & 0 & 1 \\
1 & \frac{\alpha-\varepsilon}{\beta} & 1 & 0 & 0 & 0 \\
0 & 1 & \frac{\alpha-\varepsilon}{\beta} & 1 & 0 & 0 \\
0 & 0 & 1 & \frac{\alpha-\varepsilon}{\beta} & 1 & 0 \\
0 & 0 & 0 & 1 & \frac{\alpha-\varepsilon}{\beta} & 1 \\
1 & 0 & 0 & 0 & 1 & \frac{\alpha-\varepsilon}{\beta}
\end{pmatrix}
\begin{pmatrix} c_1 \\ c_2 \\ c_3 \\ c_4 \\ c_5 \\ c_6 \end{pmatrix} = N^2 \qquad \text{A2.6}$$

Solving this equation is equivalent to diagonalizing the adjacency matrix

$$\begin{pmatrix}
0 & 1 & 0 & 0 & 0 & 1 \\
1 & 0 & 1 & 0 & 0 & 0 \\
0 & 1 & 0 & 1 & 0 & 0 \\
0 & 0 & 1 & 0 & 1 & 0 \\
0 & 0 & 0 & 1 & 0 & 1 \\
1 & 0 & 0 & 0 & 1 & 0
\end{pmatrix} \qquad \text{A2.7}$$

with eigenvalues in units of $(\alpha - \varepsilon/\beta)$ for normalized Hückel molecular orbitals.

[2] Frontline Systems Inc, www.frontsys.com/index.html.

[3] *See also* Computational Quantum Chemistry, an interactive guide to basis set theory, Charles M. Quinn [Elsevier, Academic Press, San Diego, 2002].

Thus, to perform a Hückel calculation on a spreadsheet, the adjacency matrix has to be constructed with the normalization and orthogonality constraints on the Hückel molecular orbitals satisfied by taking the sums of the products of the coefficients of the $\pi$-LCAO-MOs and requiring these to be of zero or unit values. For normalization, these products involve only one set of coefficients for a single orbital, for orthogonalization, the products are taken between the coefficients of different orbitals.

With reference to Figure 2.19, Figure A2.1[a] displays a portion of the 'Function' worksheet in the file hga.xls for the calculation of the projections of the $5H_{g[a]}$ listed in Table A1.1 of Appendix 1 on the vertex positions of the regular orbit cage of $I_h$ point symmetry. These are the coefficients of the linear combinations of equation A2.1, with the assumption that the vertices are decorated by $p_\sigma$ local radial functions to which Hückel approximations can be applied and are collected as a coefficient matrix on the 'SetUp' worksheet shown in Figure A2.1[b].

The remainder of the 'Setup' worksheet is devoted to the construction of the Distance and Adjacency matrices for the Hückel calculations set out on the other worksheets in the file. Some details are shown in Figure A2.2.

In Figure A2.2[a] the application of the powerful TABLE() and LOOKUP() functions of EXCEL® as the formula in cell \$E\$16, the uppermost top-left cell of the 121 × 121 [120 + 1 to include the *look_up values*] selected area of the worksheet

$$\begin{aligned}\$E\$16 = &\mathbf{SQRT}(\mathbf{POWER}(\mathbf{HLOOKUP}(\$C\$16, \$F\$1:\$DU\$6, 4, \text{FALSE})\\ &- \mathbf{HLOOKUP}(\$C\$17, \$F\$1:\$DU\$6, 4, \text{FALSE}), 2)\\ &+ \mathbf{POWER}(\mathbf{HLOOKUP}(\$C\$16, \$F\$1:\$DU\$6, 5, \text{FALSE})\\ &- \mathbf{HLOOKUP}(\$C\$17, \$F\$1:\$DU\$6, 5, \text{FALSE}), 2)\\ &+ \mathbf{POWER}(\mathbf{HLOOKUP}(\$C\$16, \$F\$1:\$DU\$6, 6, \text{FALSE})\\ &- \mathbf{HLOOKUP}(\$C\$17, \$F\$1:\$DU\$6, 6, \text{FALSE}), 2))\end{aligned} \qquad \text{A2.8}$$

generates the distance matrix. Note, that this is an ARRAY formula. The TABLE() macro function is implemented by selecting all the cells of the table, with the formula in the left-uppermost cell of the table, surrounded by a row and a column of *look_up values*, then choosing TABLE in the TOOLS menu and enabling the dummy indices of cells \$C\$16 and \$C\$17 and completing the ARRAY formula over all the cells by the command sequence CTL/SHIFT/ENTER.

As you can see in Figure A2.2[a] the 'bond length' of the Hückel regular orbit cage is about 0.263 units and so the Adjacency matrix is constructed, Figure A2.2[b], over the 120 × 120 matrix using, in each cell of the matrix, the conditional formula, for example, in cell \$F\$138

$$\$F\$138 = \mathbf{IF}(\mathbf{AND}(F17\langle\$D\$139, F17\rangle\$D\$138), 1, 0) \qquad \text{A2.9}$$

For the Hückel calculations, the remainder of the 'Setup' worksheet is devoted to the imposition of the orthonormality condition on the $5H_{g[a]}$ functions of Table A1.1. This condition in Hückel theory requires only matrix multiplications between the matrix of coefficients and its transpose, with stepwise imposition of, for example, Gram-Schmidt orthogonalization until

| | AB | | C | D | E | F | G | H | I | J | K | L | M | N | O | P | Q |
|---|---|---|---|---|---|---|---|---|---|---|---|---|---|---|---|---|---|
| 1 | Coordinates of Regular orbit of $I_h$ | | | | # | 1 | 2 | 3 | 4 | 5 | 6 | 7 | 8 | 9 | 10 | 11 | 12 |
| 2 | [Figure 2.19] | | | | ≅ | 25.191908 | 25.1919077 | 25.1919077 | 25.1919077 | 25.191908 | 25.191908 | 25.1919077 | 25.1919077 | 25.1919077 | 25.191908 | 39.938169 | 39.9381694 |
| 3 | | | | | ≅ | 18.02013 | 53.9798699 | 90.0201301 | 125.97987 | 162.02013 | 197.97987 | 234.02013 | 269.97987 | 306.02013 | 341.97987 | 11.836376 | 60.1636243 |
| 4 | | | | | x | 0.405 | 0.250 | 0.000 | -0.250 | -0.405 | -0.405 | -0.250 | 0.000 | 0.250 | 0.405 | 0.628 | 0.319 |
| 5 | | | | | y | 0.132 | 0.344 | 0.426 | 0.344 | 0.131 | -0.131 | -0.344 | -0.426 | -0.344 | -0.132 | 0.132 | 0.557 |
| 6 | | | | | z | 0.905 | 0.905 | 0.905 | 0.905 | 0.905 | 0.905 | 0.905 | 0.905 | 0.905 | 0.905 | 0.767 | 0.767 |
| 7 | | | | | N | 1.0000 | 1.0000 | 1.0000 | 1.0000 | 1.0000 | 1.0000 | 1.0000 | 1.0000 | 1.0000 | 1.0000 | 1.0000 | 1.0000 |
| 9 | 1hga | | | | | | | | | | | | | | | | |
| 10 | 3 | | | | | | | | | | | | | | | | |
| 11 | 0 | 0 | 2 | 2.0 | | 1.638 | 1.638 | 1.638 | 1.638 | 1.638 | 1.638 | 1.638 | 1.638 | 1.638 | 1.638 | 1.176 | 1.176 |
| 12 | 2 | 0 | 0 | -1.0 | | -0.164 | -0.063 | 0.000 | -0.063 | -0.164 | -0.164 | -0.063 | 0.000 | -0.063 | -0.164 | -0.395 | -0.102 |
| 13 | 0 | 2 | 0 | -1.0 | | -0.017 | -0.119 | -0.181 | -0.119 | -0.017 | -0.017 | -0.119 | -0.181 | -0.119 | -0.017 | -0.000 | -0.310 |
| 14 | | | | | | 0.000 | 0.000 | 0.000 | 0.000 | 0.000 | 0.000 | 0.000 | 0.000 | 0.000 | 0.000 | 0.000 | 0.000 |
| 15 | | | | | | 0.000 | 0.000 | 0.000 | 0.000 | 0.000 | 0.000 | 0.000 | 0.000 | 0.000 | 0.000 | 0.000 | 0.000 |
| 16 | | | | | | 0.000 | 0.000 | 0.000 | 0.000 | 0.000 | 0.000 | 0.000 | 0.000 | 0.000 | 0.000 | 0.000 | 0.000 |
| 17 | | | | | | 0.000 | 0.000 | 0.000 | 0.000 | 0.000 | 0.000 | 0.000 | 0.000 | 0.000 | 0.000 | 0.000 | 0.000 |
| 18 | | | | | | 0.000 | 0.000 | 0.000 | 0.000 | 0.000 | 0.000 | 0.000 | 0.000 | 0.000 | 0.000 | 0.000 | 0.000 |
| 19 | | | | | | 0.000 | 0.000 | 0.000 | 0.000 | 0.000 | 0.000 | 0.000 | 0.000 | 0.000 | 0.000 | 0.000 | 0.000 |
| 20 | | | | | | 0.000 | 0.000 | 0.000 | 0.000 | 0.000 | 0.000 | 0.000 | 0.000 | 0.000 | 0.000 | 0.000 | 0.000 |
| 92 | | | | 1 | | 1.456 | 1.456 | 1.456 | 1.456 | 1.456 | 1.456 | 1.456 | 1.456 | 1.456 | 1.456 | 0.764 | 0.764 |

[a]

| | A | B | C | D | E | F | G | H | I | J | K | L | M | N |
|---|---|---|---|---|---|---|---|---|---|---|---|---|---|---|
| 1 | O120 coordinates | | | | # | 12 | | 34 | | 56 | | 78 | | 9 |
| 2 | | | | | ≅ | 25.1919077 | 25.1919077 | 25.1919077 | 25.1919077 | 25.1919077 | 25.1919077 | 25.1919077 | 25.1919077 | 25.1919077 |
| 3 | | | | | ≅ | 18.0201301 | 53.9798699 | 90.0201301 | 125.9798699 | 162.0201301 | 197.9798699 | 234.0201301 | 269.9798699 | 306.0201301 |
| 4 | | | | | x | 0.405 | 0.250 | 0.000 | -0.250 | -0.405 | -0.405 | -0.250 | 0.000 | 0.250 |
| 5 | | | | | y | 0.132 | 0.344 | 0.426 | 0.344 | 0.131 | -0.131 | -0.344 | -0.426 | -0.344 |
| 6 | | | | | z | 0.905 | 0.905 | 0.905 | 0.905 | 0.905 | 0.905 | 0.905 | 0.905 | 0.905 |
| 7 | | | | | N | 1.0000 | 1.0000 | 1.0000 | 1.0000 | 1.0000 | 1.0000 | 1.0000 | 1.0000 | 1.0000 |
| 8 | | | | | | | | | | | | | | |
| 9 | | | | | 1hga | 1.45646 | 1.45646 | 1.45646 | 1.45646 | 1.45646 | 1.45646 | 1.45646 | 1.45646 | 1.45646 |
| 10 | | | | | 2hga | 1.90174 | 1.90174 | 1.90174 | 1.90174 | 1.90174 | 1.90174 | 1.90174 | 1.90174 | 1.90174 |
| 11 | | | | | 3hga | -3.40302 | -3.40302 | -3.40302 | -3.40302 | -3.40302 | -3.40302 | -3.40302 | -3.40302 | -3.40302 |
| 12 | | | | | 4hga | -52.15619 | -52.15619 | -52.15619 | -52.15619 | -52.15619 | -52.15619 | -52.15619 | -52.15619 | -52.15619 |
| 13 | | | | | 5hga | -0.012 | -0.012 | -0.012 | -0.012 | -0.012 | -0.012 | -0.012 | -0.012 | -0.012 |

[b]

**Figure A2.1** Screen dumps from the file Ih.xls in the BonusPack, showing portions of the 'Function'[a] and 'Setup'[b] worksheets. In [a] the function projections on the regular orbit cage are summed over the individual terms. Then in [b] the 5 × $h_{ga}$ coefficients over the p$\pi$ decorations of the regular orbit cage are displayed in matrix form.

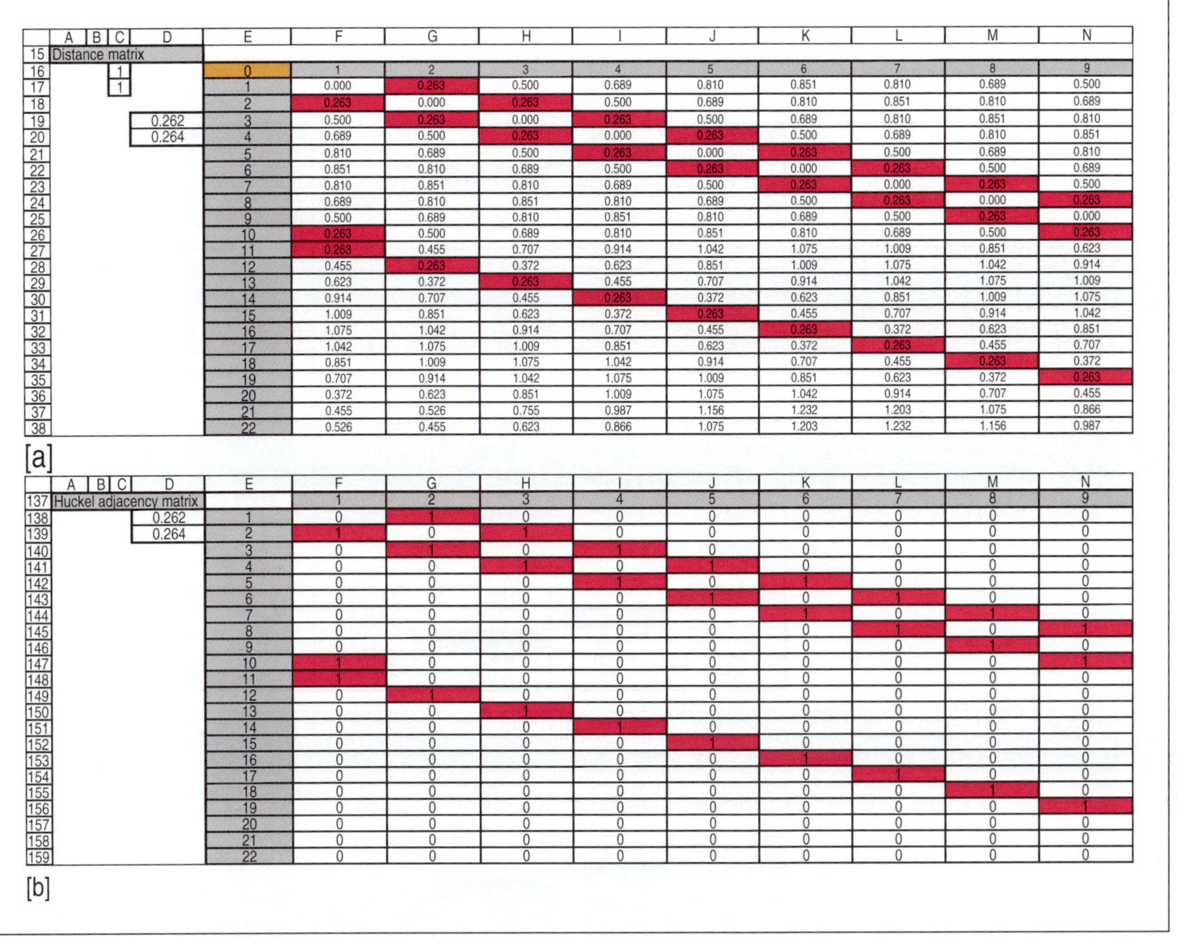

|  | A | B | C | D | E | F | G | H | I | J | K | L | M | N |
|---|---|---|---|---|---|---|---|---|---|---|---|---|---|---|
| 15 | Distance matrix |  |  |  |  |  |  |  |  |  |  |  |  |  |
| 16 |  |  | 1 |  | 0 | 1 | 2 | 3 | 4 | 5 | 6 | 7 | 8 | 9 |
| 17 |  |  | 1 |  | 1 | 0.000 | 0.263 | 0.500 | 0.689 | 0.810 | 0.851 | 0.810 | 0.689 | 0.500 |
| 18 |  |  |  |  | 2 | 0.263 | 0.000 | 0.263 | 0.500 | 0.689 | 0.810 | 0.851 | 0.810 | 0.689 |
| 19 |  |  |  | 0.262 | 3 | 0.500 | 0.263 | 0.000 | 0.263 | 0.500 | 0.689 | 0.810 | 0.851 | 0.810 |
| 20 |  |  |  | 0.264 | 4 | 0.689 | 0.500 | 0.263 | 0.000 | 0.263 | 0.500 | 0.689 | 0.810 | 0.851 |
| 21 |  |  |  |  | 5 | 0.810 | 0.689 | 0.500 | 0.263 | 0.000 | 0.263 | 0.500 | 0.689 | 0.810 |
| 22 |  |  |  |  | 6 | 0.851 | 0.810 | 0.689 | 0.500 | 0.263 | 0.000 | 0.263 | 0.500 | 0.689 |
| 23 |  |  |  |  | 7 | 0.810 | 0.851 | 0.810 | 0.689 | 0.500 | 0.263 | 0.000 | 0.263 | 0.500 |
| 24 |  |  |  |  | 8 | 0.689 | 0.810 | 0.851 | 0.810 | 0.689 | 0.500 | 0.263 | 0.000 | 0.263 |
| 25 |  |  |  |  | 9 | 0.500 | 0.689 | 0.810 | 0.851 | 0.810 | 0.689 | 0.500 | 0.263 | 0.000 |
| 26 |  |  |  |  | 10 | 0.263 | 0.500 | 0.689 | 0.810 | 0.851 | 0.810 | 0.689 | 0.500 | 0.263 |
| 27 |  |  |  |  | 11 | 0.263 | 0.455 | 0.707 | 0.914 | 1.042 | 1.075 | 1.009 | 0.851 | 0.623 |
| 28 |  |  |  |  | 12 | 0.455 | 0.263 | 0.372 | 0.623 | 0.851 | 1.009 | 1.075 | 1.042 | 0.914 |
| 29 |  |  |  |  | 13 | 0.623 | 0.372 | 0.263 | 0.455 | 0.707 | 0.914 | 1.042 | 1.075 | 1.009 |
| 30 |  |  |  |  | 14 | 0.914 | 0.707 | 0.455 | 0.263 | 0.372 | 0.623 | 0.851 | 1.009 | 1.075 |
| 31 |  |  |  |  | 15 | 1.009 | 0.851 | 0.623 | 0.372 | 0.263 | 0.455 | 0.707 | 0.914 | 1.042 |
| 32 |  |  |  |  | 16 | 1.075 | 1.042 | 0.914 | 0.707 | 0.455 | 0.263 | 0.372 | 0.623 | 0.851 |
| 33 |  |  |  |  | 17 | 1.042 | 1.075 | 1.009 | 0.851 | 0.623 | 0.372 | 0.263 | 0.455 | 0.707 |
| 34 |  |  |  |  | 18 | 0.851 | 1.009 | 1.075 | 1.042 | 0.914 | 0.707 | 0.455 | 0.263 | 0.372 |
| 35 |  |  |  |  | 19 | 0.707 | 0.914 | 1.042 | 1.075 | 1.009 | 0.851 | 0.623 | 0.372 | 0.263 |
| 36 |  |  |  |  | 20 | 0.372 | 0.623 | 0.851 | 1.009 | 1.075 | 1.042 | 0.914 | 0.707 | 0.455 |
| 37 |  |  |  |  | 21 | 0.455 | 0.526 | 0.755 | 0.987 | 1.156 | 1.232 | 1.203 | 1.075 | 0.866 |
| 38 |  |  |  |  | 22 | 0.526 | 0.455 | 0.623 | 0.866 | 1.075 | 1.203 | 1.232 | 1.156 | 0.987 |

[a]

|  | A | B | C | D | E | F | G | H | I | J | K | L | M | N |
|---|---|---|---|---|---|---|---|---|---|---|---|---|---|---|
| 137 | Huckel adjacency matrix |  |  |  |  | 1 | 2 | 3 | 4 | 5 | 6 | 7 | 8 | 9 |
| 138 |  |  |  | 0.262 | 1 | 0 | 1 | 0 | 0 | 0 | 0 | 0 | 0 | 0 |
| 139 |  |  |  | 0.264 | 2 | 1 | 0 | 1 | 0 | 0 | 0 | 0 | 0 | 0 |
| 140 |  |  |  |  | 3 | 0 | 1 | 0 | 1 | 0 | 0 | 0 | 0 | 0 |
| 141 |  |  |  |  | 4 | 0 | 0 | 1 | 0 | 1 | 0 | 0 | 0 | 0 |
| 142 |  |  |  |  | 5 | 0 | 0 | 0 | 1 | 0 | 1 | 0 | 0 | 0 |
| 143 |  |  |  |  | 6 | 0 | 0 | 0 | 0 | 1 | 0 | 1 | 0 | 0 |
| 144 |  |  |  |  | 7 | 0 | 0 | 0 | 0 | 0 | 1 | 0 | 1 | 0 |
| 145 |  |  |  |  | 8 | 0 | 0 | 0 | 0 | 0 | 0 | 1 | 0 | 1 |
| 146 |  |  |  |  | 9 | 0 | 0 | 0 | 0 | 0 | 0 | 0 | 1 | 0 |
| 147 |  |  |  |  | 10 | 1 | 0 | 0 | 0 | 0 | 0 | 0 | 0 | 1 |
| 148 |  |  |  |  | 11 | 1 | 0 | 0 | 0 | 0 | 0 | 0 | 0 | 0 |
| 149 |  |  |  |  | 12 | 0 | 1 | 0 | 0 | 0 | 0 | 0 | 0 | 0 |
| 150 |  |  |  |  | 13 | 0 | 0 | 1 | 0 | 0 | 0 | 0 | 0 | 0 |
| 151 |  |  |  |  | 14 | 0 | 0 | 0 | 1 | 0 | 0 | 0 | 0 | 0 |
| 152 |  |  |  |  | 15 | 0 | 0 | 0 | 0 | 1 | 0 | 0 | 0 | 0 |
| 153 |  |  |  |  | 16 | 0 | 0 | 0 | 0 | 0 | 1 | 0 | 0 | 0 |
| 154 |  |  |  |  | 17 | 0 | 0 | 0 | 0 | 0 | 0 | 1 | 0 | 0 |
| 155 |  |  |  |  | 18 | 0 | 0 | 0 | 0 | 0 | 0 | 0 | 1 | 0 |
| 156 |  |  |  |  | 19 | 0 | 0 | 0 | 0 | 0 | 0 | 0 | 0 | 1 |
| 157 |  |  |  |  | 20 | 0 | 0 | 0 | 0 | 0 | 0 | 0 | 0 | 0 |
| 158 |  |  |  |  | 21 | 0 | 0 | 0 | 0 | 0 | 0 | 0 | 0 | 0 |
| 159 |  |  |  |  | 22 | 0 | 0 | 0 | 0 | 0 | 0 | 0 | 0 | 0 |

[b]

**Figure A2.2** For the same spreadsheet, portions of the 'Setup' worksheet on which the orthogonalization of the original projections is performed.

<table>
<tr><th></th><th>A</th><th>B</th><th>C</th><th>D</th><th>E</th><th>F</th><th>G</th><th>H</th><th>I</th><th>J</th><th>K</th><th>L</th><th>M</th><th>N</th></tr>
<tr><td>259</td><td>Overlap matrix on starting vectors</td><td></td><td></td><td></td><td></td><td>96.004</td><td>-81.722</td><td>-163.457</td><td>-1575.405</td><td>182.818</td><td></td><td></td><td></td><td></td></tr>
<tr><td>260</td><td></td><td></td><td></td><td></td><td></td><td>-81.722</td><td>708.049</td><td>-757.886</td><td>-3095.938</td><td>-617.115</td><td></td><td></td><td></td><td></td></tr>
<tr><td>261</td><td></td><td></td><td></td><td></td><td></td><td>-163.457</td><td>-757.886</td><td>4911.082</td><td>-3705.263</td><td>-137.224</td><td></td><td></td><td></td><td></td></tr>
<tr><td>262</td><td></td><td></td><td></td><td></td><td></td><td>-1575.405</td><td>-3095.938</td><td>-3705.263</td><td>192605.464</td><td>4375.309</td><td></td><td></td><td></td><td></td></tr>
<tr><td>263</td><td></td><td></td><td></td><td></td><td></td><td>182.818</td><td>-617.115</td><td>-137.224</td><td>4375.309</td><td>6764.9</td><td></td><td></td><td></td><td></td></tr>
<tr><td>264</td><td></td><td></td><td></td><td></td><td></td><td></td><td></td><td></td><td></td><td></td><td></td><td></td><td></td><td></td></tr>
<tr><td>265</td><td>Normed starting vectors</td><td></td><td></td><td></td><td>|1></td><td>0.148646</td><td>0.148646</td><td>0.148646</td><td>0.148646</td><td>0.148646</td><td>0.148646</td><td>0.148646</td><td>0.148646</td><td>0.148646</td></tr>
<tr><td>266</td><td></td><td></td><td></td><td></td><td>|2></td><td>0.071469</td><td>0.071469</td><td>0.071469</td><td>0.071469</td><td>0.071469</td><td>0.071469</td><td>0.071469</td><td>0.071469</td><td>0.071469</td></tr>
<tr><td>267</td><td></td><td></td><td></td><td></td><td>|3></td><td>-0.048560</td><td>-0.048560</td><td>-0.048560</td><td>-0.048560</td><td>-0.048560</td><td>-0.048560</td><td>-0.048560</td><td>-0.048560</td><td>-0.048560</td></tr>
<tr><td>268</td><td></td><td></td><td></td><td></td><td>|4></td><td>-0.118842</td><td>-0.118842</td><td>-0.118842</td><td>-0.118842</td><td>-0.118842</td><td>-0.118842</td><td>-0.118842</td><td>-0.118842</td><td>-0.118842</td></tr>
<tr><td>269</td><td></td><td></td><td></td><td></td><td>|5></td><td>-0.000145</td><td>-0.000145</td><td>-0.000145</td><td>-0.000145</td><td>-0.000145</td><td>-0.000145</td><td>-0.000145</td><td>-0.000145</td><td>-0.000145</td></tr>
<tr><td>270</td><td></td><td></td><td></td><td></td><td></td><td></td><td></td><td></td><td></td><td></td><td></td><td></td><td></td><td></td></tr>
<tr><td>271</td><td></td><td></td><td></td><td></td><td></td><td>1.000</td><td>-0.313</td><td>-0.238</td><td>-0.366</td><td>0.227</td><td></td><td></td><td></td><td></td></tr>
<tr><td>272</td><td></td><td></td><td></td><td></td><td></td><td>-0.313</td><td>1.000</td><td>-0.406</td><td>-0.265</td><td>-0.282</td><td></td><td></td><td></td><td></td></tr>
<tr><td>273</td><td></td><td></td><td></td><td></td><td></td><td>-0.238</td><td>-0.406</td><td>1.000</td><td>-0.120</td><td>-0.024</td><td></td><td></td><td></td><td></td></tr>
<tr><td>274</td><td></td><td></td><td></td><td></td><td></td><td>-0.366</td><td>-0.265</td><td>-0.120</td><td>1.000</td><td>0.121</td><td></td><td></td><td></td><td></td></tr>
<tr><td>275</td><td></td><td></td><td></td><td></td><td></td><td>0.227</td><td>-0.282</td><td>-0.024</td><td>0.121</td><td>1.000</td><td></td><td></td><td></td><td></td></tr>
<tr><td>276</td><td></td><td></td><td></td><td></td><td></td><td></td><td></td><td></td><td></td><td></td><td></td><td></td><td></td><td></td></tr>
<tr><td>277</td><td></td><td></td><td></td><td></td><td></td><td></td><td></td><td></td><td></td><td></td><td></td><td></td><td></td><td></td></tr>
<tr><td>278</td><td></td><td></td><td></td><td></td><td></td><td></td><td></td><td></td><td></td><td></td><td></td><td></td><td></td><td></td></tr>
<tr><td>279</td><td>Orthogonalize |2> to |1></td><td></td><td></td><td></td><td>|1></td><td>0.149</td><td>0.149</td><td>0.149</td><td>0.149</td><td>0.149</td><td>0.149</td><td>0.149</td><td>0.149</td><td>0.149</td></tr>
<tr><td>280</td><td></td><td></td><td></td><td></td><td>|2>-I|1>/N1</td><td>0.118</td><td>0.118</td><td>0.118</td><td>0.118</td><td>0.118</td><td>0.118</td><td>0.118</td><td>0.118</td><td>0.118</td></tr>
<tr><td>281</td><td></td><td></td><td></td><td></td><td>|3></td><td>-3.403</td><td>-3.403</td><td>-3.403</td><td>-3.403</td><td>-3.403</td><td>-3.403</td><td>-3.403</td><td>-3.403</td><td>-3.403</td></tr>
<tr><td>282</td><td></td><td></td><td></td><td></td><td>|4></td><td>-52.156</td><td>-52.156</td><td>-52.156</td><td>-52.156</td><td>-52.156</td><td>-52.156</td><td>-52.156</td><td>-52.156</td><td>-52.156</td></tr>
<tr><td>283</td><td></td><td></td><td></td><td></td><td>|5></td><td>-0.012</td><td>-0.012</td><td>-0.012</td><td>-0.012</td><td>-0.012</td><td>-0.012</td><td>-0.012</td><td>-0.012</td><td>-0.012</td></tr>
</table>

[c]

<table>
<tr><th></th><th>A</th><th>B</th><th>C</th><th>D</th><th>E</th><th>F</th><th>G</th><th>H</th><th>I</th><th>J</th><th>K</th><th>L</th><th>M</th><th>N</th></tr>
<tr><td>327</td><td>Orthogonal vectors</td><td></td><td></td><td></td><td>|1></td><td>0.149</td><td>0.149</td><td>0.149</td><td>0.149</td><td>0.149</td><td>0.149</td><td>0.149</td><td>0.149</td><td>0.149</td></tr>
<tr><td>328</td><td></td><td></td><td></td><td></td><td>|2></td><td>0.124</td><td>0.124</td><td>0.124</td><td>0.124</td><td>0.124</td><td>0.124</td><td>0.124</td><td>0.124</td><td>0.124</td></tr>
<tr><td>329</td><td></td><td></td><td></td><td></td><td>|3></td><td>0.060</td><td>0.060</td><td>0.060</td><td>0.060</td><td>0.060</td><td>0.060</td><td>0.060</td><td>0.060</td><td>0.060</td></tr>
<tr><td>330</td><td></td><td></td><td></td><td></td><td>|4></td><td>0.022</td><td>0.022</td><td>0.022</td><td>0.022</td><td>0.022</td><td>0.022</td><td>0.022</td><td>0.022</td><td>0.022</td></tr>
<tr><td>331</td><td></td><td></td><td></td><td></td><td>|5></td><td>-0.003</td><td>-0.003</td><td>-0.003</td><td>-0.003</td><td>-0.003</td><td>-0.003</td><td>-0.003</td><td>-0.003</td><td>-0.003</td></tr>
<tr><td>332</td><td></td><td></td><td></td><td></td><td></td><td></td><td></td><td></td><td></td><td></td><td></td><td></td><td></td><td></td></tr>
<tr><td>333</td><td></td><td></td><td></td><td></td><td></td><td>1.0000000000</td><td>0.0000000000</td><td>0.0000000000</td><td>0.0000000000</td><td>0.0000000000</td><td></td><td></td><td></td><td></td></tr>
<tr><td>334</td><td></td><td></td><td></td><td></td><td></td><td>0.0000000000</td><td>1.0000000000</td><td>0.0000000000</td><td>0.0000000000</td><td>0.0000000000</td><td></td><td></td><td></td><td></td></tr>
<tr><td>335</td><td></td><td></td><td></td><td></td><td></td><td>0.0000000000</td><td>0.0000000000</td><td>1.0000000000</td><td>0.0000000000</td><td>0.0000000000</td><td></td><td></td><td></td><td></td></tr>
<tr><td>336</td><td></td><td></td><td></td><td></td><td></td><td>0.0000000000</td><td>0.0000000000</td><td>0.0000000000</td><td>1.0000000000</td><td>0.0000000000</td><td></td><td></td><td></td><td></td></tr>
<tr><td>337</td><td></td><td></td><td></td><td></td><td></td><td>0.0000000000</td><td>0.0000000000</td><td>0.0000000000</td><td>0.0000000000</td><td>1.0000000000</td><td></td><td></td><td></td><td></td></tr>
</table>

[d]

**Figure A2.2** Continued.

the resulting transformed linear combinations are rendered mutually orthogonal[4] as shown in Figure A2.2[d].

The calculations of the Hückel eigenvalues in the energy spectrum are set out on the worksheets 'E1' to 'E5'. All are very similar except for the imposition of extra orthogonalization constraints in the operation of the SOLVER® macro. Thus Figure A2.3[a] shows a portion of the 'E5' worksheet with its associated SOLVER® dialog window, accessible on the TOOLS menu contains the optimization conditions specified for the calculation of the eigenvalue in cell $E$26[5], displayed in Figure A2.3[b].

Because these are Hückel calculations, with the integrals simplified as in equation A2.5, the normalization and orthogonalization constraints can be written using the **SUMSQ**() and **SUMPRODUCT**() functions, and so we find

⟨5|5⟩ = $H$26 = **SUMSQ**($D$11:$D$S11) A2.10

⟨5|1⟩ = $H$27 = **SUMPRODUCT**($J$32:$J$151, $P$26:$P$145) A2.11

⟨5|2⟩ = $H$28 = **SUMPRODUCT**($K$32:$K$151, $P$26:$P$145) A2.12

⟨5|3⟩ = $H$29 = **SUMPRODUCT**($L$32:$L$151, $P$26:$P$145) A2.13

⟨5|4⟩ = $H$30 = **SUMPRODUCT**($M$32:$M$151, $P$26:$P$145) A2.14

as the relationships subject to the constraints shown in the SOLVER window, with the added conditions that the energy found [cell $E$26 value be greater than, E4, the previous one and that the solution be optimized to the *maximum* value consistent with these constraints by variation of the coefficients in $c_1$ to $c_5$ of the linear combination, cells $A$5 to $A$9.

Several other details should be mentioned. Note, that the optimization is subject to variation of the coefficients of the linear combinations over the original set of projections of the 5 × hga functions onto the regular orbit cage. It is important that only linear combinations over the projections of the distinct a, b, c, d and e sets of polynomials listed in Table A1.1 are applied. The transformation constraints that lead to these polynomials depend on this matching of the different order polynomials. This is a general observation and should be maintained with all the other sets of polynomials.

Note that Visual Basic code activated by the command buttons 'SETUP', 'SOLVE' and 'CHOOSE POLYNOMIALS' provide for concerted calculations from the final worksheet. The code controlling these actions can be viewed using the command sequence ALT/F11 which renders the Visual Basic window visible.

Extended Hückel theory calculations are the maximum level of approximation that can be employed for the calculation of the eigenvalue spectra for the regular orbit cages of the cubic point groups, since different bond lengths have to be accommodated on the $H_{1s}$ orbital

[4]Note, that the spreadsheets loaded into the BonusPack were based on an original template, designed for both Hückel and Extended Hückel calculations. Thus the extra seemingly, in this context, strange matrix multiplication of the distance matrix with the transpose of the different coefficient matrices only become relevant in the Extended Hückel theory calculations for the cubic point group regular orbits decorated by hydrogen 1s orbitals. In these calculations the distance matrix cells contain the overlap integrals based on the tables and equations published by Mulliken and his colleagues many years ago [Mulliken, R.S., Rieke, C.A., Orloff, D. and Orloff, H. 1949, Formulas and numerical tables for overlap integrals. *J. Chem. Phys.*, **17**, 1248 – 1267. This design was abandoned in later calculations for the cubic groups, see Figure A2.4.

[5]Remember that, in Hückel theory, the most stable eigenvalue corresponds to the maximum value, and so E5 is the energy of the least stable MO!

| | A | B | C | D | E | F | G | H | I | J | K | L | M | N | O | P | Q | R | S |
|---|---|---|---|---|---|---|---|---|---|---|---|---|---|---|---|---|---|---|---|
| 1 | The projection values of the functions chosen to be distinct on the regular orbit of Icosahedral symmetry. | | | | | | | | | | | | | | | | | | |
| 2 | | | x | 0.405 | 0.250 | 0.000 | -0.250 | -0.405 | 0.405 | -0.250 | 0.000 | 0.250 | 0.405 | 0.628 | 0.319 | 0.069 | -0.431 | -0.586 | 0.586 |
| 3 | | | y | 0.132 | 0.344 | 0.426 | 0.344 | 0.131 | -0.131 | -0.344 | -0.426 | -0.344 | -0.132 | 0.132 | 0.557 | 0.638 | 0.476 | 0.263 | -0.263 |
| 4 | $C_i$ | | z | 0.905 | 0.905 | 0.905 | 0.905 | 0.905 | 0.905 | 0.905 | 0.905 | 0.905 | 0.905 | 0.767 | 0.767 | 0.767 | 0.767 | 0.767 | 0.767 |
| 5 | -0.0064 | | 1hga | 0.14865 | 0.14865 | 0.14865 | 0.14865 | 0.148646 | 0.14865 | 0.14865 | 0.148646 | 0.14865 | 0.14865 | 0.07794 | 0.07794 | 0.07794 | 0.07794 | 0.07794 | 0.07794 |
| 6 | -0.0020 | | 2hga | 0.12433 | 0.12433 | 0.12433 | 0.12433 | 0.124327 | 0.12433 | 0.12433 | 0.124327 | 0.12433 | 0.12433 | -0.0748 | -0.0748 | -0.0748 | -0.0748 | -0.0748 | -0.0748 |
| 7 | -0.0129 | | 3hga | 0.0601 | 0.0601 | 0.0601 | 0.0601 | 0.060103 | 0.0601 | 0.0601 | 0.060103 | 0.0601 | 0.0601 | -0.1373 | -0.1373 | -0.1373 | -0.1373 | -0.1373 | -0.1373 |
| 8 | 0.0738 | | 4hga | 0.02229 | 0.02229 | 0.02229 | 0.02229 | 0.022288 | 0.02229 | 0.02229 | 0.022288 | 0.02229 | 0.02229 | -0.1041 | -0.1041 | -0.1041 | -0.1041 | -0.1041 | -0.1041 |
| 9 | 0.9972 | | 5hga | -0.00263 | -0.00263 | -0.00263 | -0.00263 | -0.00263 | -0.00263 | -0.00263 | -0.00263 | -0.00263 | -0.00263 | 0.01763 | 0.01763 | 0.01763 | 0.01763 | 0.01763 | 0.01763 |
| 10 | | | | | | | | | | | | | | | | | | | |
| 11 | | | ÷$_5$> | -0.0029 | -0.0029 | -0.0029 | -0.0029 | -0.00295 | -0.0029 | -0.0029 | -0.002948 | -0.0029 | -0.0029 | 0.01132 | 0.01132 | 0.01132 | 0.01132 | 0.01132 | 0.01132 |
| 12 | | | | | | | | | | | | | | | | | | | |
| 13 | | | | Orthonormality test | | | | | | | | | | | | | | | |
| 14 | | | | 1.0000 | 0.0000 | 0.0000 | 0.0000 | 0.0000 | | | | | | | | | | | |
| 15 | | | | 0.0000 | 1.0000 | 0.0000 | 0.0000 | 0.0000 | | | | | | | | | | | |
| 16 | | | | 0.0000 | 0.0000 | 1.0000 | 0.0000 | 0.0000 | | | | | | | | | | | |
| 17 | | | | 0.0000 | 0.0000 | 0.0000 | 1.0000 | 0.0000 | | | | | | | | | | | |
| 18 | | | | 0.0000 | 0.0000 | 0.0000 | 0.0000 | 1.0000 | | | | | | | | | | | |
| 19 | | | | | | | | | | hga | | | | | | | | | |
| 20 | | | | | | | | | | $E_1$ | $E_2$ | $E_3$ | $E_4$ | $E_5$ | Row | | | | |
| 21 | | | | | | | | | | 2.7214 | 1.8884 | 0.6846 | -0.4664 | -1.8280 | 427 | | | | |
| 22 | | | SETUP | | SOLVE | | CHOOSE POLYNOMIAL | | | 2.7214 | 1.8884 | 0.6846 | -0.4664 | -1.8280 | 510 | | | | |
| 23 | | | | | | | | | | | | | | | | | | | |
| 24 | | | | | | | | | | | | | | | | | | | |
| 25 | | | | | | | | | | ÷$_1$> | ÷$_2$> | ÷$_3$> | ÷$_4$> | ÷$_5$> | | ÷$_5$>$^1$ | | ÷ζ$_5$>$^1$ | |
| 26 | | | | $E_5$ | -1.82801 | | <5\|5> | 1.00000 | $E_i$ | 2.7214 | 1.8884 | 0.6846 | -0.4664 | -1.8280 | | -0.0029 | | 0.0054 | |
| 27 | | | | | | | <5\|1> | 0.00000 | $C_{1i}$ | 0.989 | 0.136 | 0.035 | 0.053 | 0.006 | | -0.0029 | | 0.0054 | |
| 28 | | | | | | | <5\|2> | 0.00000 | $C_{2i}$ | -0.145 | 0.939 | 0.297 | 0.096 | 0.002 | | -0.0029 | | 0.0054 | |
| 29 | | | | | | | <5\|3> | 0.00000 | $C_{3i}$ | -0.001 | -0.315 | 0.925 | 0.213 | 0.013 | | -0.0029 | | 0.0054 | |
| 30 | | | | | | | <5\|4> | 0.00000 | $C_{4i}$ | -0.039 | -0.031 | -0.233 | 0.968 | 0.074 | | -0.0029 | | 0.0054 | |
| 31 | | | | | | | | | $C_{5i}$ | 0.009 | 0.001 | 0.030 | -0.068 | 0.997 | | -0.0029 | | 0.0054 | |
| 32 | | | | | | | | | ÷$_{1i}$> | 0.128 | 0.11729 | 0.09251 | 0.05441 | -0.00295 | | 0.0029 | | 0.0054 | |
| 33 | | | | | | | | | | 0.128 | 0.11729 | 0.09251 | 0.05441 | -0.00295 | | 0.0029 | | 0.0054 | |
| 34 | | | | | | | | | | 0.128 | 0.11729 | 0.09251 | 0.05441 | -0.00295 | | 0.0029 | | 0.0054 | |
| 35 | | | | | | | | | | 0.128 | 0.11729 | 0.09251 | 0.05441 | -0.00295 | | 0.0029 | | 0.0054 | |

[a]

**Figure A2.3** The final worksheet of the file hga.xls upon which all the results are summarized and the fifth energy of the regular orbit cage in Hückel is determined using the SOLVER® macro.

**Figure A2.3** Continued.

decorations of the vertices. For hydrogen 1s atomic orbitals the simple formula[4] for the overlap integral between two [Slater-type] hydrogen 1s atomic operated separated at a distance R is

$$S_{12} = e^{\zeta R}\left(1 + \zeta R + \frac{\zeta^2 R^2}{3}\right) \qquad \text{A2.15}$$

wherein $\zeta$ is the Slater exponent, in these calculations set at the usual value 1.3 $a_0^{-1}$. In the Extended Hückel calculations, the Hamiltonian diagonal matrix elements are the Valence State Ionization Energies, while the off-diagonal elements between different atomic orbitals are the means of the VSIE for the appropriate orbitals, weighted by their overlap integrals and an extra factor, for example, the Cusack coefficient of value 1.75.

There are two template EXCEL® files for the orthogonality tests using the Extended Hückel approximation in the BonusPack. The first is a similar template to the one used for the Hückel calculations. It is applied to check the 2-dimensional basis functions of the groups, which are taken in pairs, with the totally symmetric basis function included in the calculation to make up the $5 \times 5$ structure of the matrix multiplications. In the second template, for the 3-dimensional basis functions, a $9 \times 9$ structure is employed so that a full set of, for example, $1t_{1u}$, $2t_{1u}$ and $3t_{1u}$ functions can be checked at the same time.

In both templates, the areas of the 'Setup' worksheets contain the overlap matrix elements defined by equation A2.11, with the distance matrix for the different regular orbit cages being located below all the orthogonalization transformations. Figure A2.4 displays portions of the 9 x 9 template based spreadsheet containing the overlap and Extended Hückel Hamiltonian matrices. Both are constructed by writing the appropriate formula in the top-left cells of the arrays and then filling right and down to complete the matrices. Thus in cell \$F\$20, we find

$$\$F\$20 = \text{EXP}(-(0.5^*(\$D\$19 + \$D\$20)^*F465/\$D\$21))^*(1 + (0.5^*(\$D\$19 + \$D\$20) {}^*F465/\$D\$21) + (1/3)^*\textbf{POWER}((0.5^*(\$D\$19 + \$D\$20)^*F465/\$D\$21), 2)) \qquad \text{A2.16}$$

which is equation A2.11, while in \$F\$72 the Hamiltonian matrix elements follow from the propagation of the formula,

$$\$F\$72 = \textbf{IF}(F\$71 = \$E72, \$D\$72, \$D\$74^*F20^*(\$D\$72 + \$D\$73)/2) \qquad \text{A2.17}$$

Note, that in both the EXCEL® equations only absolute notation is used to distinguish cells containing fixed values, from those which changes during the propagation stages.

The aim of these exercises, in the BonusPack, is to demonstrate that the polynomial functions of Table A1.1, which are mutually orthonormal on the unit sphere, can also be made to transform appropriate as basis functions for the irreducible spaces of a Hamiltonian with icosahedral point symmetry.

Figure A2.5 displays the results of carrying out the similarity transform of the final five orthonormal functions of hu irreducible symmetry on the Hückel Hamiltonian for the decorated regular orbit cage. Given the accuracy level applied in the calculations, with no precautions to ensure that notoriously unreliable functions are being calculated appropriately, the 3 figure accuracy in the overall diagonalization is acceptable.

| | A | B | C | D | E | F | G | H | I | J | K | L | M | N | O | P | Q |
|---|---|---|---|---|---|---|---|---|---|---|---|---|---|---|---|---|---|
| 1 | **Oh regular orbit** | | | | # | 1 | 2 | 3 | 4 | 5 | 6 | 7 | 8 | 9 | 10 | 11 | 12 |
| 2 | | | | | θ | 34.31583 | 58.61132 | 77.54109 | 145.68417 | 77.54109 | 58.61132 | 121.38868 | 34.31583 | 145.68417 | 102.45891 | 145.68417 | 102.45891 |
| 3 | | | | | ϕ | 67.50000 | 14.63881 | 147.76439 | 292.50000 | 57.76439 | 255.36119 | 345.36119 | 157.50000 | 112.50000 | 212.23561 | 22.50000 | 302.23561 |
| 4 | | | | | x | 0.21574 | 0.82594 | -0.82594 | 0.21574 | 0.52084 | -0.21574 | 0.82594 | -0.52084 | -0.21574 | -0.82594 | 0.52084 | 0.52084 |
| 5 | | | | | y | 0.52084 | 0.21574 | 0.52084 | -0.52084 | 0.82594 | -0.82594 | -0.21574 | 0.21574 | 0.52084 | -0.52084 | 0.21574 | -0.82594 |
| 6 | | | | | z | 0.82594 | 0.52084 | 0.21574 | -0.82594 | 0.21574 | 0.52084 | -0.52084 | 0.82594 | -0.82594 | -0.21574 | -0.82594 | -0.21574 |
| 7 | | | | | N | 1.0000 | 1.0000 | 1.0000 | 1.0000 | 1.0000 | 1.0000 | 1.0000 | 1.0000 | 1.0000 | 1.0000 | 1.0000 | 1.0000 |
| 8 | | | | | | | | | | | | | | | | | |
| 9 | | | | | 1t1ua | 0.8259 | 0.5208 | 0.2157 | -0.8259 | 0.2157 | 0.5208 | -0.5208 | 0.8259 | -0.8259 | -0.2157 | -0.8259 | -0.2157 |
| 10 | | | | | 1t1ub | 0.5208 | 0.2157 | 0.5208 | -0.5208 | 0.8259 | -0.8259 | -0.2157 | 0.2157 | 0.5208 | -0.5208 | 0.2157 | -0.8259 |
| 11 | | | | | 1t1uc | 0.2157 | 0.8259 | -0.8259 | 0.2157 | 0.5208 | -0.2157 | 0.8259 | -0.5208 | -0.2157 | -0.8259 | 0.5208 | 0.5208 |
| 12 | | | | | 2t1ua | 0.3394 | -0.8561 | -0.5970 | -0.3394 | -0.5970 | -0.8561 | 0.8561 | 0.3394 | -0.3394 | 0.5970 | -0.3394 | 0.5970 |
| 13 | | | | | 2t1ub | -0.8561 | -0.5970 | -0.8561 | 0.8561 | 0.3394 | -0.3394 | 0.5970 | -0.5970 | -0.8561 | 0.8561 | -0.5970 | -0.3394 |
| 14 | | | | | 2t1uc | -0.5970 | 0.3394 | -0.3394 | -0.5970 | -0.8561 | 0.5970 | 0.3394 | 0.8561 | 0.5970 | -0.3394 | -0.8561 | -0.8561 |
| 15 | | | | | 3t1ua | 0.1791 | 0.1030 | 0.0096 | -0.1791 | 0.0096 | 0.1030 | -0.1030 | 0.1791 | -0.1791 | -0.0096 | -0.1791 | -0.0096 |
| 16 | | | | | 3t1ub | 0.1030 | 0.0096 | 0.1030 | -0.1030 | 0.1791 | -0.1791 | -0.0096 | 0.0096 | 0.1030 | -0.1030 | 0.0096 | -0.1791 |
| 17 | | | | | 3t1uc | 0.0096 | 0.1791 | -0.1791 | 0.0096 | 0.1030 | -0.0096 | 0.1791 | -0.1030 | -0.0096 | -0.1791 | 0.1030 | 0.1030 |
| 18 | Slater H-1s Overlap Matrix | | | | | | | | | | | | | | | | |
| 19 | $\mu_a$ = = > | | | 1.3 | | 1 | 2 | 3 | 4 | 5 | 6 | 7 | 8 | 9 | 10 | 11 | 12 |
| 20 | $\mu_b$ = = > | | | 1.3 | 1 | 1.000 | 0.631 | 0.355 | 0.111 | 0.631 | 0.251 | 0.182 | 0.598 | 0.167 | 0.142 | 0.167 | 0.161 |
| 21 | **$a_0$ = = >** | | | **0.529** | 2 | 0.631 | 1.000 | 0.167 | 0.182 | 0.631 | 0.241 | 0.397 | 0.277 | 0.161 | 0.111 | 0.277 | 0.306 |
| 22 | | | | | 3 | 0.355 | 0.167 | 1.000 | 0.142 | 0.277 | 0.228 | 0.111 | 0.631 | 0.355 | 0.397 | 0.161 | 0.111 |
| 23 | | | | | 4 | 0.111 | 0.182 | 0.142 | 1.000 | 0.161 | 0.251 | 0.631 | 0.111 | 0.397 | 0.355 | 0.598 | 0.631 |
| 24 | | | | | 5 | 0.631 | 0.631 | 0.277 | 0.161 | 1.000 | 0.136 | 0.306 | 0.288 | 0.306 | 0.111 | 0.355 | 0.167 |
| 25 | | | | | 6 | 0.251 | 0.241 | 0.228 | 0.251 | 0.136 | 1.000 | 0.200 | 0.397 | 0.121 | 0.466 | 0.131 | 0.444 |
| 26 | | | | | 7 | 0.182 | 0.397 | 0.111 | 0.631 | 0.306 | 0.200 | 1.000 | 0.111 | 0.306 | 0.167 | 0.725 | 0.631 |
| 27 | | | | | 8 | 0.598 | 0.277 | 0.631 | 0.111 | 0.288 | 0.397 | 0.111 | 1.000 | 0.167 | 0.306 | 0.111 | 0.142 |

[a]

| | A | B | C | D | E | F | G | H | I | J | K | L | M | N | O | P | Q |
|---|---|---|---|---|---|---|---|---|---|---|---|---|---|---|---|---|---|
| 71 | EHT Hamiltonian | | | | | 1 | 2 | 3 | 4 | 5 | 6 | 7 | 8 | 9 | 10 | 11 | 12 |
| 72 | | | **H11** | **-13.6** | 1 | -13.6000 | -15.0269 | -8.4577 | -2.6446 | -15.0269 | -5.9670 | -4.3381 | -14.2250 | -3.9692 | -3.3892 | -3.9692 | -3.8278 |
| 73 | | | **H22** | **-13.6** | 2 | -15.0269 | -13.6000 | -3.9692 | -4.3381 | -15.0269 | -5.7326 | -9.4371 | -6.5854 | -3.8278 | -2.6446 | -6.5854 | -7.2883 |
| 74 | | | **K** | **1.75** | 3 | -8.4577 | -3.9692 | -13.6000 | -3.3892 | -6.5854 | -5.4204 | -2.6446 | -15.0269 | -8.4577 | -9.4371 | -3.8278 | -2.6446 |
| 75 | | | | | 4 | -2.6446 | -4.3381 | -3.3892 | -13.6000 | -3.8278 | -5.9670 | -15.0269 | -2.6446 | -9.4371 | -8.4577 | -14.2250 | -15.0269 |
| 76 | | | | | 5 | -15.0269 | -15.0269 | -6.5854 | -3.8278 | -13.6000 | -3.2256 | -7.2883 | -6.8655 | -7.2883 | -2.6446 | -8.4577 | -3.9692 |
| 77 | | | | | 6 | -5.9670 | -5.7326 | -5.4204 | -5.9670 | -3.2256 | -13.6000 | -4.7504 | -9.4371 | -2.8685 | -11.0998 | -3.1157 | -10.5743 |
| 78 | | | | | 7 | -4.3381 | -9.4371 | -2.6446 | -15.0269 | -7.2883 | -4.7504 | -13.6000 | -2.6446 | -7.2883 | -3.9692 | -17.2653 | -15.0269 |
| 79 | | | | | 8 | -14.2250 | -6.5854 | -15.0269 | -2.6446 | -6.8655 | -9.4371 | -2.6446 | -13.6000 | -3.9692 | -7.2883 | -2.6446 | -3.3892 |
| 80 | | | | | 9 | -3.9692 | -3.8278 | -8.4577 | -9.4371 | -7.2883 | -2.8685 | -7.2883 | -3.9692 | -13.6000 | -6.8655 | -14.2250 | -4.3381 |
| 81 | | | | | 10 | -3.3892 | -2.6446 | -9.4371 | -8.4577 | -2.6446 | -11.0998 | -3.9692 | -7.2883 | -6.8655 | -13.6000 | -4.3381 | -6.5854 |
| 82 | | | | | 11 | -3.9692 | -6.5854 | -3.8278 | -14.2250 | -8.4577 | -3.1157 | -17.2653 | -2.6446 | -14.2250 | -4.3381 | -13.6000 | -8.4577 |
| 83 | | | | | 12 | -3.8278 | -7.2883 | -2.6446 | -15.0269 | -3.9692 | -10.5743 | -15.0269 | -3.3892 | -4.3381 | -6.5854 | -8.4577 | -13.6000 |
| 84 | | | | | 13 | -4.7504 | -3.2256 | -10.5743 | -5.7326 | -2.8685 | -15.0269 | -3.2256 | -11.0998 | -4.7504 | -20.0743 | -3.1157 | -5.9670 |
| 85 | | | | | 14 | -11.0998 | -20.0743 | -3.2256 | -5.4204 | -9.4371 | -8.4577 | -10.5743 | -5.9670 | -3.1157 | -3.2256 | -5.9670 | -11.0998 |
| 86 | | | | | 15 | -5.9670 | -4.7504 | -11.0998 | -5.9670 | -10.5743 | -2.6446 | -5.7326 | -5.4204 | -20.0743 | -5.4204 | -11.0998 | -3.2256 |

[b]

**Figure A2.4** The final summary worksheet of the file EHTO48-T1u.xls. The display is similar to that for the summary sheet in Figure A2.3 except that a 9 × 9 matrix problem has been analysed.

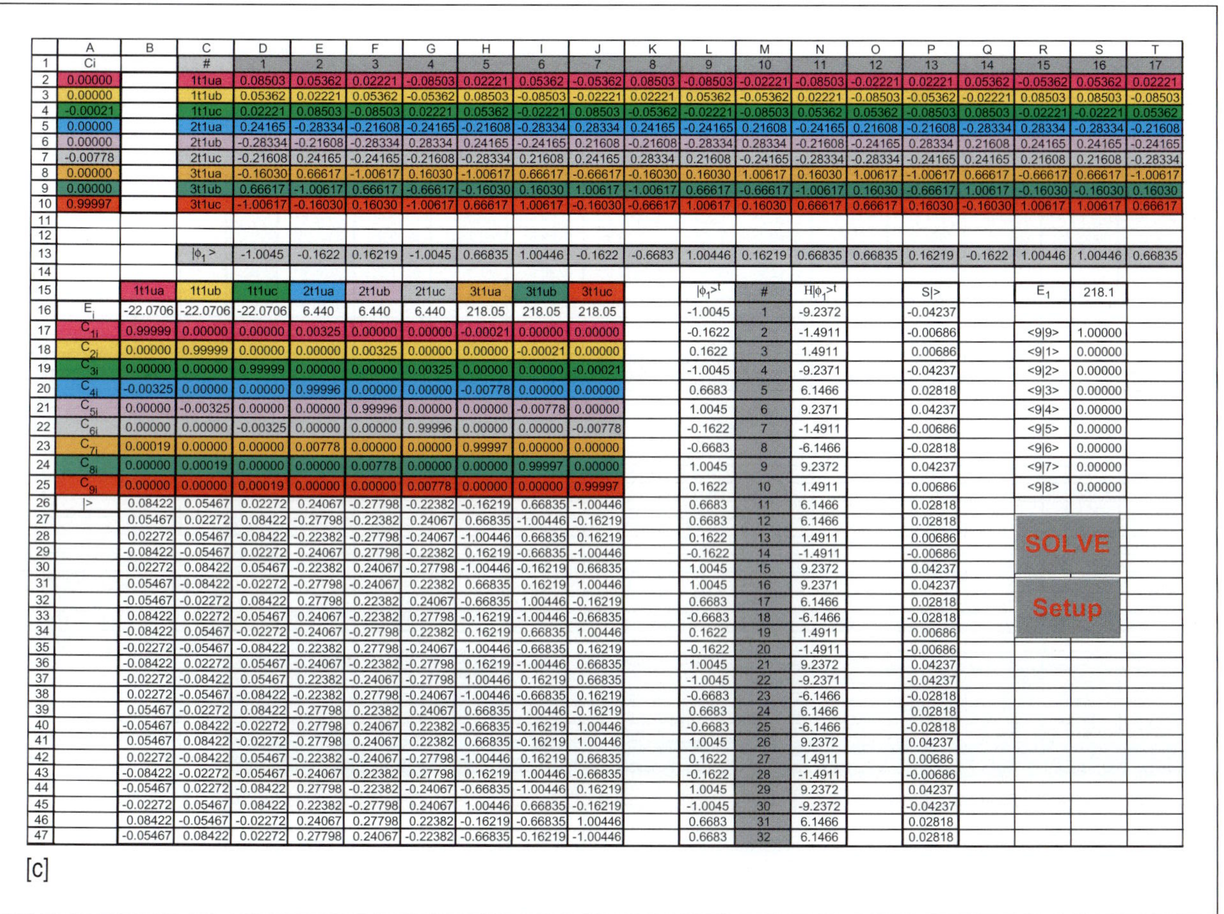

| | A | B | C | D | E | F | G | H | I | J | K | L | M | N | O | P | Q | R | S | T |
|---|---|---|---|---|---|---|---|---|---|---|---|---|---|---|---|---|---|---|---|---|
| 1 | Ci | | # | 1 | 2 | 3 | 4 | 5 | 6 | 7 | 8 | 9 | 10 | 11 | 12 | 13 | 14 | 15 | 16 | 17 |
| 2 | 0.00000 | | 1t1ua | 0.08503 | 0.05362 | 0.02221 | -0.08503 | 0.02221 | 0.05362 | -0.05362 | 0.08503 | -0.08503 | -0.02221 | -0.08503 | -0.02221 | 0.02221 | 0.05362 | -0.05362 | 0.05362 | 0.02221 |
| 3 | 0.00000 | | 1t1ub | 0.05362 | 0.02221 | 0.05362 | -0.05362 | 0.08503 | -0.08503 | -0.02221 | 0.02221 | 0.05362 | -0.05362 | 0.02221 | -0.08503 | -0.05362 | -0.02221 | 0.08503 | 0.08503 | -0.08503 |
| 4 | -0.00021 | | 1t1uc | 0.02221 | 0.08503 | -0.08503 | 0.02221 | 0.05362 | -0.02221 | 0.08503 | -0.05362 | -0.02221 | -0.08503 | 0.05362 | 0.05362 | -0.08503 | 0.08503 | -0.02221 | -0.02221 | 0.05362 |
| 5 | 0.00000 | | 2t1ua | 0.24165 | -0.28334 | -0.21608 | -0.24165 | -0.21608 | -0.28334 | 0.28334 | 0.24165 | -0.24165 | 0.21608 | -0.24165 | 0.21608 | -0.21608 | -0.28334 | 0.28334 | -0.28334 | -0.21608 |
| 6 | 0.00000 | | 2t1ub | -0.28334 | -0.21608 | -0.28334 | 0.28334 | 0.24165 | -0.24165 | 0.21608 | -0.21608 | -0.28334 | 0.28334 | -0.21608 | -0.24165 | 0.28334 | 0.21608 | 0.24165 | 0.24165 | -0.24165 |
| 7 | -0.00778 | | 2t1uc | -0.21608 | 0.24165 | -0.24165 | -0.21608 | -0.28334 | 0.21608 | 0.24165 | 0.28334 | 0.21608 | -0.24165 | -0.28334 | -0.28334 | -0.24165 | 0.24165 | 0.21608 | 0.21608 | -0.28334 |
| 8 | 0.00000 | | 3t1ua | -0.16030 | 0.66617 | -1.00617 | 0.16030 | -1.00617 | 0.66617 | -0.66617 | -0.16030 | 0.16030 | 1.00617 | 0.16030 | 1.00617 | -1.00617 | 0.66617 | -0.66617 | 0.66617 | -1.00617 |
| 9 | 0.00000 | | 3t1ub | 0.66617 | -1.00617 | 0.66617 | -0.66617 | -0.16030 | 0.16030 | 1.00617 | -1.00617 | 0.66617 | -0.66617 | -1.00617 | 0.16030 | -0.66617 | 1.00617 | -0.16030 | -0.16030 | 0.16030 |
| 10 | 0.99997 | | 3t1uc | -1.00617 | -0.16030 | 0.16030 | -1.00617 | 0.66617 | 1.00617 | -0.16030 | -0.66617 | 1.00617 | 0.16030 | 0.66617 | 0.66617 | 0.16030 | -0.16030 | 1.00617 | 1.00617 | 0.66617 |
| 11 | | | | | | | | | | | | | | | | | | | | |
| 12 | | | | | | | | | | | | | | | | | | | | |
| 13 | | | $\vert\phi_1>$ | -1.0045 | -0.1622 | 0.16219 | -1.0045 | 0.66835 | 1.00446 | -0.1622 | -0.6683 | 1.00446 | 0.16219 | 0.66835 | 0.66835 | 0.16219 | -0.1622 | 1.00446 | 1.00446 | 0.66835 |
| 14 | | | | | | | | | | | | | | | | | | | | |
| 15 | | 1t1ua | 1t1ub | 1t1uc | 2t1ua | 2t1ub | 2t1uc | 3t1ua | 3t1ub | 3t1uc | | $\vert\phi_1>^t$ | # | $H\vert\phi_1>^t$ | | S\|> | | $E_1$ | 218.1 | |
| 16 | $E_i$ | -22.0706 | -22.0706 | -22.0706 | 6.440 | 6.440 | 6.440 | 218.05 | 218.05 | 218.05 | | -1.0045 | 1 | -9.2372 | | -0.04237 | | | | |
| 17 | $C_{1i}$ | 0.99999 | 0.00000 | 0.00000 | 0.00325 | 0.00000 | 0.00000 | -0.00021 | 0.00000 | 0.00000 | | -0.1622 | 2 | -1.4911 | | -0.00686 | | <9\|9> | 1.00000 | |
| 18 | $C_{2i}$ | 0.00000 | 0.99999 | 0.00000 | 0.00000 | 0.00325 | 0.00000 | 0.00000 | -0.00021 | 0.00000 | | 0.1622 | 3 | 1.4911 | | 0.00686 | | <9\|1> | 0.00000 | |
| 19 | $C_{3i}$ | 0.00000 | 0.00000 | 0.99999 | 0.00000 | 0.00000 | 0.00325 | 0.00000 | 0.00000 | -0.00021 | | -1.0045 | 4 | -9.2371 | | -0.04237 | | <9\|2> | 0.00000 | |
| 20 | $C_{4i}$ | -0.00325 | 0.00000 | 0.00000 | 0.99996 | 0.00000 | 0.00000 | -0.00778 | 0.00000 | 0.00000 | | 0.6683 | 5 | 6.1466 | | 0.02818 | | <9\|3> | 0.00000 | |
| 21 | $C_{5i}$ | 0.00000 | -0.00325 | 0.00000 | 0.00000 | 0.99996 | 0.00000 | 0.00000 | -0.00778 | 0.00000 | | 1.0045 | 6 | 9.2371 | | 0.04237 | | <9\|4> | 0.00000 | |
| 22 | $C_{6i}$ | 0.00000 | 0.00000 | -0.00325 | 0.00000 | 0.00000 | 0.99996 | 0.00000 | 0.00000 | -0.00778 | | -0.1622 | 7 | -1.4911 | | -0.00686 | | <9\|5> | 0.00000 | |
| 23 | $C_{7i}$ | 0.00019 | 0.00000 | 0.00000 | 0.00778 | 0.00000 | 0.00000 | 0.99997 | 0.00000 | 0.00000 | | -0.6683 | 8 | -6.1466 | | -0.02818 | | <9\|6> | 0.00000 | |
| 24 | $C_{8i}$ | 0.00000 | 0.00019 | 0.00000 | 0.00000 | 0.00778 | 0.00000 | 0.00000 | 0.99997 | 0.00000 | | 1.0045 | 9 | 9.2372 | | 0.04237 | | <9\|7> | 0.00000 | |
| 25 | $C_{9i}$ | 0.00000 | 0.00000 | 0.00019 | 0.00000 | 0.00000 | 0.00778 | 0.00000 | 0.00000 | 0.99997 | | 0.1622 | 10 | 1.4911 | | 0.00686 | | <9\|8> | 0.00000 | |
| 26 | \|> | 0.08422 | 0.05467 | 0.02272 | 0.24067 | -0.27798 | -0.22382 | -0.16219 | 0.66835 | -1.00446 | | 0.6683 | 11 | 6.1466 | | 0.02818 | | | | |
| 27 | | 0.05467 | 0.02272 | 0.08422 | -0.27798 | -0.22382 | 0.24067 | 0.66835 | -1.00446 | -0.16219 | | 0.6683 | 12 | 6.1466 | | 0.02818 | | | | |
| 28 | | 0.02272 | 0.05467 | -0.08422 | -0.22382 | -0.27798 | -0.24067 | -1.00446 | 0.66835 | 0.16219 | | 0.1622 | 13 | 1.4911 | | 0.00686 | | SOLVE | | |
| 29 | | -0.08422 | -0.05467 | 0.02272 | -0.24067 | 0.27798 | -0.22382 | 0.16219 | -0.66835 | -1.00446 | | -0.1622 | 14 | -1.4911 | | -0.00686 | | | | |
| 30 | | 0.02272 | 0.08422 | 0.05467 | -0.22382 | 0.24067 | -0.27798 | -1.00446 | -0.16219 | 0.66835 | | 1.0045 | 15 | 9.2372 | | 0.04237 | | | | |
| 31 | | 0.05467 | -0.08422 | -0.02272 | -0.27798 | -0.24067 | 0.22382 | 0.66835 | 0.16219 | 1.00446 | | 1.0045 | 16 | 9.2371 | | 0.04237 | | | | |
| 32 | | -0.05467 | -0.02272 | 0.08422 | 0.27798 | 0.22382 | 0.24067 | -0.66835 | 1.00446 | -0.16219 | | 0.6683 | 17 | 6.1466 | | 0.02818 | | Setup | | |
| 33 | | 0.08422 | 0.02272 | -0.05467 | 0.24067 | -0.22382 | 0.27798 | -0.16219 | -1.00446 | -0.66835 | | -0.6683 | 18 | -6.1466 | | -0.02818 | | | | |
| 34 | | -0.08422 | 0.05467 | -0.02272 | -0.24067 | -0.27798 | 0.22382 | 0.16219 | 0.66835 | 1.00446 | | 0.1622 | 19 | 1.4911 | | 0.00686 | | | | |
| 35 | | -0.02272 | -0.05467 | -0.08422 | 0.22382 | 0.27798 | -0.24067 | 1.00446 | -0.66835 | 0.16219 | | -0.1622 | 20 | -1.4911 | | -0.00686 | | | | |
| 36 | | -0.08422 | 0.02272 | 0.05467 | -0.24067 | -0.22382 | -0.27798 | 0.16219 | -1.00446 | 0.66835 | | 1.0045 | 21 | 9.2372 | | 0.04237 | | | | |
| 37 | | -0.02272 | -0.08422 | 0.05467 | 0.22382 | -0.24067 | -0.27798 | 1.00446 | 0.16219 | 0.66835 | | -1.0045 | 22 | -9.2371 | | -0.04237 | | | | |
| 38 | | 0.02272 | -0.05467 | -0.08422 | -0.22382 | 0.27798 | -0.24067 | -1.00446 | -0.66835 | 0.16219 | | -0.6683 | 23 | -6.1466 | | -0.02818 | | | | |
| 39 | | 0.05467 | -0.02272 | 0.08422 | -0.27798 | 0.22382 | 0.24067 | 0.66835 | 1.00446 | -0.16219 | | 0.6683 | 24 | 6.1466 | | 0.02818 | | | | |
| 40 | | -0.05467 | 0.08422 | -0.02272 | 0.27798 | 0.24067 | 0.22382 | -0.66835 | -0.16219 | 1.00446 | | -0.6683 | 25 | -6.1466 | | -0.02818 | | | | |
| 41 | | 0.05467 | 0.08422 | -0.02272 | -0.27798 | 0.24067 | 0.22382 | 0.66835 | -0.16219 | 1.00446 | | 1.0045 | 26 | 9.2372 | | 0.04237 | | | | |
| 42 | | 0.02272 | -0.08422 | 0.05467 | -0.22382 | -0.24067 | -0.27798 | -1.00446 | 0.16219 | 0.66835 | | 0.1622 | 27 | 1.4911 | | 0.00686 | | | | |
| 43 | | -0.08422 | -0.02272 | -0.05467 | -0.24067 | 0.22382 | 0.27798 | 0.16219 | 1.00446 | -0.66835 | | -0.1622 | 28 | -1.4911 | | -0.00686 | | | | |
| 44 | | -0.05467 | 0.02272 | -0.08422 | 0.27798 | -0.22382 | -0.24067 | -0.66835 | -1.00446 | 0.16219 | | 1.0045 | 29 | 9.2372 | | 0.04237 | | | | |
| 45 | | -0.02272 | 0.05467 | 0.08422 | 0.22382 | -0.27798 | 0.24067 | 1.00446 | 0.66835 | -0.16219 | | -1.0045 | 30 | -9.2372 | | -0.04237 | | | | |
| 46 | | 0.08422 | -0.05467 | -0.02272 | 0.24067 | 0.27798 | 0.22382 | -0.16219 | -0.66835 | 1.00446 | | 0.6683 | 31 | 6.1466 | | 0.02818 | | | | |
| 47 | | -0.05467 | 0.08422 | 0.02272 | 0.27798 | 0.24067 | -0.22382 | -0.66835 | -0.16219 | -1.00446 | | 0.6683 | 32 | 6.1466 | | 0.02818 | | | | |

[c]

**Figure A2.4** Continued.

| | A | B | C | D | E | F | G | H | I | J | K | L | M | N | O |
|---|---|---|---|---|---|---|---|---|---|---|---|---|---|---|---|
| 1 | Regular orbit cage of Ih point symmetry - test of the diagonalization of the Huckel Hamiltonian by the hu basis functions after orthonormalization. | | | | | | | | | | | | | | |
| 2 | | | | | | | | | | | | | | | |
| 3 | \|> | 1 | 2 | 3 | 4 | 5 | 6 | 7 | 8 | 9 | 10 | 11 | 12 | 13 | 14 |
| 4 | $\phi_1$ | 0.15407 | 0.083206 | 0.012014 | -0.10265 | -0.14664 | -0.14664 | -0.10265 | 0.012014 | 0.083206 | 0.15407 | 0.04439 | -0.01403 | 0.04146 | -0.05306 |
| 5 | $\phi_2$ | 0.092762 | 0.09623 | -0.0389 | -0.03329 | -0.1168 | -0.1168 | -0.03329 | -0.0389 | 0.09623 | 0.092762 | -0.14572 | -0.00894 | -0.08112 | 0.140191 |
| 6 | $\phi_3$ | 0.080074 | -0.00307 | 0.052557 | -0.08197 | -0.04759 | -0.04759 | -0.08197 | 0.052557 | -0.00307 | 0.080074 | -0.1318 | -0.13052 | 0.049065 | 0.05113 |
| 7 | $\phi_4$ | 0.031619 | -0.09261 | 0.112147 | -0.08885 | 0.037692 | 0.037692 | -0.08885 | 0.112147 | -0.09261 | 0.031619 | 0.001379 | 0.031181 | -0.03033 | 0.017892 |
| 8 | $\phi_5$ | 0.043813 | -0.13 | 0.157077 | -0.12416 | 0.053266 | 0.053266 | -0.12416 | 0.157077 | -0.13 | 0.043813 | -0.03302 | 0.152909 | -0.17332 | 0.127527 |
| 9 | | | | | | | | | | | | | | | |
| 10 | | | | | | | | | | | | | | | |
| 11 | | | | | | | | | | | | | | | |
| 12 | | | | | | | | H\|> | | | | | | | |
| 13 | | 1.8280 | 0.0000 | 0.0000 | 0.0005 | 0.0000 | | 0.281665 | 0.043275 | -0.05479 | -0.05961 | -0.11921 | | | |
| 14 | | 0.0000 | 0.4664 | 0.0000 | 0.0000 | -0.0002 | | 0.152058 | 0.044919 | 0.002113 | 0.174946 | 0.353798 | | | |
| 15 | | 0.0000 | 0.0000 | -0.6846 | 0.0001 | 0.0000 | | 0.02202 | -0.01817 | -0.03597 | -0.21179 | -0.42747 | | | |
| 16 | | 0.0005 | 0.0000 | 0.0001 | -1.8884 | -0.0002 | | -0.18769 | -0.01551 | 0.056095 | 0.167731 | 0.337869 | | | |
| 17 | | 0.0000 | -0.0002 | 0.0000 | -0.0002 | -2.7214 | | -0.26806 | -0.05451 | 0.032556 | -0.07128 | -0.14498 | | | |
| 18 | | | | | | | | -0.26806 | -0.05451 | 0.032556 | -0.07128 | -0.14498 | | | |
| 19 | | | | | | | | -0.18769 | -0.01551 | 0.056095 | 0.167731 | 0.337869 | | | |

**Figure A2.5** Test of the overall diagonalization of the hu subspace of the Hückel Hamiltonian by the hua polynomial functions after orthonormalization. The results are satisfactory to 3 figure, suggesting that not sufficient precautions have been taken in the calculations of intermediate transcendental functions.

# Appendix 3

## Sample input files for the ApianusII.exe and FunctionPlot.exe programmes on the CDROM

Most of the diagrams in Chapters 2 and 3 have been generated using computer programmes written originally in FORTRAN applied as DOS executables. Over the years these have been transformed by the merging of POSTSCRIPT code onto the originals and then, more recently, the generation of Console executables for the MS-Windows environment. The files in the folder\Programmes\ApianusII.exe and FunctionPlot.exe can be used to draw the ApianusII projections of structure orbits and to display the linear combinations of decorations on these orbits, which correspond to the LCAO-MOs over the atomic orbitals used in the decoration.

The following instructions permit the reader to apply the programmes to check the various results displayed in the diagrams in the main manuscript and to generate their own projections for other examples.

### ApianusII.exe

Two input control files are required to generate a POSTSCRIPT output file [*.ps], for example, Figure2.19f, which is shown in Figure A3.1.

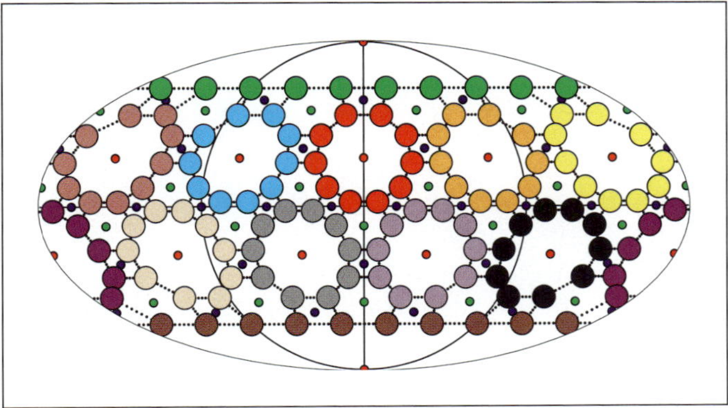

**Figure A3.1** Reproduction of Figure2.19f.ps as an example with which to illustrate the construction of the input control files Figure2.19f.in and Figure 2.19fsu.in.

The figure displays the 120-vertex equilateral regular orbit cage of $I_h$ point symmetry. The locations of the poles of the rotational axes are marked on the projection and the vertices are divided into sets of decagons of different colours. The orbit cage is 3-valent, which means that there are 3 connections about each vertex contributing to the net of the structure.

The file Figure2.19f.in contains a list of the polar coordinates of the vertices on the unit sphere, arranged so that the $(\theta, \phi)$ values of each decagon are read in during execution of the programme in the sequence that all the $\theta$ – values are read before the $\phi$ – values. This order is convenient because most of the coordinate data for the structures were calculated on spreadsheets and so could be copied and then pasted [PASTESPECIAL command] into a NOTEPAD file as 'unformatted' text.

**Table A3.1 Sample control input file, Figure2.19fsu.in, detailing the orbit-by-orbit properties of a structure, here the regular orbit of $I_h$ divided into sets of decagons of vertices about the $C_5$ axes, colours and locations of the rotational poles on the unit sphere. Note that the lines in red that start with an exclamation mark are comments and do not appear in the input file.**

```
! Positioning of diagram on drawing plane, in points [1/72"] wrt top-left origin, and scale
!factor (3f5.2)
300. 155. 1.4
! # of rotational poles, # of vertices, control orbits, structure orbits, net?,
!extraconnections?, size of vertices, #of extra connections
!(6i4,f10.4,i4)
62 120 03 12 1 0 14.0000 00
! Sizes of circles identifying poles of rotational axes, each one (f5.2)
5.01 5.01 5.01
!List of all orbits to be displayed, including orbits of poles of rotation axes, up to 20 on first
!line, then add line (20i3)
12 20 30 10 10 10 10 10 10 10 10 10 10 10 10
!Control of net, RGB colours, size of net dots and whether 3-, 4, 5- connected (4f5.2,2i3)
0.0 0.0 0.0 1.0 10 3
!RGB colours for orbit, here 1st three are the colours for the rotational poles
! then the control [0/1] to place a black dot in the centre of coloured circle (3f5.2,i3)
1.0000.0000.000 0
0.0001.0000.000 0
0.0000.0001.000 0
0.0001.0000.000 0
1.0000.0000.000 0
1.0000.6270.000 0
1.0001.0000.000 0
0.7840.3730.392 0
0.0001.0001.000 0
0.6270.4710.686 0
0.0000.0000.000 0
0.5020.0000.502 0
0.9800.8430.643 0
0.5020.5020.502 0
0.5020.2000.204 0
!Here follows the list of (θ, φ) coordinates for the locations of the rotation axes poles,
!θ first then φ. It is simplest to leave these data as a universal endlist to each control file
!for a particular symmetry.
```

All the FORTRAN programmes in the BonusPack require formatted input. For the coordinate data input file, for example, Figure2.19f.in, the formal requirement is that the $(\theta, \phi)$ values be in F18.4 format[1].

Tables A3.1 displays the structure dependent part of the formatted input control file for the construction of Figure2.19f shown in Figure A3.1. The remainder of this input file is the list of $(\theta, \phi)$ coordinates identifying the positions of the poles of the $I_h$ rotational axes using the same format as in the structure input file, Figure2.19f.in, in this example. In the table comments are identified in red text and marked by an '!' and are not part of the input data.

Because the structure in Figure A3.1 is equilateral, there is no need to specify extra connections to join particular vertices of the net. This is not true for cases in which there are major differences in the distances between adjacent vertices, for example, Figure 2.15b, which is reproduced in Figure A3.2a.

For this and similar cases, a connected list is specified in the input data for all the coordinate positions in the projected net of the structure.

## FunctionPlot.exe

This programme was used to generate the diagrams of Chapter 3. Figure 3.18 provides a good example with which to describe the input details for the FunctionPlot.exe programme. Only one input file is required for each diagram drawn. The first diagram in the Figure A3.3, is Figure 3.18e, the dependent linear combination over the $O_6$ orbit of $D_{3h}$, which is fashioned into the orthogonal linear combination displayed in the second diagram superimposed on the polynomial that is found on rendering the $2e''(1)$ function orthogonal to the $1e''(1)$ function.

The input files for these examples are shown in Table A3.3, and the formatting and other details, as in the previous example, are comments identified by '!' in red.

Note, that, as in the third diagram of Figure A3.3, it is possible to generate high contrast diagrams as the output files from FunctionPlot.exe. The setting of the definition parameter to the value 1 means that the phase of the central polynomial function is evaluated on a very fine $(\theta, \phi)$ mesh with the virtually continuous shading result shown. The penalty in this option is that the resulting output file sizes become large.

[1]Since *formatted* input files are not common nowadays, it is appropriate to make this note. 'f**. *' indicates a floating point number, with the number of decimal places indicated by the integer corresponding to the 3rd, e.g. F10.6 is a floating point number occupying 10 spaces, one of which is the decimal point, after which 6 figures of decimal can be specified. The decimal point and the total number of spaces are critical: there is freedom to write fewer entries in the 10 spaces. Integer numbers are identified in input data by the descriptor 'i*', the '*' being the number of spaces devoted to this information. Integer descriptors are *right-justified* within the integer space. This incomplete account is sufficient to operate all the programmes in the BonusPack.

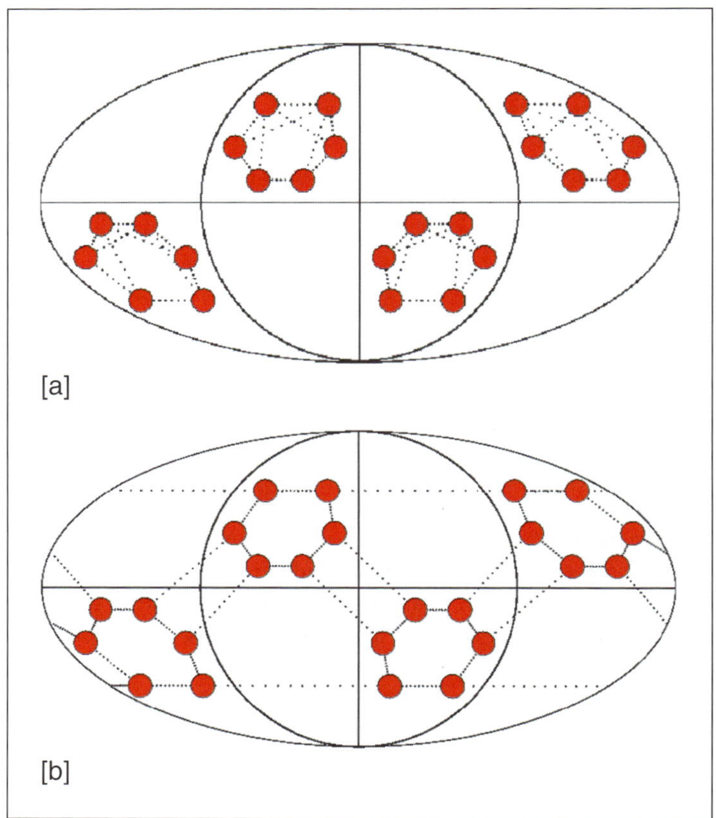


**Figure A3.2** Demonstration of the need to use a connected-list input format to reproduce Figure 2.15b. Figure A3.2a results if it is assumed that the 3-connection sorting is sufficient to construct the correct diagram. Because the sort is to find the first 3 lowest 'bond lengths', only the hexagonal rings are connected and cross links are present. The correct projection [with some distortion of the structure!] is found using the connection list input of Table A3.2.

**Table A3.2 Example of 'connection list' input files for the ApianusII.exe programmes, using the example of the non-equilateral regular orbit of $T_d$ [Figure 2.15b].**

Control File [Figure 2.15b.in]

300. 170. 1.4
00 24 00 01 0 1 15.0 36
24
0.0 0.0 0.0 1.0 10 0
1.0 0.0 0.0 0
0.0 1.0 0.0
!Coordinates of $T_d$ rotational poles
!on unit sphere, in this case!

Coordinate data file starts with list of values as before.

| # | $\theta$ | $\phi$ |
|---|---|---|
| 1 | 34.916 | 157.500 |
| 2 | 34.316 | 112.500 |
| 3 | 34.316 | 292.500 |
| 4 | 34.316 | 337.500 |
| 5 | 58.611 | 104.639 |
| 6 | 58.611 | 165.361 |
| 7 | 58.611 | 345.361 |
| 8 | 58.611 | 284.639 |
| 9 | 77.541 | 122.236 |
| 10 | 77.541 | 147.764 |
| 11 | 77.541 | 302.236 |
| 12 | 77.541 | 327.764 |
| 13 | 102.459 | 32.236 |
| 14 | 102.459 | 57.764 |
| 15 | 102.459 | 212.236 |
| 16 | 102.459 | 237.764 |
| 17 | 121.389 | 14.639 |
| 18 | 121.389 | 75.361 |
| 19 | 121.389 | 255.361 |
| 20 | 121.389 | 194.639 |
| 21 | 145.684 | 22.500 |
| 22 | 145.684 | 67.500 |
| 23 | 145.684 | 247.500 |
| 24 | 145.684 | 202.500 |

Coordinate data file
[Connections list]

!Connections and **special case** 1 option to
!connect through $\phi = 0$,
!Apportionment of number of dots in net
!between such links (5i4).
1 2
1 6
1 3
2 5
2 4 **1** 8 2
5 14
5 9
9 18
9 10
10 6
10 20
6 15
4 7
4 3
7 12
7 13
12 17
12 11
11 8
11 19
8 3
8 16
16 15
16 19
15 20
20 24
24 22
24 23
23 19
23 21 **1** 8 2
21 22
22 18
18 14
13 14
13 17
17 21

**Table A3.3 Sample input files to run FunctionPlot.exe and construct the first 2 diagrams of Figure A3.3. Note the importance of the 'definition' parameter. Its function is to determine the fineness of the shading of the regions of positive phase of the central functions. Set to 1 the output is as in Figure A3.3c.**

| Figure A3.3a.in | Figure A3.3b.in |
|---|---|
| ! Position on canvas<br>!Scale Factor<br>!Definition (3f6.4,i3)<br>225.0 175.0 1.000 **3**<br>! Plot zero contour,<br>! n title [#lines] (2i3)<br>1 1<br>!If n title > 0 then<br>!position on canvas<br>!and font<br>−130 20 18<br>!If n title > 0 input title lines<br>2e″(1) = (201) − (021)<br>!# terms in function (i3)<br>2<br>!Function terms, Elert<br>!form (3i2,f18.4)<br>2 0 1 1.00<br>0 2 1−1.00<br>!Orbital type [Capitals]<br>!# of vertices<br>!orbital size (a1,i4,f10.4)<br>S 12 100.0<br>! $\theta$ values of orbit vertices<br>45.00<br>45.00<br>45.00<br>45.00<br>......<br>135.00<br>135.00<br>135.00<br>135.00<br>! $\phi$ values of orbit vertices<br>15.00<br>105.00<br>135.00<br>225.00<br>......<br>135.00<br>225.00<br>255.00<br>345.00 | ! Position on canvas<br>!Scale Factor<br>!Definition (3f6.4,i3)<br>225.0 175.0 1.000 **3**<br>! Now, 2 title lines on output<br>1 2<br>−130 20 18<br>! Title on 2 lines<br>2e″(1) = (201) − (021) − 0.5611(110)<br>Slater exponent = 1.625, R = 3.0 a.u.<br>! Now, 3 terms to include<br>! orthogonalizing term<br>3<br>2 0 1 1.00<br>0 2 1−1.00<br>1 1 0−0.54517<br>! Input as before, but full list of<br>! $(\theta, \phi)$ values<br>S 12 0 100.000000<br>45.00<br>45.00<br>45.00<br>45.00<br>45.00<br>45.00<br>135.00<br>135.00<br>135.00<br>135.00<br>135.00<br>135.00<br>15.00<br>105.00<br>135.00<br>225.00<br>255.00<br>345.00<br>15.00<br>105.00<br>135.00<br>225.00<br>255.0<br>345.00 |

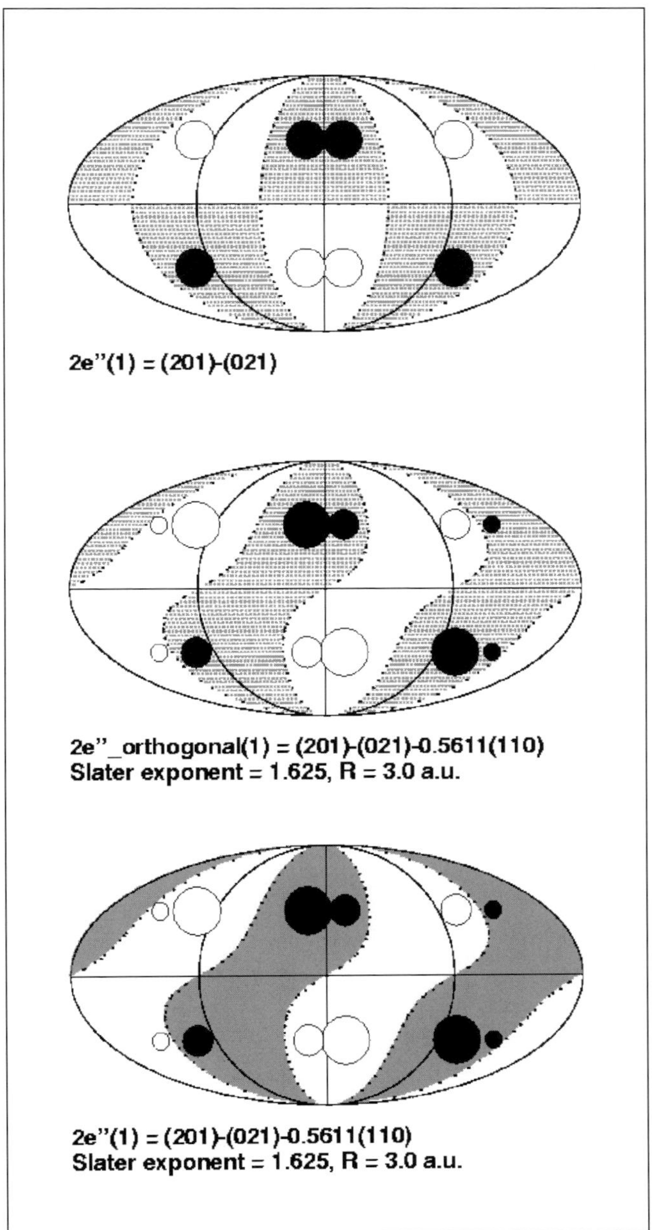


**Figure A3.3** Reproductions of Figure 3.18 diagrams generated using the input files of Table A3.3 to run FunctionPlot.exe. Note, in the 3rd diagram the possibility to display high contrast shading is controlled by the definition parameter at the end of the first line of input.

# Bibliography


### *Group Theory in Chemistry*

D.M. Bishop, *Group Theory and Chemistry* [Dover Publications, 1993].

F.A. Cotton, *Chemical Applications of Group Theory* [Wiley, 1990].

### *Bonding in Clusters*

P.W. Fowler and D.E. Manolopoulos, *An Atlas of Fullerenes* [Oxford University Press, 1995].

R.L. Johnston, Mathematical Cluster Bonding, in *Structure and Bonding,* **87** [Springer-Verlag, Berlin, 1997].

D.M.P. Mingos and R.L. Johnston, Theoretical Models of Cluster Bonding, in *Structure and Bonding,* **68** [Springer-Verlag, Berlin, 1987].

D.M.P. Mingos and D.J. Wales, *Introduction to Cluster Chemistry (Prentice Hall Advanced Reference Series)* [Prentice-Hall, 1990].

A. Streitwieser, *Molecular Orbital Theory for Organic Chemists* [Wiley, 1961].

J.G. Verkade, *A Pictorial Approach to Molecular Bonding* [Springer, 1986].

### *Spherical-Shell Technique, Tensor Surface Harmonic Theory*

C.M. Quinn, J.G. McKiernan and D.B. Redmond, *Inorg. Chem.*, **22** (1983) 2310.

C.M. Quinn, J.G. McKiernan and D.B. Redmond, *J. Chem. Ed.*, **61** (1984) 569.

C.M. Quinn, J.G. McKiernan and D.B. Redmond, *J. Chem. Ed.*, **61** (1984) 572.

D.B. Redmond, C.M. Quinn and J.G. McKiernan, *J. Chem. Soc.*, *Faraday Trans. II*, **79** (1983) 1791.

A.J. Stone, *Mol. Phys.*, **41** (1980) 1339.

A.J. Stone, *Inorg. Chem.*, **20** (1981) 563.

A.J. Stone and M.J. Alderton, *Inorg. Chem.*, **21** (1982) 2297.

A.J. Stone, *Polyhedra*, **3** (1984) 2051.

### *Projections*

K.C. Clarke, Flattening the earth: Two thousand years of map projections, *American Geographical Society*, **84** (1994) 490.

D.H. Maling, *Coordinate Systems and Map Projections*, 2nd ed [Butterworth-Heinemann, 1993].

### *Polyhedra*

H.S.M. Coxeter, *Regular Polytopes* [Dover Publications, 1974].

P.R. Cromwell, *Polyhedra* [Cambridge University Press, 1999].

# Index